Networks-on-Chip:
From Implementations
to Programming
Paradigms

Networks-on-Chip: From Implementations to Programming Paradigms

Editor-in-Chief

Zhiying Wang

State Key Laboratory of High Performance Computing
National University of Defense Technology, China

Authors

Sheng Ma, Libo Huang, Mingche Lai and Wei Shi

State Key Laboratory of High Performance Computing
National University of Defense Technology, China

AMSTERDAM • BOSTON • HEIDELBERG • LONDON
NEW YORK • OXFORD • PARIS • SAN DIEGO
SAN FRANCISCO • SINGAPORE • SYDNEY • TOKYO

Morgan Kaufmann is an imprint of Elsevier

Morgan Kaufmann is an imprint of Elsevier
225 Wyman Street, Waltham, MA 02451, USA

Notices

Knowledge and best practice in this field are constantly changing. As new research and
experience broaden our understanding, changes in research methods, professional practices,
or medical treatment may become necessary.

Practitioners and researchers must always rely on their own experience and knowledge in
evaluating and using any information, methods, compounds, or experiments described herein.
In using such information or methods they should be mindful of their own safety and the
safety of others, including parties for whom they have a professional responsibility.

To the fullest extent of the law, neither the Publisher nor the authors, contributors, or editors,
assume any liability for any injury and/or damage to persons or property as a matter of
products liability, negligence or otherwise, or from any use or operation of any methods,
products, instructions, or ideas contained in the material herein.

Library of Congress Cataloging-in-Publication Data
A catalog record for this book is available from the Library of Congress

British Library Cataloguing in Publication Data
A catalogue record for this book is available from the British Library

For information on all Morgan Kaufmann publications
visit our web site at store.elsevier.com

Printed and bound in USA
15 16 17 18   10 9 8 7 6 5 4 3 2 1

ISBN: 978-0-12-800979-6

Contents in Brief

Contents

PART V   EPILOGUE

Preface

The continuing advance of semiconductor technology has led to dramatic increases in transistor densities. Designing complex single-core processors brings about a sharp increase in power consumption, but this is accompanied by diminishing performance returns. Instead, computer architects are actively pursuing multicore designs to efficiently use the billions of transistors on a single chip. Current processors have already integrated tens or hundreds of cores; the community is entering the many-core era. Although great breakthroughs have already been made in the design of many-core processors, there are still many critical challenges to solve, ranging from high-level parallel programming paradigms to low-level logic implementations. The correctness, performance, and efficiency of many-core processors rely heavily on communication mechanisms; addressing these critical challenges requires communication-centric cross-layer optimizations.

The traditional on-chip bus communication mechanisms face several limitations, such as low bandwidth, large latency, high power consumption, and poor scalability. To address these limitations, the network-on-chip (NoC) is proposed as an efficient and scalable communication structure for many-core platforms. Owing to its many desirable properties, the NoC has rapidly crystallized into a significant research domain of the computer architecture community. Although the NoC has some similarities with off-chip networks, its physical fabric in the latency, power, and area is fundamentally different. The NoC competes with processing cores for the scarce on-chip power and area. Thus, it can only leverage limited resources. To support high performance within tight power and area budgets, more attention should be paid to NoC optimizations, including low-level logic implementations, network-level routing and flow control, and support for high-level programming paradigms.

Zhiying Wang's research group at the National University of Defense Technology has been studying the frontier subjects of computer architecture. His group has been conducting NoC research for about 10 years, and has presented or published tens of peer-reviewed papers at several prestigious and influential conferences and in several journals. This book reviews and summarizes the research progress and outcomes reported in these publications, including three papers from the top-tier ISCA-2011 and HPCA-2012 conferences, three papers from the prestigious DAC-2008, ICCD-2011, and ASAP-2012 conferences, four papers from the flagship *IEEE Transactions on Computers*, *IEEE Transactions on Parallel and Distributed Systems*, and *ACM Transactions on Architecture and Code Optimization* journals, and two papers from the influential *Microprocessors and Microsystems* journal.

This book is not a simple overview or collection of research ideas and design experiences. The editor, Zhiying Wang, and the authors wrote this volume with the purpose of exploring the NoC design space in a coherent, uniform, and bottom-up fashion, from low-level router, buffer, and topology implementations, to network-level routing and flow control designs, to co-optimizations of the NoC and high-level programming paradigms. The book is composed of five parts. Parts I and IV introduce and conclude this book respectively. Part V also presents future work.

Part II covers low-level logic implementations, and consists of Chapters 2–1. Chapter 2 discusses a wing-channel-based single-cycle router architecture, which significantly reduces communication delays with low hardware costs. Chapter 1 studies two dynamic virtual channel structures with congestion awareness. The designs dynamically share buffers among virtual channels or ports to reduce buffer amount requirements, while maintaining or improving performance. Chapter 4 introduces an NoC topology enhanced with virtual bus structures. The topology efficiently supports unicast as well as multicast/broadcast communications, and the latter is critical for parallel application executions.

Part III investigates routing and flow control, and includes Chapters 5–7. Chapter 5 delves into routing algorithms for workload consolidation. The routing algorithms provide high adaptivity and dynamic isolation for concurrent applications. Chapter 6 proposes efficient flow control for fully adaptive routing. It maximizes the utilization of limited buffer resources to scale the performance without inducing network deadlock. Chapter 7 explores deadlock-free flow control theories and designs for torus NoCs. Conventional deadlock avoidance designs for torus NoCs either negatively affect the router complexity and frequency, or reduce the buffer utilization. The proposed flit bubble flow control achieves low complexity, high frequency, and efficient buffer utilization for high performance.

Part IV explores co-optimizations of the NoC and programming paradigms. It contains Chapters 8–10. Chapter 8 addresses the NoC optimization for shared memory paradigms. It provides hardware implementations for cache-coherent collective communications to prevent these communications from becoming system bottlenecks. Chapter 9 customizes NoC designs for message passing paradigms. The NoC presented offers special and low-cost hardware for basic message passing interface (MPI) primitives, upon which other MPI functions can be built, to effectively boost the performance of MPI functions. Chapter 10 studies an adaptive MPI communication protocol, which combines the buffered and synchronous communication modes to provide robust and high communication performance.

Overall, on the basis of the communication-centric cross-layer optimization methods, the work presented in this book has significantly advanced the state of the art in the NoC design and the many-core processor research domain. Based on a bottom-up and thorough exploration of the NoC design space, the research presented here has addressed a multitude of pressing concerns spanning a wide spectrum of design topics. It not only greatly improves the performance and reduces the overhead for the communication layer of many-core processors, but also significantly mitigates the challenges in the logic implementation layer and the parallel programming layer.

The research described in this book has been supported by several grants from various organizations, including the 863 Program of China (2012AA010905), the National Natural Science Foundation of China (61272144, 61303065, 61103016, 61202481, 61202123, 61202122), the Hunan Provincial Natural Science Foundation of China (12JJ4070, 14JJ3002), the Doctoral Fund of the Ministry of Education of China (20134307120028, 20114307120010), and the Research Project of the National University of Defense Technology (JC12-06-01, JC13-06-02). Sheng Ma's research has also been supported by the University of Toronto and the Natural Sciences and Engineering Research Council of Canada when he visited the University of Toronto.

About the Editor-in-Chief and Authors

EDITOR-IN-CHIEF

Zhiying Wang received his Ph.D. degree in electrical engineering from the National University of Defense Technology in 1988. He is currently a professor in the School of Computer Science at the National University of Defense Technology. He has contributed more than 10 invited chapters to book volumes, published 240 papers in archival journals, refereed conference proceedings, and delivered over 30 keynote speeches. His main research fields include computer architecture, computer security, VLSI design, reliable architecture, multicore memory systems, and asynchronous circuits. He is a member of the IEEE and the Association for Computing Machinery.

AUTHORS

Sheng Ma received his B.S. and Ph.D. degrees in computer science and technology from the National University of Defense Technology in 2007 and 2012 respectively. He visited the University of Toronto from September 2010 to September 2012. He is currently an assistant professor in the School of Computer Science at the National University of Defense Technology. His research interests include on-chip networks, single instruction, multiple data architectures, and arithmetic unit designs.

Libo Huang received his B.S. and Ph.D. degrees in computer engineering from the National University of Defense Technology in 2005 and 2010 respectively. From 2010, he has been a lecturer in the School of Computer Science. His research interests include computer architecture, hardware/software co-design, VLSI design, and on-chip communication. He has served as a technical reviewer for several conferences and journals, for example *IEEE Transactions on Computers, IEEE Transactions on Computer-Aided Design of Integrated Circuits and Systems,* and DAC 2012. Since 2004, he has authored more than 20 papers published in internationally recognized journals and presented at international conferences.

Mingche Lai received his Ph.D. degree in computer engineering from the National University of Defense Technology in 2008. Currently, he is an associate professor in the School of Computer Science at the National University of Defense Technology, and is employed to develop high-performance computer interconnection systems. Since 2008, he has also been a faculty member of the National Key Laboratory for Parallel and Distributed Processing of China. His research interests include on-chip networks, optical communication, many-core processor architecture, and hardware/software co-design. He is a member of the IEEE and the Association for Computing Machinery.

Wei Shi received his Ph.D. degree in computer science from the National University of Defense Technology in 2010. Currently, he is an assistant professor in the School of Computer Science at the National University of Defense Technology, and is employed to develop high-performance processors. His research interests include computer architecture, VLSI design, on-chip communication, and asynchronous circuit techniques.

Prologue

I

Introduction

1

CHAPTER OUTLINE

1.1 THE DAWN OF THE MANY-CORE ERA

The development of the semiconductor industry has followed the "self-fulfilling" Moore's law [108] in the past half century: to maintain a competitive advantage

Networks-on-Chip. http://dx.doi.org/10.1016/B978-0-12-800979-6.00001-9
Copyright © 2015 China Machine Press/Beijing Huazhang Graphics & Information Co., Ltd.
Published by Elsevier Inc. All rights reserved.

for products, the number of transistors per unit area on integrated circuits doubles every 2 years. Currently, a single chip is able to integrate several billion transistors [114, 128], and Moore's law is expected to continue to be valid in the next decade [131]. The steadily increasing number of transistors, on the one hand, offers the computer architecture community enormous opportunities to design and discover innovative techniques to scale the processor's performance [54]. On the other hand, the huge number of transistors exacerbates several issues, such as power consumption [42, 111], resource utilization [38], and reliability problems [14], which bring about severe design challenges for computer architects.

Until the beginning of this century, the primary performance scaling methods focused on improving the sequential computation ability for single-core processors [110]. Since the birth of the first processor, the processor's sequential performance-price ratio has been improved by over 10 billion times [117]. Some of the most efficient performance enhancement techniques include leveraging deep pipelines to boost the processor frequency [56, 60], utilizing sophisticated superscalar techniques to squeeze the instruction-level parallelism [73, 83], and deploying large on-chip caches to exploit the temporal and spatial locality [70, 121]. Yet, since the switching power is proportional to the processor frequency, sophisticated superscalar techniques involve complex logic, and large caches configure huge memory arrays, these traditional performance-scaling techniques induce a sharp increase in processor complexity [3] and also power consumption [134]. In addition, there are diminishing performance returns as the already exploited instruction-level parallelism and temporal/spatial locality have nearly reached the intrinsic limitations of sequential programs [120, 143]. The unconquerable obstacles, including the drastically rising power consumption and diminishing performance gains for single-core processors, have triggered a profound revolution in computer architecture.

To efficiently reap the benefits of billions of transistors, computer architects are actively pursuing the multicore design to maintain sustained performance growth [50]. Instead of an increase in frequency or the use of sophisticated superscalar techniques to improve the sequential performance, the multicore designs scale the computation throughput by packing multiple cores into a single chip. Different threads of a single application or different concurrent applications can be assigned to the multiple cores to efficiently exploit the thread-level and task-level parallelism respectively. This can achieve significantly higher performance than using a single-core processor. Also, multicore processors reduce engineering efforts for the producers, since different amounts of the same core can be stamped down to form a family of processors. Since Intel's cancellation of the single-core processors Tejas and Jayhawk in 2004 and its switching to developing dual-core processors to compete with AMD, the community has been rapidly embracing the multicore design method [45].

Generally speaking, when the processor integrates eight or fewer cores, it is called a multicore processor. A processor with more cores is called a many-core processor. Current processors have already integrated several tens or hundreds of cores, and a new version of Moore's law uses the core count as the exponentially increasing parameter [91, 135]. Intel and IBM announced their own 80-core and 75-core prototype processors, the Teraflops [141] and Cyclops-64 [29], in 2006. Tilera released the 64-core TILE64 [11] and the 100-core TILE-Gx100 [125] processors in 2007 and 2009 respectively. In 2008, ClearSpeed launched the 192-core CSX700 processor [105]. In 2011, Intel announced its 61-core Knights Corner co-processor [66], which is largely and successfully deployed in the currently most powerful supercomputer, the TianHe-2 [138]. Intel plans to release its next-generation acceleration co-processor, the 72-core Knights Landing co-processor, in 2015 [65]. Other typical many-core processors include Intel's Single-chip Cloud Computer (SCC) [58], Intel's Larrabee [130], and AMD's and NVIDIA's general purpose computing on graphics processing units (GPGPUs) [41, 90, 114, 146]. The many-core processors are now extensively utilized in several computation fields, ranging from scientific supercomputing to desktop graphic applications; the computer architecture and its associated computation techniques are entering the many-core era.

1.2 COMMUNICATION-CENTRIC CROSS-LAYER OPTIMIZATIONS

Although there have already been great breakthroughs in the design of many-core processors in the theoretical and practical fields, there are still many critical problems to solve. Unlike in single-core processors, the large number of cores increases the design and deployment difficulties for many-core processors; the design of efficient many-core architecture faces several challenges ranging from high-level parallel programming paradigms to low-level logic implementations [2, 7]. These challenges are also opportunities. Whether they can be successfully and efficiently solved may largely determine the future development for the entire computer architecture community. Figure 1.1 shows the three key challenges for the design of a many-core processor.

The first challenge lies in the parallel programming paradigm layer [7, 31]. For many years, only a small number of researchers used parallel programming to conduct large-scale scientific computations. Nowadays, the widely available multicore and many-core processors make parallel programming essential even for desktop users [103]. The programming paradigm acts as a bridge to connect the parallel applications with the parallel hardware; it is one of the most significant challenges for the development of many-core processors [7, 31]. Application developers desire an opaque paradigm which hides the underlying architecture to ease programming efforts and improve application portability. Architecture designers hope for a visible paradigm to utilize the hardware features for high performance. An efficient parallel programming paradigm should be an excellent tradeoff between these two requirements.

The second challenge is the interconnection layer. The traditional bus and crossbar structures have several shortcomings, including poor scalability, low bandwidth, large

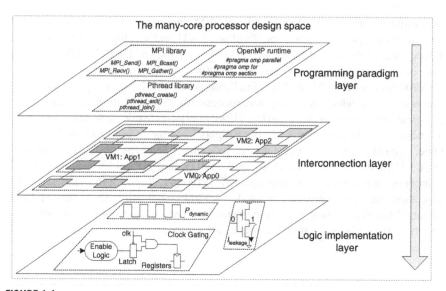

The many-core processor design space

FIGURE 1.1

Key challenges for the design of a many-core processor.

latency, and high power consumption. To address these limitations, the network-on-chip (NoC) introduces a packet-switched fabric for the on-chip communication [25], and it becomes the de facto many-core interconnection mechanism. Although significant progress has been made in NoC research [102, 115], most works have focused on the optimizations of pure network-layer performance, such as the zero-load latency and saturation throughput. These works do not consider sufficiently the upper programming paradigms and the lower logic implementations. Indeed, the cross-layer optimization is regarded as an efficient way to extend Moore's law for the whole information industry [2]. The future NoC design should meet the requirements of both upper programming paradigms and lower logic implementations.

The third challenge appears in the logic implementation layer. Although multicore and many-core processors temporarily mitigate the problem of the sharply rising power consumption, the computer architecture will again face the power consumption challenge with the steadily increasing transistor count driven by Moore's law. The evaluation results from Esmaeilzadeh *et al.* [38] show that the power limitation will force approximately 50% of the transistors to be off for processors based on 8 nm technology in 2018, and thus the emergence of "dark silicon." The "dark silicon" phenomenon strongly calls for innovative low-power logic designs and implementations for many-core processors.

The aforementioned three key challenges directly determine whether many-core processors can be successfully developed and widely used. Among them, the design of an efficient interconnection layer is one of the most critical challenges, because of the following issues. First, the many-core processor design has already evolved from the "computation-centric" method into the "communication-centric" method [12, 53]. With the abundant computation resources, the efficiency of the interconnection layers largely and heavily determines the performance of the many-core processor. Second, and more importantly, mitigating the challenges for the programming paradigm layer and the logic implementation layer requires optimizing the design of the interconnection layer. The opaque programming paradigm requires the interconnection layer to intelligently manage the application traffic and hide the communication details. The visible programming paradigm requires the interconnection layer to leverage the hardware features for low communication latencies and high network throughput. In addition, the interconnection layers induce significant power consumption. In the Intel SCC [59], Sun Niagara [81], Intel Teraflops [57] and MIT Raw [137] processors, the interconnection layer's power consumption accounts for 10%, 17%, 28%, and 36% of the processor's overall power consumption respectively. The design of low-power processors must optimize the power consumption for the interconnection layer.

On the basis of the "communication-centric cross-layer optimization" method, this book explores the NoC design space in a coherent and uniform fashion, from the low-level logic implementations of the router, buffer, and topology, to network-level routing and flow control schemes, to the co-optimizations of the NoC and high-level programming paradigms. In Section 1.3, we first conduct a baseline design space exploration for NoCs, and then in Section 1.4 we review the current NoC research status. Section 1.5 summarizes NoC design trends for several real processors, including academia prototypes and commercial products. Section 1.6 briefly introduces the main content of each chapter and gives an overview of this book.

1.3 A BASELINE DESIGN SPACE EXPLORATION OF NoCs

Figure 1.2 illustrates a many-core platform with an NoC interconnection layer. The 16 processing cores are organized in a 4 × 4 mesh network. Each network node includes a processing node and a network router. The processing node may be replaced

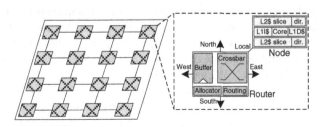

FIGURE 1.2

A many-core platform with an NoC.

by other hardware units, such as special accelerators or memory controllers. The network routers and the associated links form the NoC. The router is composed of the buffer, crossbar, allocator, and routing unit. The baseline NoC design exploration mainly consists of the design of the network topology, routing algorithm, flow control mechanism, and router microarchitecture [26, 35]. In addition, we also discuss the baseline performance metrics for NoC evaluations.

1.3.1 TOPOLOGY

The topology is the first fundamental aspect of NoC design, and it has a profound effect on the overall network cost and performance. The topology determines the physical layout and connections between nodes and channels. Also, the message traverse hops and each hop's channel length depend on the topology. Thus, the topology significantly influences the latency and power consumption. Furthermore, since the topology determines the number of alternative paths between nodes, it affects the network traffic distribution, and hence the network bandwidth and performance achieved.

Since most current integrated circuits use the 2D integrated technology, an NoC topology must match well onto a 2D circuit surface. This requirement excludes several excellent topologies proposed for high-performance parallel computers; the hypercube is one such example. Therefore, NoCs generally apply regular and simple topologies. Figure 1.3 shows the structure of three typical NoC topologies: a ring, a 2D mesh, and a 2D torus.

The ring topology is widely used with a small number of network nodes, including the six-node Ivy Bridge [28] and the 12-node Cell [71] processors. The ring network has a very simple structure, and it is easy to implement the routing and flow control, or even apply a centralized control mechanism [28, 71]. Also, since Intel has much experience in implementing the cache coherence protocols on the ring networks, it prefers to utilize the ring topology, even in the current 61-core Knights Corner processor [66]. Yet, owing to the high average hop count, the scalability of a ring network is limited. Multiple rings may mitigate this limitation. For example, to improve the scalability, Intel's Knights Corner processor leverages ten rings, which is about two times more rings than the eight-core Nehalem-EX processor [22, 30, 116]. Another method to improve the scalability is to use 2D mesh or torus networks.

The 2D mesh is quite suitable for the wire routing in a multimetal layer CMOS technology, and it is also very easy to achieve deadlock freedom. Moreover, its scalability is better than that of the ring. These factors make the 2D mesh topology widely used in several academia and industrial chips, including the U.T. Austin TRIPS [48], Intel Teraflops [140], and Tilera TILE64 [145] chips. Also, the 2D mesh is the most widely assumed topology for the research community. A limitation of this topology is that its center portion is easily congested and may become a bottleneck with increasing core counts. Designing asymmetric topologies, which assign more wiring resources to the center portion, may be an appropriate way to address the limitation [107].

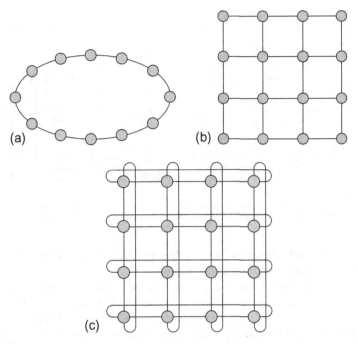

FIGURE 1.3

Three typical NoC topologies. (a) Ring. (b) 2D mesh. (c) 2D torus.

The node symmetry of a torus network helps to balance the network utilization. And the torus network supports a much higher bisection bandwidth than the mesh network. Thus, the 2D or higher-dimensional torus topology is widely used in the off-chip networks for several supercomputers, including the Cray T3E [129], Fujitsu K Computer [6], and IBM Blue Gene series [1, 16]. Also, the torus's wraparound links convert plentiful on-chip wires into bandwidth, and reduce the hop count and latency. Yet, the torus needs additional effort to avoid deadlock. Utilizing two virtual channels (VCs) for the dateline scheme [26] and utilizing bubble flow control (BFC) [18, 19, 100, 123] are two main deadlock avoidance methods for torus networks.

1.3.2 ROUTING ALGORITHM

Once the NoC topology has been determined, the routing algorithm calculates the traverse paths for packets from the source nodes to the destination nodes. This section leverages the 2D mesh as the platform to introduce routing algorithms.

There are several classification criteria for routing algorithms. As shown in Figure 1.4, according to the traverse path length, the routing algorithms can be classified as minimal routing or nonminimal routing. The paths of nonminimal routing may be outside the minimal quadrant defined by the source node and the destination node. The Valiant routing algorithm is a typical nonminimal routing procedure [139]. It firstly routes the packet to a random intermediate destination

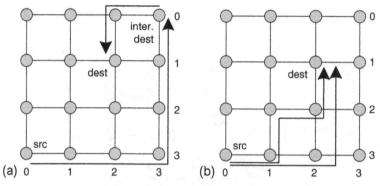

FIGURE 1.4

Nonminimal routing and minimal routing. (a) Nonminimal routing. (b) Minimal routing.

(the "inter. dest" node shown in Figure 1.4a), and then routes the packet from this intermediate destination to the final destination.

In contrast, the packet traverse path of minimal routing is always inside the minimal quadrant defined by the source node and the destination node. Figure 1.4b shows two minimal routing paths for one source and destination node pair. Since minimal routing's packet traverse hops are fewer than those of nonminimal routing, its power consumption is generally lower. Yet, nonminimal routing is more appropriate to achieve global load balance and fault tolerance.

The routing algorithms can be divided into deterministic routing and nonde-terministic routing on the basis of the path count provided. Deterministic routing offers only one path for each source and destination node pair. Dimensional order routing (DOR) is a typical deterministic routing algorithm. As shown in Figure 1.5a, the *XY* DOR offers only one path between nodes (3,0) and (1,2); the packet is firstly forwarded along the *X* direction to the destination's *X* position, and then it is forwarded along the *Y* direction to the destination node.

In contrast, nondeterministic routing may offer more than one path. According to whether it considers the network status, nondeterministic routing can be classified as oblivious routing or adaptive routing. Oblivious routing does not consider the network status when calculating the routing paths. As shown in Figure 1.5b, O1Turn routing is an oblivious routing algorithm [132]. It offers two paths for each source and destination pair: one is the *XY* DOR path, and the other one is the *YX* DOR path. The injected packet randomly chooses one of them. Adaptive routing considers the network status when calculating the routing paths. As shown in Figure 1.5c, if the algorithm detects there is congestion between nodes (3,1) and (3,2), adaptive routing tries to avoid the congestion by forwarding the packet to node (2,1) rather than node (3,2). Adaptive routing has some abilities to avoid network congestion; thus, it generally supports higher saturation throughput than deterministic routing.

One critical issue of the routing algorithm design is to achieve deadlock freedom. Deadlock freedom requires that there is no cyclic dependency among the network

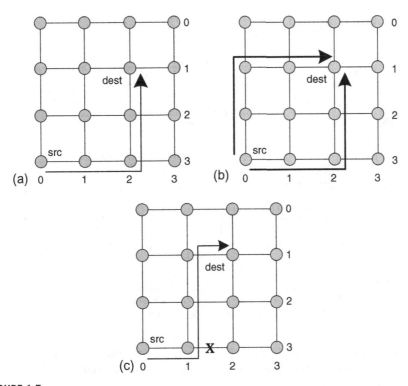

FIGURE 1.5

DOR, O1Turn routing, and adaptive routing. (a) *XY* DOR. (b) O1Turn routing. (c) Adaptive routing.

resources, typically the buffers and channels [24, 32]. Some proposals eliminate the cyclic resource dependency by forbidding certain turns for packet routing [20, 43, 46]. These routing algorithms are generally classified as partially adaptive routing since they do not allow the packets to utilize all the paths between the source and the destination. Other proposals [32, 33, 97, 99], named fully adaptive routing, allow the packet to utilize all minimal paths between the source and the destination. They achieve deadlock freedom by utilizing the VCs [24] to form acyclic dependency graphs for routing subfunctions, which act as the backup deadlock-free guarantee.

1.3.3 FLOW CONTROL

Flow control allocates the network resources, such as the channel bandwidth, buffer capacity, and control state, to traversing packets. Store-and-forward (SAF) [26], virtual cut-through (VCT) [72], and wormhole [27] flow control are three primary flow control types.

Only after receiving the entire packet does the SAF flow control apply the buffers and channels of the next hop. Figure 1.6a illustrates an SAF example, where a five-flit

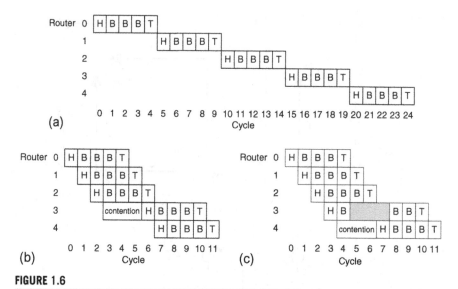

FIGURE 1.6

Examples of SAF, VCT, and wormhole flow control. H, head flit; B, body flit; T, tail flit. (a) An SAF example. (b) A VCT example. (c) A wormhole example.

packet is routed from router 0 to router 4. At each hop, the packet is forwarded only after receiving all five flits. Thus, SAF flow control induces serialization latency at each packet-forwarding hop.

To reduce serialization latency, VCT flow control applies the buffers and channels of the next hop after receiving the header flit of a packet, as shown in Figure 1.6b. There is no contention at router 1, router 2, and router 3; thus, these routers forward the packets immediately after receiving the header flit. VCT flow control allocates the buffers and channels at the packet granularity; it needs the downstream router to have enough buffers for the entire packet before forwarding the header flit. In cycle 3 in Figure 1.6b, there are only two buffer slots at router 3, which is not enough for the five-flit packet. Thus, the packet waits at router 3 for three more cycles to be forwarded. VCT flow control requires the input port to have enough buffers for an entire packet.

In contrast to the SAF and VCT mechanisms, the wormhole flow control allocates the buffers at the flit granularity. It can send out the packet even if there is only one buffer slot downstream. When there is no network congestion, wormhole flow control performs the same as VCT flow control, as illustrated for routers 0, 1, and 2 in Figure 1.6b and c. However, when congestion occurs, VCT and wormhole flow control perform differently. Wormhole flow control requires only one empty downstream slot. Thus, as shown in Figure 1.6c, router 2 can send out the head flit and the first body flit at cycle 3, even if the two free buffer slots at router 3 are not enough for the entire packet. Wormhole flow control allocates the channels at the packet granularity. Although the channel between routers 2 and 3 is free from cycle 5

Table 1.1 Summary of Flow Control Mechanisms

Flow Control	Channel	Buffer	Descriptions
SAF	Packet	Packet	The head flit must wait for the arrival of the entire packet before proceeding to the next link
VCT	Packet	Packet	The head flit can begin the next link traversal before the tail arrives at the current node
Wormhole	Packet	Flit	The channel is allocated on the packet granularity
VC	Flit	Flit	Can interleave flits of different packets on links

to cycle 7 in Figure 1.6c, this channel cannot be allocated to other packets until the tail flit of the last packet is sent out at cycle 10.

The VC techniques [24] can be combined with all three flow control types. A VC is an independent FIFO buffer queue. The VC can achieve many design purposes, including mitigating the head-of-line blocking and improving the physical channel utilization for wormhole flow control, avoiding the deadlock and supporting the quality of service (QoS). For example, to improve the physical channel utilization, multiple VCs can be configured to share the same physical channel to allow packets to bypass the current blocked packet. The router allocates the link bandwidth to different VCs and packets at the flit granularity. When the downstream VC for one packet does not have enough buffers, other packets can still use the physical channel through other VCs. Table 1.1 summarizes the properties of these flow control techniques [35].

1.3.4 ROUTER MICROARCHITECTURE

The router microarchitecture directly determines the router delay, area overhead, and power consumption. Figure 1.7 illustrates the structure of a canonical VC NoC router. The router has five ports, with one port for each direction of a 2D mesh network and another port for the local node. A canonical NoC router is mainly composed of the input units, routing computation (RC) logic, VC allocator, switch allocator, crossbar, and output units [26]. The features and functionality of these components are as follows:

- Input unit. The input unit consists of the input buffer and the related link control logic. The input buffer is built by SRAM cells or registers. The buffer slots are organized into several queues, and each queue is one VC. The queue has a circular or linked structure, with dynamically managed head and tail pointers.
- RC unit. If the flit currently at the VC head is a head flit, the RC logic calculates the output path for this packet. The calculation is performed on the basis of the

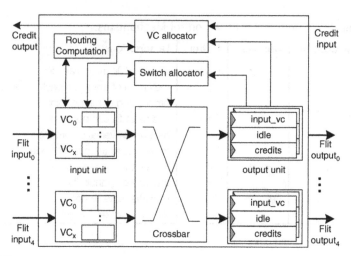

FIGURE 1.7

The structure of a typical virtual channel NoC router.

destination location carried by the head flit, and it produces the output port(s) and output VC(s) for the packet. Different routing algorithms can provide different numbers of ports and VCs. The RC results may be one output VC of a particular port, or multiple or all VCs of a particular port, or multiple VCs of multiple ports. The RC result affects the implementation complexity of the VC allocator and switch allocator.

- VC allocator. After the RC has finished, the head flit requests the output VCs. The VC allocator collects the requests from all input VCs, and then allocates the output VCs to the requesting input VCs. It guarantees that one output VC is allocated at most to one input VC, and each input VC is granted at most one output VC. The requesting input VCs without getting the grants will request the VC allocation (VA) again in the next cycle. The RC and VA are performed at the packet granularity. Only the head flit conducts these two stages.
- Switch allocator. When the input VC has been successfully allocated an output VC, the router will check whether the granted downstream VC has buffers available. If it does, the flit requests the switch traversal (ST) from the switch allocator. Similarly to the VC allocator, the switch allocator guarantees each switch output port is allocated to at most one flit. The switch allocator also generates control signals to connect the input port and the output port of the crossbar. Both the VC allocator and the switch allocator consist of several arbiters. To reduce the area and power consumption, these arbiters are generally simple round-robin arbiters [26, 35] or matrix arbiters [26, 35].
- Crossbar. The crossbar is generally implemented with multiple multiplexers. The control signals for these multiplexers are generated by the switch allocator.

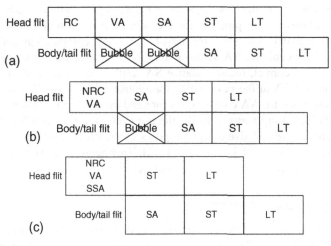

FIGURE 1.8

The baseline router pipeline and its optimizations. NRC, next-hop routing computation; SSA, speculative SA. (a) The baseline four-stage pipeline. (b) The three-stage pipeline. (c) The two-stage pipeline.

The area and power consumption of the crossbar can be roughly modeled as $O((pw)^2)$, where p is the port count and w is the port width. Once a flit has passed the ST stage, the crossbar generates a credit to inform the upstream router of the release of a buffer slot.

- Output unit. The output unit tracks the status of downstream VCs with several registers. The "input_vc" register records the input VC that a downstream VC is allocated to. When the value of the 1-bit "idle" register is "TRUE," the downstream VC receives the tail flit of the last allocated packet, and it is now available for reallocation to other packets. The "credits" register records the credit amount.

The canonical NoC router has a four-stage pipeline, as shown in Figure 1.8a. When the head flit arrives, it firstly performs the RC, and then applies the VA. After the head flit has been granted an output VC, it performs the switch allocation (SA). Finally, the flit passes the ST stage to move out of the router. The flit needs an additional link traversal (LT) stage before arriving at the next router. The body and tail flits do not pass through the RC and VA stages, and they inherit the output port and VC from the head flit.

Several optimization techniques are proposed to reduce router latency. One example is the lookahead routing computation [44] shown in Figure 1.8b. Lookahead routing calculates the output port for the next router at the current router. This pipeline optimization technique can be combined with both deterministic routing and adaptive routing [78]. The head flit undergoes the next-hop routing computation in parallel with the current-hop VA, thus reducing the number of pipeline stages by one.

The speculative design is another pipeline optimization method [119]. Figure 1.8c illustrates the pipeline for the speculative SA router. The head flit undergoes speculative SA when the next-hop routing computation and VA are performed. The speculative SA assumes the head flit can get the VA grant. If this assumption is not correct, the speculative SA result is nullified, and the head flit undergoes normal SA at the next cycle. Speculative SA is very efficient with low network loads, since the VA contention is rare and most VA result assumptions are correct. With increasing network loads, the accuracy of speculative SA decreases. A more aggressive speculative design proposes speculatively performing the ST; it performs the VA, SA, and ST in parallel [112]. This aggressive speculative design achieves performance improvement only with very low network loads.

1.3.5 PERFORMANCE METRIC

The network capacity and latency are two common performance metrics used for NoC evaluation [26, 35]. The network capacity is defined as the injection bandwidth offered to each of the N terminal nodes for uniform random traffic [9, 26]. It can be calculated from the bisection bandwidth (B_B) or the bisection channel count (B_C) and the channel bandwidth (b) as follows:

$$\text{Capacity} = \frac{2B_B}{N} = \frac{2bB_C}{N}. \tag{1.1}$$

Since global wiring requirements depend on the channel bisection, Equation (1.1) expresses a tradeoff between NoC performance and the implementation overhead and complexity. Typically, designers desire an NoC with high network capacity. The modern CMOS technology supports a high bisection bandwidth or a large bisection channel count.

The average contention-free latency T_0 from source s to destination d depends on several parameters, including the average hop count (H) from s to d and the router traversal latency (t_r), which determine the routing latency; the channel traversal latency (T_c); and the packet length (L) and channel bandwidth (b), which determine the serialization latency $\left(T_s = \frac{L}{b}\right)$ [9]. The contention-free latency T_0 is calculated as

$$T_0(s, d) = H(s, d)t_r + T_c(s, d) + \frac{L}{b}. \tag{1.2}$$

As the network load increases, greater resource contention contributes to longer router traversal latencies [26, 35]. The network latency versus injection bandwidth relationship generally exhibits a trend similar to the exponential curve, as shown in Figure 1.9. The zero-load latency is the average contention-free latency shown in Equation (1.2). The saturation throughput is defined as the injection bandwidth when the network is saturated. As shown in Figure 1.9, an obvious phenomenon for the saturated network is the sharply increased average network latency. Thus, the saturation throughput is generally measured as the injection bandwidth at which the average latency is three times the zero-load latency [95–97, 100].

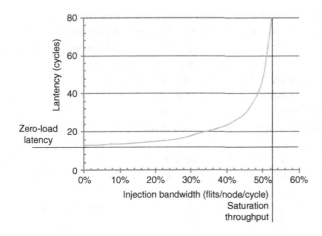

FIGURE 1.9

The latency versus injection bandwidth curve for an NoC.

1.4 REVIEW OF NoC RESEARCH

In this section, we review the current research progress on several aspects of NoC design, including the topology, unicast routing, supporting collective communications, flow control, and router microarchitecture.

1.4.1 RESEARCH ON TOPOLOGIES

Reducing the network latency is the first important research direction for topology design. Balfour and Dally [9] evaluated several topologies for on-chip communications. They point out that concentrating multiple cores into one router is an efficient way to reduce the network latency. Adding physical express channels to connect nonneighboring routers can further reduce the network latency. The concentration and physical express channels add ports to the router, improving the implementation complexity and may reduce the frequency [86]. Kim *et al.* [77] leveraged the off-chip flattened butterfly topology into the design of on-chip networks (OCNs). The flattened butterfly provides full connection along each dimension, and it efficiently utilizes the abundant on-chip wiring resources to reduce the network latency. The long channels in the flattened butterfly add complexities to the flow control scheme and the wiring procedure.

Improving the network saturation throughput is the second important research aspect for topology design. Mishra *et al.* [107] observed that the center portions of mesh networks are easily congested, which limits the saturation throughput. Thus, they propose a heterogeneous mesh network, which assigns higher bandwidth to center routers. This heterogeneous network needs to combine several small packets on narrow channels into big packets for wide channels. And it also needs to divide one large packet into several small packets for traversing narrow channels. Manevich *et al.* [101] combined the bus structure and the mesh network to improve the saturation throughput; the bus structure is utilized for local communication and the mesh network is leveraged for global communication.

Reducing the area and power overhead is the third important research field for topology design. Cho *et al.* [21] suggested adaptively adjusting the direction for each channel to improve the network throughput. Hesse *et al.* [55] presented a fine-grained bandwidth adaptivity mechanism based on bidirectional channels. Compared with a baseline design, their design achieves similar performance with half of the channel resources. By combining ring and mesh networks, Zafar *et al.* [148] proposed the cubic ring topology, which supports dynamically shutting off 30% of routers to reduce power consumption. Chen and Pinkston [17] added a ring into the baseline mesh network. The ring acts as a backup connected network when mesh routers are powered off. Their design allows all mesh routers to be shut off.

Our group conducts research on topology design. Huang *et al.* [64] proposed a virtual bus network which dynamically reconfigures links between routers to form the bus structure to reduce network latency. On the basis of the virtual bus structure, Huang *et al.* [63] further designed a hierarchical cache coherence protocol; the virtual bus delivers the local communication generated by a snoopy protocol, and the mesh network delivers the global communication generated by a directory protocol.

1.4.2 RESEARCH ON UNICAST ROUTING

Guaranteeing deadlock freedom and achieving high performance are two primary objectives for unicast routing designs. Designing the routing algorithm based on some deadlock avoidance theories is the general way to guarantee deadlock freedom. Dally and Seitz [24] proposed the seminal deadlock avoidance theory, which declares that an acyclic channel dependency graph is the sufficient condition for deadlock freedom. This theory can be used to design deterministic or partially adaptive routing algorithms.

Duato [32–34] introduced the concept of the routing subfunction, and indicated that a routing subfunction with an acyclic channel dependency graph is the sufficient condition for deadlock-free fully adaptive routing. Duato's theory was a great breakthrough; it allows the cyclic channel dependency for high routing flexibility. This theory has been widely used for about 20 years in several industrial products, including the Cray T3E [129] and IBM Blue Gene series [1, 16]. Duato's theory requires each VC to hold at most one packet for wormhole fully adaptive routing. Since most NoC packets are short, this requirement strongly limits the performance.

Our group conducts research to address the limitation of Duato's theory. Ma *et al.* [97] proposed the whole packet forwarding (WPF) flow control scheme, which extends Duato's theory to allow multiple packets to be stored in one VC. Ma *et al.* proved the deadlock-free property of the WPF scheme, which was also verified by Verbeek and Schmaltz [142] with the decision procedure method. Ma *et al.* [99] further proved that the escape VCs in Duato's theory can apply aggressive VC reallocation without inducing deadlock, which can further improve the performance.

Achieving high performance requires the routing algorithm to provide high path diversity and to avoid network congestion. The fully adaptive routing allows the packet to use all minimal paths between the source and destination nodes; thus, it supports higher path diversity than partially adaptive and deterministic routing [98]. Kim et al. [78] proposed a fully adaptive routing algorithm which utilizes lookahead routing and the port preselection scheme for low latency. Lookahead routing calculates at most two available output ports one hop ahead, and then the preselection scheme selects the appropriate port one cycle before the arrival of the packet. Hu and Marculescu [61] proposed the DyAD design, which adaptively applies deterministic or adaptive routing according to the network loads.

Avoiding network congestion requires routing algorithms to be efficiently aware of network status. The DyXY design proposed by Li et al. [89] uses dedicated wires to convey neighboring router status, and then forwards packets to either the X or the Y dimension according to the congestion status. The neighbors-on-path (NoP) design proposed by Ascia et al. [8] selected the output port on the basis of the status of routers next to neighboring routers. The NoP design leverages a sideband network to deliver congestion status for two-hop-away routers. Gratz et al. [47] proposed the regional congestion awareness (RCA) design to select the output port on the basis of the status of all routers along the same column or row. Similarly to the NoP design, a dedicated congestion information network is leveraged. When multiple applications run concurrently on a many-core platform, RCA considers the status of nodes belonging to other applications, which couples the behaviors of different applications, and negatively affects the application performance [95].

Our group conducts research to address RCA's limitations. Ma et al. [95] proposed destination-based adaptive routing (DBAR) to integrate destination locations into port selection procedures; DBAR provides dynamic isolation for concurrent applications. Ma et al. [98] further extended the DBAR design to other topologies, such as the concentrated mesh.

1.4.3 RESEARCH ON SUPPORTING COLLECTIVE COMMUNICATIONS

Collective communications, including multicast and reduction communications, easily become the system bottleneck. There has been much research on supporting multicast communication. Lu et al. [93] proposed a path-based multicast routing design. The design first sends out the path establishment request; after receiving acknowledgment messages (ACKs) from destinations, it then sends out multicast packets. The path-based multicast routing induces path establishment and acknowledgment delays, reducing the performance.

Tree-based multicast routing eliminates path establishment and acknowledgment delays. The virtual circuit tree multicasting proposed by Enright Jerger et al. [37] introduces the concept of a virtual multicast tree. It sends out multicast packets after building up a virtual multicast tree. The building up procedure results in some overheads. To mitigate these overheads, the recursive partitioning multicast (RPM) proposed by Wang *et al.* [144] builds up the tree during transmission. To avoid deadlock, RPM leverages two dedicated virtual networks for upward and downward multicast packets. These two virtual networks cause unbalanced buffer utilization and reduce saturation throughput. The bLBDR design proposed by Rodrigo *et al.* [126] uses broadcasting in a small region for multicast communications. The broadcasting causes redundant communications. The Whirl multicast proposed by Krishna *et al.* [82] can efficiently support broadcasting and dense multicasting with optimized one-cycle-delay routers.

Our group researches supporting multicast communication. The virtual bus structure proposed by Huang *et al.* can support multicast/broadcast operations in the message passing interface (MPI) [62] and cache-coherent communications [63]. To address the unbalanced buffer utilization in the RPM design, Ma *et al.* [96] proposed balanced, adaptive multicast (BAM) routing. On the basis of the observation that multicast packets in cache-coherent NoCs are single-flit packets, BAM extends Duato's theory [32] for unicast routing to the design of deadlock-free multicast routing. It achieves balanced buffer utilization between different dimensions, and supports efficient bandwidth utilization through a heuristic port selection strategy.

Currently, there is little research on the support of reduction communication in NoCs. Bolotin *et al.* [13] noticed that combining ACKs of the cache line invalidation message may improve the performance, but they do not offer a detailed design or evaluation. Krishna *et al.* [82] proposed combining the ACKs at the NoCs to reduce the network load. The design keeps the earlier-arriving ACKs at the input VCs, and then combines them with the later-arriving ACKs. The earlier-arriving ACKs occupy the input VCs, which limits the performance. To mitigate this effect, Krishna *et al.* employed a dedicated virtual network with 12 VCs to deliver ACKs. Our group researches the support of reduction communication. The message combination framework proposed by Ma *et al.* [96] adds a message combination framework table to each router to support ACK combination. This design does not keep earlier-arriving ACKs inside input VCs, thus reducing the number of VCs required.

1.4.4 RESEARCH ON FLOW CONTROL

Bufferless flow control is the first research aspect for flow control designs. The BLESS design proposed by Moscibroda *et al.* [109] performs RC at the flit granularity. When multiple flits contend for one port, only one flit is granted the port, and the other flits are misrouted. In contrast, the SCARAB design proposed by Hayenga *et al.* [51] drops the nongranted flits, and sends a negative ACK back to the source node to inform the retransmission. Jafri *et al.* [68] presented an adaptive flow control combining buffered and bufferless schemes that is based on a normal buffered router. The low-load routers turn off buffers to save power and apply bufferless flow control, while high-load routers apply buffered flow control. Fallin *et al.* [39] optimized the BLESS design in three aspects. First, they used a permutation network to implement allocators, reducing critical path delays. Second, they introduced a golden packet to avoid the livelock. The golden packet is the packet with the highest priority, and it cannot be misrouted. Third, they leveraged miss status holding registers as reordering buffers, thus reducing the required buffer amount. On the basis of this work, Fallin *et al.* [40] further proposed the MinBD design, which adds a side buffer to baseline bufferless routers, thus improving the network saturation throughput.

Reducing the packet transmission latency is the second research aspect for flow control designs. The flit reservation flow control proposed by Peh and Dally [118] eliminates the buffer usage limitation induced by credit round-trip delays. The express VC proposed by Kumar *et al.* [85] reduces latencies by preallocating buffers of multiple routers. On the basis of this work, Kumar *et al.* [84] further proposed token flow control, which broadcasts the current buffer occupation status to nearby routers. Then, the packet forwarding can bypass the pipeline of several routers. The layered switching design proposed by Lu *et al.* [92] divides the long packet into several flit groups. It maintains SA results for an entire flit group to reduce contention and latencies. Enright Jerger *et al.* [36] proposed hybrid circuit switching to combine the circuit switching and packet switching mechanisms. The wormhole cut-through design presented by Samman *et al.* [127] allows several packets to be mixed in one VC at the flit granularity.

Avoiding deadlock for ring/torus networks is the third research aspect of flow control designs. Bubble flow control (BFC) avoids deadlock by prohibiting occupation of the last free packet-size buffers [15, 122, 123]. On the basis of the BFC scheme, Chen *et al.* [19] proposed the critical bubble scheme (CBS); this scheme achieves deadlock freedom by marking one bubble inside the ring as the critical one, and this critical bubble can be occupied only by packets traveling inside the ring. On the basis of the CBS design, Chen and Pinkston [18] further proposed worm bubble flow control (WBFC), which extends the BFC idea to wormhole networks. It assigns to the buffers three different colors to convey the global buffer utilization status into local information and to avoid starvation. WBFC allows the ring to leave only one free flit-size buffer to avoid deadlock for wormhole networks. The prevention flow control proposed by Joshi *et al.* [69] combines the priority arbitration and prevention slot cycling scheme to avoid deadlock. The priority arbitration gives higher priority to already injected packets, and the prevention slot cycling scheme keeps one free packet-size buffer inside the ring. Luo and Xiang [94] proposed a fully adaptive routing with only two VCs for torus networks. Their design allocates VCs according to whether the packet needs to cross wraparound links in the future.

Our group researches flow control designs. On the basis of the observation of the inefficiency of BFC, CBS, and prevention flow control in handling variable-sized packets, Ma *et al.* [100] proposed flit bubble flow control (FBFC). The kernel idea of FBFC is similar to that of WBFC. By maintaining one free flit-size buffer inside the ring, FBFC achieves deadlock freedom for wormhole ring/torus networks. FBFC implementations can be regarded as a hybrid flow control; the packet injection is similar to VCT flow control, and the packet movement follows wormhole flow control.

1.4.5 RESEARCH ON ROUTER MICROARCHITECTURE

Reducing the number of pipeline stages is the first aspect of router design. Peh and Dally [119] proposed speculative SA to reduce the number of pipeline stages by one; their design speculatively performs SA in parallel with VA. Mullins *et al.* [112] further used the speculative ST to reduce the number of stages by one more. Their design allows the packet to go straight into the ST under low loads. Matsutani *et al.* [104] evaluated the performance of several different prediction algorithms. The speculative SA and ST need an additional switch allocator, increasing the router complexity. On the basis of priority allocation, Kim [76] proposed a low-cost router for DOR. It simplifies the router into two multiplexers and a few buffers. To support low-latency transmission, the router assigns higher priorities to straight-traveling packets than to turning ones. Recently, Kim *et al.* [75] extended this low-cost router to hierarchical ring networks. Hayenga and Lipasti [52] observed that most NoC packets are short packets, and they creatively leveraged the property of XOR logic into NoC router designs. Their proposed NoX router eliminates one pipeline stage, while it does not induce additional switch allocators. The NoX router needs additional effort to handle long packets. Our group researches low-latency routers. By adding directly connecting channels between input and output ports, Lai *et al.* [87] implemented a low-cost single-cycle router.

Reducing the buffer overhead is the second aspect for router design. The dynamically allocated multiqueue (DAMQ) buffer [136] is an efficient way to reduce buffer amount requirements. The ViChaR design proposed by Nicopoulos *et al.* [113] introduced the DAMQ buffer into the NoC buffer design, and it achieves obvious area and power improvements. Xu *et al.* [147] noticed the ViChaR structure needs too many VCs, which limits the frequencies. Thus, they proposed a DAMQ structure with limited and constant VCs. Ramanujam *et al.* [124] implemented a distributed buffer to achieve the performance of the output buffer, while avoiding the overhead of the output buffer. Their design configures both input buffers and DAMQ-style center buffers in NoC routers. Ahmadinia *et al.* [4] proposed a DAMQ buffer to share buffers among different physical ports. Becker *et al.* [10] observed that previous DAMQ buffer designs cannot support QoS efficiently; thus, they proposed a DAMQ buffer design which assigns a minimum number of buffers to each VC to provide QoS support. The elastic buffer proposed by Michelogiannakis *et al.* [106] utilizes flip-flops inside long channels and routers as buffers, thus removing router input buffers. Kodi *et al.* [80] used repeaters inside channels as buffers, which reduces the input buffer amount. Kim *et al.* [74] observed that most buffers are idle during the application running period; thus, they applied fine-grained power-gating techniques to control the status of each buffer slot to reduce power consumption.

Our group researches buffer design. Lai *et al.* [88] proposed a DAMQ buffer design which can efficiently avoid network congestion. Their research supports sharing buffers among different VCs of the same physical port. Shi *et al.* [133] further proposed a DAMQ buffer to support sharing buffers among different physical ports.

1.5 TRENDS OF REAL PROCESSORS

In this section, we discuss the NoC design trends for several commercial or prototype processors, including the MIT Raw, Tilera TILE64, Sony/Toshiba/IBM Cell, U.T. Austin TRIPS, Intel Teraflops, Intel SCC, Intel Larrabee, and Intel Knights Corner processors.

1.5.1 THE MIT RAW PROCESSOR

MIT has developed the Raw processor since the late 1990s to address several emerging design challenges, including internal wire lengths being increasingly limited by the scaling frequency, market constraints for quickly verifying new designs, and applications emphasizing stream-based multimedia computations [137]. The kernel design idea of the Raw processor is to keep the hardware simple and leave most performance-scaling jobs to the software. The Raw processor has 16 tiles, and each tile can execute instructions independently. There is an eight-stage MIPS-like processing core, a 32 kB data cache, and a 96 kB instruction cache on each tile. It also has static and dynamic routers for static and dynamic communications respectively. The Raw processor is manufactured with IBM's SA-27E, 0.15-μm, six-level, copper, application-specific integrated circuit (ASIC) process. Its design target is a 225 MHz worst-case frequency.

As illustrated in Figure 1.10, the 16 tiles are organized in a 4 × 4 mesh topology. The Raw processor has two static networks and two dynamic networks. The static and dynamic networks both have a 32-bit physical channel width. Both static and dynamic networks are register mapped directly into the bypass path of the processing core; the core loads or stores registers R24 and R25 to access two static networks. Dynamic networks are mapped to registers R26 and R27. Static network configurations are scheduled at compile time. On the basis of the compiled instructions, static routers can collectively reconfigure the entire communication pattern of the static network on a cycle-by-cycle basis. Since the static router knows the packet destination long before the packet arrives, it performs preconfigurations to allow packets that have arrived to be delivered immediately. The per-hop latency of static networks is one cycle. This low-latency property is critical for efficiently bypassing scalar operands to exploit instruction-level parallelism.

In contrast, the packets in dynamic networks are generated and routed according to computational results during the run time. Thus, dynamic routers have more latency than static routers as they have to wait for the arrival of the header flit to initiate router configurations. The dynamic network applies DOR. Its per-hop latency is one cycle for straight-traveling packets and two cycles for turning packets. The dynamic network applies credit-based wormhole flow control. Although DOR eliminates the network-level deadlock, dynamic networks still need to consider the protocol-level deadlock. The two dynamic networks apply different schemes to handle the protocol-level deadlock. The memory network applies the deadlock avoidance scheme on the

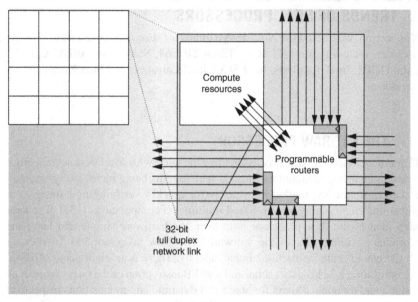

FIGURE 1.10

The NoC in the Raw processor. The Raw processor comprises 16 tiles, and each tile has computational resources and four networks. From Ref. [137], copyright 2002 IEEE.

basis of resource usage limitations. The general network applies the deadlock recovery scheme, which uses the memory network as backup deadlock-free delivery paths.

1.5.2 THE TILERA TILE64 PROCESSOR

Tilera's TILE64 processor is a commercial product inspired by the MIT Raw processor [145]. The TILE64 processor is a general-purpose many-core processor, which mainly focuses on multimedia, networking, and wireless applications. It has a multiple instruction, multiple data (MIMD) architecture with 64 cores arranged in an 8 × 8 tiled mesh topology. Each tile has a three-way very long instruction word core with an independent program counter, a two-level cache hierarchy with 768 kB capacity, a 2D direct memory access (DMA) subsystem to exchange data among cores and memory controllers, and a switch to interconnect neighboring tiles. Each tile is a fully featured computing system which can independently run the Linux operating system. Also, multiple tiles can be combined to run a multiprocessor operating system, such as SMP Linux. To optimize the power consumption, the TILE64 processor employs extensive clock gating and applies power gating on the tile granularity.

The TILE64 processor is implemented in 90 nm technology, with a frequency up to 1 GHz. It can perform 192 billion 32-bit operations per second. The TILE64 processor supports subword arithmetic and can achieve 256 billion 16-bit operations

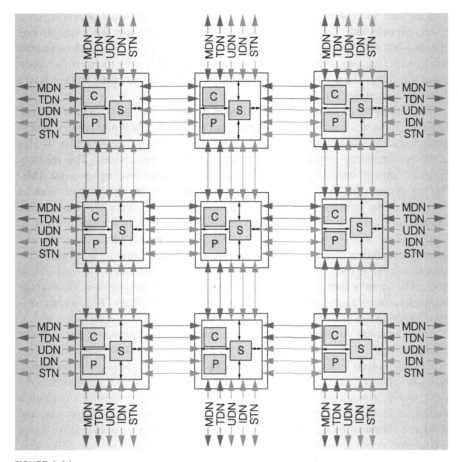

FIGURE 1.11

The NoC of the TILE64 processor for a 3 × 3 array. MDN, memory dynamic network; TDN, tile dynamic network; UDN, user dynamic network; IDN, input/output dynamic network; STN, static network. From Ref. [145], copyright 2007 IEEE.

per second, or half a teraops for 8-bit operations. To facilitate programming, the TILE64 processor provides a communication library, iLib, based on the C programming language. The iLib library supports two communication types, the socket-like streaming channel and the MPI-like communication API. The socket channel supports higher performance than the MPI, while its flexibility is poorer than that of the MPI.

The TILE64 processor employs five physical networks, including one static network and four dynamic networks, as shown in Figure 1.11. Similarly to the Raw processor, the connections of the static and dynamic networks are configured at compile time and run time respectively. Each of the five physical networks has a 32-bit channel width. The static network applies circuit switching flow control, with a per-hop latency of one cycle. The dynamic network leverages DOR with wormhole flow control. Each input port has the minimal three buffer slots to cover the credit round-trip delay. The per-hop latency for straight packets is one cycle, while it is two cycles for turning packets.

The four dynamic networks serve different purposes. The user dynamic network delivers communications for userland processes or threads. The input/output dynamic network transfers traffic between input/output devices and processing tiles. It also serves for operating-system and hypervisor-level communications. The memory dynamic network exchanges data between on-tile caches and off-chip DRAMs. The tile dynamic network works in concert with the memory dynamic network as a portion of the memory system. The request packets of direct tile-to-tile cache transfers transit the tile dynamic network, and responses transit the memory dynamic network. This avoids protocol-level deadlock for the tile dynamic network. The memory dynamic network uses the buffer preallocation scheme to avoid protocol-level deadlock. It reserves certain buffers at memory controllers for each cache block. The input/output dynamic network and the user dynamic network leverage deadlock recovery mechanisms to handle protocol-level deadlock. They use the off-chip DRAM as the backup memory space in case of deadlock.

To support the virtualization and protection among concurrent applications or supervisors, the TILE64 processor provides Multicore Hardwall technology. This technology protects links on the user-accessible user dynamic network, input/output dynamic network, and static network, while the memory dynamic network and the tile dynamic network are protected by memory protection mechanisms through a translation lookaside buffer. If a packet attempts to cross a hardwalled link, the Multicore Hardwall blocks the packet and triggers an interrupt to system software to take appropriate actions.

1.5.3 **THE SONY/TOSHIBA/IBM CELL PROCESSOR**

The Sony/Toshiba/IBM Cell Broadband Engine is developed primarily for multimedia applications and scientific computations. It has a heterogeneous architecture with 12 core elements connected through a multiring bus called an element interconnect bus (EIB) [5, 71, 79], as shown in Figure 1.12. The elements include one controlling core called a power processing element (PPE), eight computation cores called synergistic processing elements (SPEs), one memory interface controller (MIC), and two bus controllers (IOIF0, IOIF1).

The dual-threaded PPE is a complete 64-bit IBM Power Architecture processor featuring a 32 kB first-level (L1) cache and a 512 kB second-level (L2) cache, and it can run the operating system. The SPEs are 128-bit single instruction, multiple data (SIMD) processing units optimized for data-rich workloads allocated to them by the PPE. Each SPE contains 256 kB of local store memory. The PPE supports both DMA and memory-mapped input/output, while DMA is the only method for moving

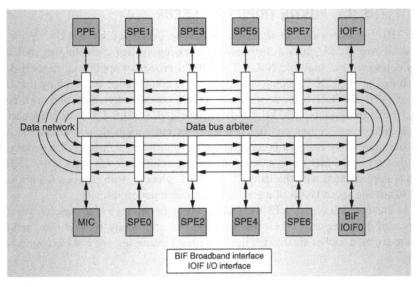

FIGURE 1.12

The EIB of the Cell processor. From Ref. [79], copyright 2006 IEEE.

data between the SPE's local store and the system memory. The Cell processor is implemented in 90 nm silicon-on-insulator technology. The SPE and the PPE run at a frequency of 3.2 GHz, and the EIB operates at 1.6 GHz.

The EIB supports communications among 12 elements. All elements contain bus interface units to access the EIB, and each element is capable of an aggregate throughput of 51.2 GB/s. Thus, the data rate requirement of the EIB is quite high. Also, the EIB needs to efficiently handle coherent data transfers whose packet size may vary. The EIB architecture is optimized to satisfy the above requirements.

The EIB has independent networks for commands (requests for data from other sources) and for the data being moved. It consists of a shared command bus, four 128-bit data rings (two in each direction), and a central data arbiter. The command bus distributes commands, sets end-to-end transactions, and handles memory coherence. It consists of five distributed address concentrators arranged in a treelike structure that lets multiple commands be outstanding simultaneously across the network. The address concentrators handle collision detection and prevention, and they provide fair access to the command bus by a round-robin mechanism. All address concentrators forward commands to a single serial command reflection point.

The data transfer is elaborate. Each of the four data rings can handle up to three concurrent nonoverlapping transfers, allowing the EIB to support 12 concurrent data transfers. The data ring applies slotted ring flow control and shortest path routing. The central data arbiter coordinates the accesses for data rings on a per transaction basis. It implements round-robin arbitration with two priority levels. The MIC has the highest priority for the data arbiter, while all other elements have a lower priority.

1.5.4 **THE U.T. AUSTIN TRIPS PROCESSOR**

The TRIPS processor is an academia prototype developed by the University of Texas at Austin [48]. It is a distributed processing system consisting of multiple tiles connected via multiple NoCs. The TRIPS processor applies an explicit data graph execution (EDGE) architecture, where the computation is driven by the data arrival [49]. The TRIPS processor contains two out-of-order cores, and each core has 16 execution units, a 32 kB L1 data cache and a 64 kB L1 instruction cache. Since each execution unit can independently execute instructions, the TRIPS core can simultaneously issue 16 instructions in the out-of-order fashion. The processor also integrates a 1 MB static nonuniform cache architecture L2 cache. The TRIPS chip is implemented with the IBM CU-11 ASIC process, with a drawn feature size of 130 nm and seven layers of metal. Its work frequency is 366 MHz.

As illustrated in Figure 1.13, the TRIPS processor applies a tiled structure; the L2 cache and the processing cores, including the execution units, register files, and L1 data/instruction caches are divided into 106 tiles. Leveraging one NoC to connect all

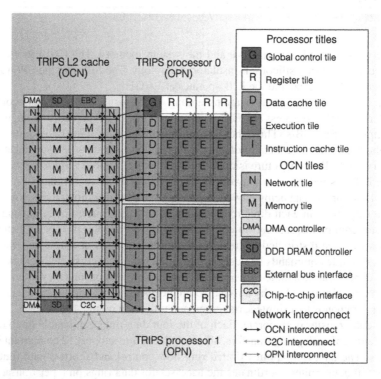

FIGURE 1.13

Block diagram of the TRIPS chip. From Ref. [48], copyright 2007 IEEE.

106 tiles is a possible design choice. Yet, to achieve high performance and resource efficiency, the TRIPS processor leverages two customized types of NoCs to connect core elements and the L2 memory system respectively. The operand network (OPN) connects 25 core element tiles with a 5×5 mesh topology, and it replaces the traditional operand bypass bus and L1 cache bus. The OPN-connected nodes include 16 execution tiles, four register file tiles, four L1 data cache tiles, and one global control tile. The remaining five L1 instruction tiles are connected with special control networks and managed by the global control tile to implement the block execution of the EDGE architecture [49].

The block execution model preallocates reservation stations for operands. Thus, the OPN does not need to handle protocol-level deadlock. It uses only one VC with four buffer slots. The OPN has a 142-bit physical channel width, including 4 bits for flow control, 29 bits for control physical digits, and 109 bits for data payloads. This physical channel width is chosen to make all operand packets delivered in the OPN be single-flit packets. The OPN utilizes *YX* DOR, with wormhole flow control based on the on/off mechanism. There are two 5×5 crossbars inside each OPN router: one crossbar is for data payloads and the other one is for the control physical digits. The control physical digits, including the destination node and instructions, are sent one cycle before the data payload to preconfigure the data payload crossbar. The per-hop latency of the OPN is one cycle.

The on-chip network (OCN) connects the 40 on-chip memory system tiles with a 4×10 mesh topology, and it replaces the traditional memory bus. The 40 tiles include 16 memory tiles and 24 network tiles. In addition to the four-way, 64 kB memory bank, each memory tile also includes an OCN router and a single-entry miss status holding register. The memory bank can be configured as an L2 cache bank or as a scratch-pad memory, with configuration commands from cores. The network tiles surrounding the memory tiles act as translation agents for determining where to route memory requests. Each network tile contains a programmable routing table to determine the request destinations.

The OCN leverages four virtual networks to avoid protocol-level deadlock, and each virtual network has one VC with two buffer slots. The OCN has a 138-bit physical channel width, with 128-bit payloads and 10-bit flow control signals. This channel width is chosen as it is the closest even divisor of a cache line size that would allow the small memory request packets to fit in one flit. The OPN packet lengths range between one flit and five flits. The OCN applies the *YX* DOR with credit-based wormhole flow control. It has a one-cycle per-hop latency. The OCN router has a 6×6 crossbar. In addition to the regular five ports to connect neighboring routers and a local tile, the other port is used for configuring the local memory address translation table.

1.5.5 THE INTEL TERAFLOPS PROCESSOR

The Teraflops processor chip is the first-generation silicon prototype of the terascale computing research initiative at Intel [140, 141]. It consists of 80 homogeneous tiles arranged in an 8×10 mesh topology operating at 5 GHz and 1.2 V. Each tile has a simple processing engine, with two independent, fully pipelined single-precision floating point multiply accumulators. A tile also has a 3 kB single-cycle instruction memory, a 2 kB data memory, and a five-port router. The Teraflops processor has an eight-way very long instruction word instruction set architecture. The design target for the power consumption is under 100 W when achieving a performance of more than 1 Tflop. It applies fine-grained clock gating and power gating techniques to optimize the power consumption. Each tile is partitioned into 21 sleep regions, and each of them can be dynamically controlled by special sleep/wake up instructions. The router is divided into 10 sleep regions with control of each individual port. The Teraflops chip is implemented in 65 nm technology. Its total area is $275 \, mm^2$, and each tile occupies $3 \, mm^2$.

The 8×10 mesh network has one 39-bit physical channel, with 32 bits for payloads, 6 bits for flow control, and 1 bit for the mesochronous communication [57]. The mesochronous communication allows neighboring routers to communicate with different clock edges, which mitigates the difficulty to design a global clock distribution network. The network has two logic lanes, and each logic lane consists of one VC with 16 buffer slots. The two logic lanes are used for the instruction and data, respectively, to avoid short instruction packets being blocked by long data packets.

The per-hop latency of the mesh network is five cycles, including the buffer write, buffer read, RC, arbitration, and ST pipeline stages, as shown in Figure 1.14. The network applies DOR with on/off-based wormhole flow control. The Teraflops processor uses source routing; the entire routing path is encoded in the head flit. Since a head flit can at most encode output ports for 10 hops, the possible remaining output ports are encoded by a chained head flit. The router arbitration utilizes a two-stage arbitration structure. The first stage performs the physical port arbitration at the packet granularity, and the second stage conducts the logic lane arbitration at the flit granularity. To reduce the area overhead, the Teraflops processor uses double-pumped crossbars. This type of crossbar leverages one crossbar wire to deliver the signals from two input wires at the high and low periods of a cycle respectively. Thus, it eliminates half of the wiring overhead for the crossbar. The entire router is implemented in $0.34 \, mm^2$.

1.5.6 THE INTEL SCC PROCESSOR

The Intel SCC processor is a prototype processor with 48 IA-32 architecture cores connected by a 6×4 mesh network [59]. As shown in Figure 1.15, every two cores are integrated into one tile, and they are concentrated to one router. A total of four memory controllers, two on each side, are attached to two opposite sides of the mesh. The IA-32 cores are the second-generation in-order Pentium P54C cores, except that the L1 ICache/DCache capacity is upgraded from 8 to 16 kB. The L1 ICache and DCache both support four-way set associativity. They can be configured in write-through or write-back modes. Each core has a 256 kB, four-way set associative, write-back L2 cache. The L1 and L2 cache line sizes are both 32 bytes. The L2 cache is

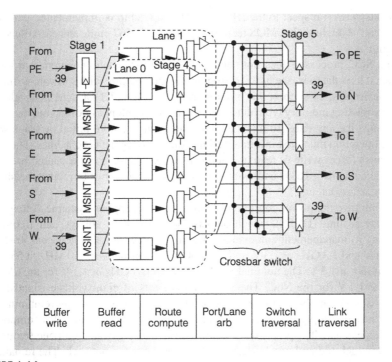

FIGURE 1.14

The router architecture and pipeline stages of the Teraflops NoC. From Ref. [57], copyright 2007 IEEE.

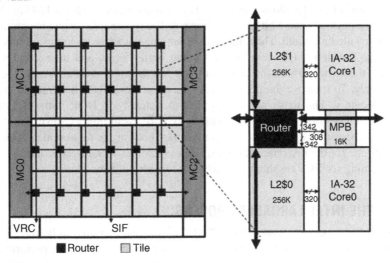

FIGURE 1.15

Block diagram and tile architecture of the SCC processor. From Ref. [59]. copyright 2011 IEEE.

not inclusive with respect to the L1 cache. The SCC chip is implemented in 45 nm, nine-metal-layer, high-K CMOS technology. Its area is 567 mm^2, and each tile's area is 18 mm^2.

The SCC processor does not support the hardware cache coherence protocol. Instead, it uses the message passing programming paradigm, with software-managed data consistency. To support this programming paradigm, L1 DCache lines add one message passing memory type bit to identify the line content as normal memory data or message passing data. To maintain data consistency, a core initial message write or read operation must first invalidate all message passing data cache lines. Then, the cache miss forces write or read operations to access the data in the shared memory. Each tile contains a 16 kB addressable message passing buffer to reduce the shared memory access latency.

To optimize the power consumption, the SCC chip applies dynamic voltage and frequency scaling (DVFS), which adjusts the processor's voltage and frequency according to management commands. The tile's DVFS range is between 300 MHz at 700 mV and 1.3 GHz at 1.3 V, and the NoC's range is between 60 MHz at 550 mV and 2.6 GHz at 1.3 V. The nominal usage condition is 1 GHz at 1.1 V for the tile and 2 GHz at 1.1 V for the NoC. The SCC chip consists of eight voltage islands: two voltage islands supply the mesh network and die periphery, and the remaining six voltage islands supply the processing cores, with four neighboring tiles composing one voltage island. The processor contains 28 frequency islands, with one frequency island for each tile and one frequency island for the mesh network. The remaining three frequency islands serve as the system interface, voltage regulator, and memory controllers respectively.

The mesh network of the SCC processor achieves a significant improvement over the NoC of the Teraflops processor. Its physical channel width is 144 bits, with 136 bits for payloads and 8 bits for flow control. The router architecture is optimized for such a wide link width. The router has a four-stage pipeline targeting a frequency of 2 GHz. The first stage includes the LT and buffer write, and the second stage performs the switch arbitration. The third and fourth stages are the VA and the ST respectively. To improve the performance, the router leverages the wrapped wavefront allocators. The mesh network applies lookahead XY DOR with VCT flow control. Each physical port configures eight VCs. To avoid protocol-level deadlock, the SCC processor employs two virtual networks: one for the request messages and the other one for the response messages. Each virtual network reserves one VC, and the remaining six VCs are shared between the two virtual networks.

1.5.7 THE INTEL LARRABEE PROCESSOR

The Larrabee processor is an Intel GPGPU, which was expected to be released in 2010 [130], but it was cancelled owing to delays and disappointing early performance figures. The Larrabee processor can be considered as a combination of a multicore CPU and a graphics processing unit (GPU), and has similarities to both. Its coherent

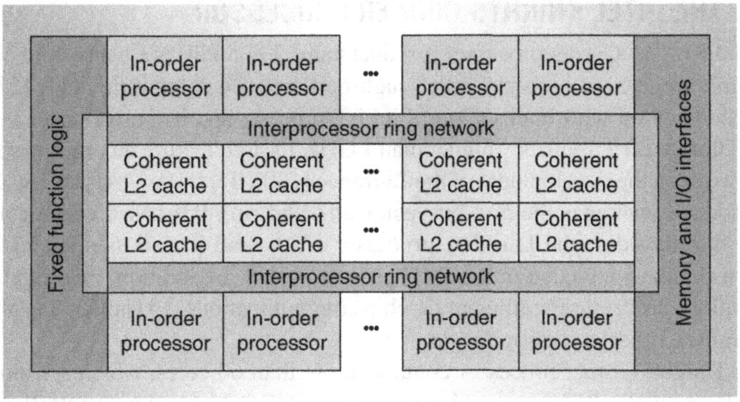

FIGURE 1.16

Schematic of the Larrabee processor. The number of processor cores is implementation dependent. From Ref. [130], copyright 2009 IEEE.

cache hierarchy and x86 architecture compatibility are CPU-like, while its wide SIMD vector units and texture sampling hardware are GPU-like.

The Larrabee processor is based on simple in-order P54C Pentium cores, and each core supports four-way interleaved multithreading, with four copies of each processor register. The original P54C core is extended with a 512-bit vector processing unit to provide high graphic processing performance. This vector unit can perform 16 single-precision floating point operations simultaneously. Also, the processor integrates one major fixed-function graphics hardware feature: the texture sampling unit. But to provide more flexibility than current GPUs, the Larrabee processor has less specialized graphic hardware than the GPGPUs from NVIDIA and AMD.

As interconnection features, the Larrabee processor has a 1024-bit (512 bits in each direction) data ring bus for communication between cores and the memory, as shown in Figure 1.16. The network structure is similar to the ring utilized in the Nehalem-EX processor [116]. The packet routing decisions are made before injection, by selecting the ring with the minimal distance. The ring applies the polarity stop scheme. Each ring stop has an odd or even polarity, and can only pull from the appropriate ring that matches that polarity. The ring changes polarity once per clock, so which stop can read from which ring on a given cycle is easy to figure out. An injecting ring stop knows the destination ring stop and the destination stop's polarity. It also knows the hop count to the destination ring stop. Thus, the source can figure out when to send a packet so that the destination can actually read it. By introducing a delay of at most one cycle, a source can ensure the destination can read the injected packet, and the case of two packets arriving at the same time never occurs. The routing and polarity stop flow control simplify the network: since injected packets are never blocked, no storage is required in the routers.

1.5.8 THE INTEL KNIGHTS CORNER PROCESSOR

The Intel Knights Corner processor (product name Xeon Phi) is a many-core processor mainly targeted at highly parallel, high-performance computing fields [22, 67].
It is manufactured with Intel's 22 nm FinFET process, and its performance is more
than 1 Tflops with a frequency higher than 1 GHz. Its architecture incorporates Intel's
earlier work on the development of the Larrabee [130], Teraflops [141] and SCC [59]
processors. It leverages the 512-bit vector unit and ring-bus-based cache-coherent
system introduced in the Larrabee processor [130], and it also uses sophisticated
DVFS techniques included in the SCC processor [59]. In addition, investigation of
the Teraflops processor for efficient on-chip communications [141] offers insights for
the Knights Corner processor design.

The Knights Corner processor contains more than 60 cores, which are modified
on the basis of the P54C core. The cores, eight GDDR5 memory controllers, and
the PCI Express interfaces are connected by a ring bus. Similarly to the Larrabee
core, each Knights Corner core includes a scalar pipeline and a vector pipeline.
Compared with the Larrabee core, in the Knights Corner core, the vector pipeline
is enhanced for high double-precision floating point performance. Each core supports four-way interleaved multithreading to alleviate memory access bottlenecks
for in-order instruction executions. Each core contains a 32 kB L1 DCache and a
32 kB L1 ICache, and a 512 kB unified private L2 cache. The L1 and L2 caches
both support eight-way set associativity, with 64-byte cache line sizes. The cache
hierarchy maintains the inclusive property to mitigate coherent overheads. The
memory system is optimized for high performance. The L1 DCache supports 64-
byte read/write per cycle. Each core contains 16 stream prefetchers and a 64-
entry translation lookaside buffer for the L2 cache. The L1 and L2 caches can
both support up to about 38 outstanding requests per core. The streaming store
instructions, which allow the cores to write an entire cache line without reading
it first, are added to reduce memory bandwidth consumption. Although the L2
cache is private to each core, the tag directory on each core is shared by the whole
processor.

To optimize power consumption, the Knights Corner processor implements a
bunch of power management modes (the C-states in Intel's terminology). The C0
mode is full power like in CPUs, the core C1 mode clock gates one core, and the core
C6 mode power gates that core but not the L2 cache. Since other cores can snoop the
L2 cache, it cannot be powered down until all cores are in at least the C1 mode. The
ring has the same problem: it cannot go to sleep until everything else does. When all
the cores are power gated and the uncore detects no activity, the tag directories, ring
bus, L2 caches, and memory controllers are clock gated. Then, the Knights Corner
processor enters the Package C6 state, which powers down everything but the PCI
Express interface to wait for wake-up commands from the host PC, and the GDDR5
is put into self-refresh mode.

In principle, the ring structure of the Knights Corner processor is similar to
those in the Nehalem-EX [116] and Larrabee [130] processors. As illustrated in
Figure 1.17, the cache coherence system leverages three kinds of rings. The main

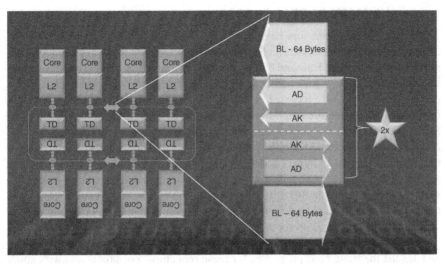

FIGURE 1.17

The NoC of the Knights Corner processor with two address/acknowledgment rings per direction. BL, data block ring; AD, address ring; AK, acknowledgment ring; TD, tag directory. From Ref. [22], copyright 2012 Intel.

ring is the 64-byte-wide data block ring, which is able to carry a full cache line. The other two types of rings are the narrower address ring and acknowledgment ring. The address ring delivers the address and read/write commands, and the acknowledgment ring carries flow control and coherence messages. Unlike the Nehalem-EX processor, the Knights Corner processor utilizes a complex controller to allow each stop to inject a packet into the ring every clock. After injection, the packets are deterministically delivered to the destination. The Knights Corner processor allows packet bouncing; in some cases, the destination may not have space to accept the packet, and so it is left on the ring and then picked up the next time the packet goes by.

Intel has found that the address and acknowledgment rings become performance bottlenecks and exhibit poor scalability beyond 32 cores [67]. Since these rings carry almost constant-sized packets, doubling the ring widths to deliver larger packets cannot alleviate the problem. Instead, Intel doubles the number of both address and acknowledgment rings to send more packets. Thus, there are now ten rings in total, or five in each direction, as shown in Figure 1.17. This allows scaling to 60+ cores, eight memory controllers, and many other supporting ring stops like the PCI Express interface. In the physical layout of the Knights Corner processor each GDDR5 memory controller is physically associated with a fixed number of cores. Although all cores can access all memory controllers, leveraging this physical proximity to reduce ring hops is helpful for performance.

1.5.9 SUMMARY OF REAL PROCESSORS

The NoCs of real processors have a significant design space as illustrated by the examples in the last sections. The functionalities and requirements of these NoCs are rich and different. The Raw and TRIPS processors mainly use the NoCs for operand bypassing; the NoC's low-latency property is critical for the whole processor's performance. The Cell processor primarily uses the NoC to deliver data for DMA operations, which requires high network bandwidth. The NoCs in the TILE64 and SCC processors mostly deliver data for the message passing paradigm, while the NoCs in the Larrabee and Knights Corner processors transfer cache-coherent data for the shared memory paradigm. Two independent physical networks or virtual networks are enough to avoid protocol-level deadlock for the message passing paradigm, such as the design in the SCC processor. In contrast, the complex cache coherence transaction causes the shared memory paradigm to use more physical networks or virtual networks to avoid protocol-level deadlock.

The design choices for each aspect of these NoCs span a continuum. Tables 1.2 and 1.3 list the basic NoC properties for these real chips. For processors with small numbers of cores, such as the Cell processor, the ring topology is appropriate for connection. Large numbers of cores require the more scalable 2D mesh topology, as in the TILE64, Teraflops, and SCC processors. In the Knights Corner processor a notable topology choice is made by using the ring network to connect more than 60 cores and other additional memory controllers and input/output interfaces. This processor leverages 10 rings to eliminate the performance bottleneck caused by the NoC congestion. Most real processors use the simple *XY* or *YX* DOR for 2D mesh NoCs, or shortest path routing for ring NoCs. These simple routing algorithms easily avoid network-level deadlock and require only a few hardware resources. Moreover, their low complexities are more convenient to achieve high frequencies.

The flow control of these NoCs includes the circuit switching, slotted ring, wormhole, and VCT mechanisms. The circuit switching flow control needs to configure a route path before sending packets; the Cell processor uses the shared command bus and data bus arbiter to build up the route path, while the TILE64 processor preconfigures the routers on the basis of compiled instructions. The Larrabee and Knights Corner processors both use slotted ring flow control, which requires all messages to be single-flit packets to fill into the ring slot. Thus, both processors have large link widths for the data ring to encode a full-size cache line. The wormhole flow control can be implemented as credit-based or on/off-based flow control. These two implementations require different minimum buffer amounts for correctness or for avoidance of buffer utilization bubbles [23]. The Raw and TILE64 processors and the OCN in the TRIPS processor use the credit-based wormhole flow control, while the Teraflops and the OPN in the TRIPS processor use the on/off-based design. Although previous research suggests that VCT flow control may not be appropriate for NoC design as it consumes large buffers [25], the SCC processor attempts to use this flow control mechanism in OCNs. Compared with wormhole flow control, VCT flow control can simplify the allocator design, thus supporting higher frequencies [59].

Table 1.2 The Node Count, Topology, and Routing Algorithms of the NoCs for Real Chips

Processor	No. of Nodes	Topology	Routing Algorithm
Raw (dynamic)	16	4 × 4 mesh, two PNs. Each PN is 32 bits wide	*XY* DOR
TILE64 (dynamic)	64	8 × 8 mesh, four PNs. Each PN is 32 bits wide	*XY* DOR
TILE64 (static)	64	8 × 8 mesh, one PN with a 32-bit width	*XY* DOR
Cell	12	Four rings, two for each direction. Each ring is 128 bits wide	Shortest path
TRIPS (OPN)	25	5 × 5 mesh, one PN with a 142-bit width	*YX* DOR
TRIPS (OCN)	40	4 × 10 mesh, four VNs. 1 PN with a 138-bit width	*YX* DOR
Teraflops	80	8 × 10 mesh, two VNs. 1 PN with a 39-bit width	*YX* DOR Source routing
SCC	24	6 × 4 mesh, two VNs. 1 PN with a 144-bit width	*XY* DOR
Larrabee	16+	Two data rings, one for each direction. Each data ring is 512 bits wide	Shortest path
Knights Corner	60+	Two data rings and eight control rings. Each direction has five rings. Each data ring is 512 bits wide	Shortest path Allow packet bouncing

PN, physical network; VN, virtual network.

These NoCs are manufactured with processes ranging from the 150 nm IBM ASIC for the academia prototype Raw processor to the 22 nm trigate for the commercial Knights Corner processor. The academia prototypes, Raw and TRIPS, target relatively lower frequencies of hundreds of megahertz. The Teraflops processor is used to demonstrate highly optimized OCN design. Thus, it targets a very high frequency of 5 GHz. Its NoC's power consumption is more than 10% of the whole processor's power consumption, which exceeds Intel's expectation [57]. Intel's following processors, including the SCC, Larrabee, and Knights Corner processors, have lower frequency goals. A moderate frequency is more efficient for throughput-oriented many-core power computation [67].

Table 1.3 The Flow Control and Per-Hop Latency of the NoCs for Real Chips

Processor	Flow control	Per-hop latency
Raw (dynamic)	Wormhole, credit based	Straight: one cycle, turn: two cycles at 225 MHz (worst case), 150 nm process
TILE64 (dynamic)	Wormhole, credit based	Straight: one cycle, turn: two cycles at 1 GHz, 90 nm process
TILE64 (static)	Circuit switching	One cycle at 1 GHz, 90 nm process
Cell	Circuit switching	One cycle at 1.6 GHz, 90 nm process
TRIPS (OPN)	Wormhole, on/off based	One cycle at 366 MHz, 130 nm process
TRIPS (OCN)	Wormhole, credit based	One cycle at 366 MHz, 130 nm process
Teraflops	Wormhole, on/off based	Five cycles at 5 GHz, 65 nm process
SCC	Virtual cut-through	Four cycles at 2 GHz, 45 nm process
Larrabee	Slotted ring, polarity stops	One cycle
Knights Corner	Slotted ring	One cycle at 1.05 GHz, 22 nm process

1.6 OVERVIEW OF THE BOOK

On the basis of the communication-centric cross-layer optimization method presented in Section 1.2, we have organized this book in five parts and 11 chapters, whose structure is illustrated in Figure 1.18. The Prologue and the Epilogue introduce and conclude this book respectively. Future work is also proposed in the Epilogue. The intermediate three parts, which consist of nine chapters, form the main content of this book. These parts describe communication-centric cross-layer optimizations for the NoC design space, in a thorough and bottom-up way.

Part II, on logic implementations, focuses on the optimization of NoC logic implementations, including efficient router architecture, buffer structure, and network topology designs. It contains three chapters, Chapters 2–4. Chapter 2 proposes a low-cost single-cycle router architecture with wing channels. The wing channels forward incoming packets to free ports immediately with the inspection of SA results, thus achieving a single-cycle router delay. Moreover, the incoming packets granted wing channels can fill in the free time slots of the crossbar and reduce contention with subsequent packets, thereby improving throughput effectively. Chapter 3 discusses the design of dynamic NoC VC structures. It first introduces a DAMQ buffer structure, which can efficiently avoid network congestion. This structure adaptively adjusts the VC depth and VC amount on the basis of network loads to mitigate network blocking. Then, it presents a novel NoC router with a shared-buffer structure named a hierarchical bit-line buffer. The hierarchical bit-line buffer supports sharing buffers among different physical ports, and it assigns higher priorities to packets heading to lightly loaded ports to avoid network congestion. Chapter 4 studies a new NoC topology of incorporating buses into NoCs in order to take advantage of both NOCs and buses in a hierarchical way. The proposed virtual bus OCN dynamically

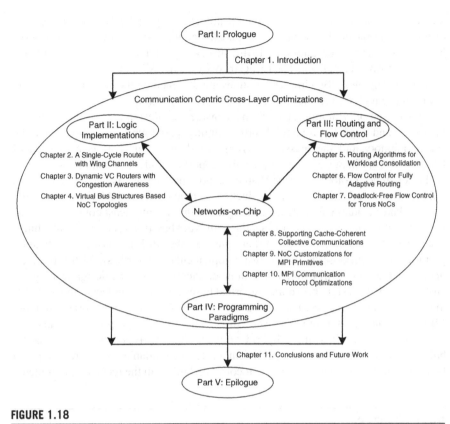

FIGURE 1.18

The book structure.

uses point-to-point links of conventional NoC designs as bus transaction links to support high-performance multicast and broadcast communications.

Part III, on routing and flow control, delves into the design of high-performance NoC routing algorithms and flow control mechanisms on the basis of the traffic characteristics. There are three chapters, Chapters 5–7. In Chapter 5 a holistic approach is used to design efficient routing algorithms for workload consolidation scenarios. It proposes a novel selection strategy, the destination-based selection strategy, for adaptive routing algorithms in many-core systems running multiple concurrent applications. Its exploration considers several critical aspects of routing algorithms, including adaptivity, path selection strategy, VA, isolation, and hardware implementation cost, to ensure an efficient design. Chapter 6 mainly focuses on efficient flow control designs for deadlock-free fully adaptive routing algorithms. It presents two novel flow control designs. First, WPF reallocates a nonempty VC if the VC has enough free buffers for an entire packet. This chapter proves that WPF does not induce deadlock, and it is an important extension to several deadlock avoidance theories. Second, the chapter extends Duato's theory to apply aggressive

VC reallocation on escape VCs without deadlock. Chapter 7 delves into the design of high-performance, low-cost, and deadlock-free flow control for wormhole torus NoCs. This chapter introduces the FBFC theory, which maintains one free flit-size buffer slot to avoid deadlock. The FBFC theory uses one VC, and does not treat short packets as long ones. Two implementations are designed on the basis of this theory; both achieve high frequencies and efficient buffer utilization.

Part IV, on programming paradigms, considers co-design and co-optimizations of NoCs and high-level parallel programming paradigms, including both the shared memory and the message passing paradigms. This part is composed of Chapters 8–10. Chapter 8 studies hardware support for collective communications in cache coherence protocols for the shared memory paradigms. Its design includes an efficient framework to support the reduction communication of ACK packets and a novel balanced, adaptive multicast routing algorithm for multicast communications. By considering hardware features available in NoC-based many-core architectures, Chapter 9 presents an NoC design that optimizes the MPI to boost the application performance. It proposes novel hardware implementations for basic MPI primitives, upon which all other MPI functions can be efficiently built. The design leverages a customized NoC design incorporating virtual buses into NoCs and an optimized MPI unit to efficiently execute MPI-related transactions. Chapter 10 explores designing MPI communication protocols over NoCs. This chapter proposes the adaptive communication mechanism, a hybrid protocol that involves behavior similar to buffered communication when sufficient buffer is available in the receiver and behavior similar to a synchronous protocol when buffers in the receiver are limited.

REFERENCES

[1] N.R. Adiga, M.A. Blumrich, D. Chen, P. Coteus, A. Gara, M.E. Giampapa, P. Heidel-berger, S. Singh, B.D. Steinmacher-Burow, T. Takken, M. Tsao, P. Vranas, Blue Gene/L torus interconnection network, IBM J. Res. Dev. 49 (2.3) (2005) 265–276. 9, 18

[2] S. Adve, D.H. Albonesi, D. Brooks, L. Ceze, S. Dwarkadas, J. Emer, B. Falsafi, A. Gonzalez, M.D. Hill, M.J. Irwin, D. Kaeli, S.W. Keckler, C. Kozyrakis, A. Lebeck, M. Martin, J.F. MartÃ-nez, M. Martonosi, K. Olukotun, M. Oskin, L.-S. Peh, M. Prvulovic, S.K. Reinhardt, M. Schulte, S. Sethumadhavan, G. Sohi, D. Sorin, J. Torrellas, T.F. Wenisch, D. Wood, K. Yelick, 21st Century computer architecture: a community white paper. Technical report, Computing Community Consortium, 2012. 5, 6

[3] V. Agarwal, M.S. Hrishikesh, S.W. Keckler, D. Burger, Clock rate versus IPC: the end of the road for conventional microarchitectures, in: Proceedings of the International Symposium on Computer Architecture (ISCA), 2000, pp. 248–259. 4

[4] A. Ahmadinia, A. Shahrabi, A highly adaptive and efficient router architecture for network-on-chip, Comput. J. 54 (8) (2011) 1295–1307. 22

[5] T.W. Ainsworth, T.M. Pinkston, Characterizing the cell EIB on-chip network, IEEE Micro 27 (5) (2007) 6–14. 26

[6] Y. Ajima, S. Sumimoto, T. Shimizu, Tofu: a 6D mesh/torus interconnect for exascale computers, Computer 42 (11) (2009) 36–40. 9

[7] K. Asanovic, R. Bodik, B.C. Catanzaro, J.J. Gebis, P. Husbands, K. Keutzer, D.A. Patterson, W.L. Plishker, J. Shalf, S.W. Williams, K.A. Yelick, The landscape of parallel computing research: a view from Berkeley, Technical report, Electrical Engineering and Computer Sciences University of California at Berkeley, 2006. 5

[8] G. Ascia, V. Catania, M. Palesi, D. Patti, Implementation and analysis of a new selection strategy for adaptive routing in networks-on-chip, IEEE Trans. Comput. 57 (6) (2008) 809–820. 19

[9] J. Balfour, W. Dally, Design tradeoffs for tiled CMP on-chip networks, in: Proceedings of the International Conference on Supercomputing (ICS), 2006, pp. 187–198. 16, 17

[10] D.U. Becker, N. Jiang, G. Michelogiannakis, W.J. Dally, Adaptive backpressure: efficient buffer management for on-chip networks, in: Proceedings of the International Conference on Computer Design (ICCD), 2012, pp. 419–426. 22

[11] S. Bell, B. Edwards, J. Amann, R. Conlin, K. Joyce, V. Leung, J. MacKay, M. Reif, L. Bao, J. Brown, et al., TILE64 - Processor: a 64-core soc with mesh interconnect, in: Proceedings of the International Solid-State Circuits Conference Digest of Technical Papers (ISSCC), 2008, pp. 88–598. 5

[12] T. Bjerregaard, S. Mahadevan, A survey of research and practices of network-on-chip, ACM Comput. Surv. 38 (1) (2006) 1-51. 7

[13] E. Bolotin, Z. Guz, I. Cidon, R. Ginosar, A. Kolodny, The power of priority: NoC based distributed cache coherency, in: Proceedings of the International Symposium on Networks-on-Chip (NOCS), 2007, pp. 117–126. 20

[14] S. Borkar, Designing reliable systems from unreliable components: the challenges of transistor variability and degradation, IEEE Micro 25 (6) (2005) 10–16. 4

[15] C. Carrion, R. Beivide, J. Gregorio, F. Vallejo, A flow control mechanism to avoid message deadlock in k-ary n-cube networks, in: Proceedings of the International Conference on High-Performance Computing (HiPC), 1997, pp. 322–329. 21

[16] D. Chen, N. Eisley, P. Heidelberger, R. Senger, Y. Sugawara, S. Kumar, V. Salapura, D. Satterfield, B. Steinmacher-Burow, J. Parker, The IBM Blue Gene/Q interconnection fabric, IEEE Micro 32 (1) (2012) 32–43. 9, 18

[17] L. Chen, T.M. Pinkston, NoRD: node-router decoupling for effective power-gating of on-chip routers, in: Proceedings of the International Symposium on Microarchitecture (MICRO), 2012, pp. 270–281. 18

[18] L. Chen, T.M. Pinkston, Worm-bubble flow control, in: Proceedings of the International Symposium on High-Performance Computer Architecture (HPCA), 2013, pp. 366–377. 9, 21

[19] L. Chen, R. Wang, T.M. Pinkston, Critical bubble scheme: an efficient implementation of globally aware network flow control, in: Proceedings of the International Parallel & Distributed Processing Symposium (IPDPS), 2011, pp. 592–603. 9, 21

[20] G.-M. Chiu, The odd-even turn model for adaptive routing, IEEE Trans. Parallel Distributed Syst. 11 (7) (2000) 729–738. 11

[21] M.H. Cho, M. Lis, K.S. Shim, M. Kinsy, T. Wen, S. Devada, Oblivious routing in on-chip bandwidth-adaptive networks, in: Proceedings of the International Conference on Parallel Architectures and Compilation Techniques (PACT), 2009, pp. 181–190. 18

[22] G. Chrysos, Intel Xeon Phi coprocessor (codename Knights Corner), in: Invited Presentation to Hot Chips 2012, 2012. 8, 34, 35

[23] N. Concer, M. Petracca, L.P. Carloni, Distributed flit-buffer flow control for networks-on-chip, in: Proceedings of the International Conference on Hardware/Software Codesign and System Synthesis (CODES+ISSS), 2008, pp. 215–220. 36

[24] W. Dally, C. Seitz, Deadlock-free message routing in multiprocessor interconnection networks, IEEE Trans. Comput. C-36 (5) (1987) 547–553. 11, 13, 18

[25] W. Dally, B. Towles, Route packets, not wires: on-chip interconnection networks, in: Proceedings of the Design Automation Conference (DAC), 2001, pp. 684–689. 6, 36

[26] W. Dally, B. Towles, Principles and Practices of Interconnection Networks, first ed., Morgan Kaufmann Publishers Inc., San Francisco, CA, USA, 2003. 8, 9, 11, 13, 14, 16

[27] W.J. Dally, C.L. Seitz, The torus routing chip, Distributed Comput. 1 (4) (1986) 187–196. 11

[28] S. Damaraju, V. George, S. Jahagirdar, T. Khondker, R. Milstrey, S. Sarkar, S. Siers, I. Stolero, A. Subbiah, A 22nm IA multi-CPU and GPU system-on-chip, in: Proceedings of the International Solid-State Circuits Conference Digest of Technical Papers (ISSCC), 2012, pp. 56–57. 8

[29] J. del Cuvillo, W. Zhu, Z. Hu, G. Gao, Toward a software infrastructure for the cyclops-64 cellular architecture, in: Proceedings of the International Symposium on High-Performance Computing in an Advanced Collaborative Environment (HPCS), 2006. 5

[30] C. Demerjian, Intel details Knights Corner architecture at long last. http://semiaccurate. com/2012/08/28/intel-details-knights-corner-architecture-at-long-last/, Cited April 2014. 8

[31] J. Diaz, C. Munoz-Caro, A. Nino, A survey of parallel programming models and tools in the multi and many-core era, IEEE Trans. Parallel Distributed Syst. 23 (8) (2011) 1369–1386. 5

[32] J. Duato, A new theory of deadlock-free adaptive routing in wormhole networks, IEEE Trans. Parallel Distributed Syst. 4 (12) (1993) 1320–1331. 11, 18, 20

[33] J. Duato, A necessary and sufficient condition for deadlock-free adaptive routing in wormhole networks, IEEE Trans. Parallel Distributed Syst. 6 (10) (1995) 1055-1067. 11

[34] J. Duato, A necessary and sufficient condition for deadlock-free routing in cut-through and store-and-forward networks, IEEE Trans. Parallel Distributed Syst. 7 (8) (1996) 841–854. 18

[35] N. Enright Jerger, L. Peh, On-Chip Networks, first ed., Morgan & Claypool, San Rafael, CA, 2009. 8, 13, 14, 16

[36] N. Enright Jerger, L. Peh, M. Lipasti, Circuit-switched coherence, in: Proceedings of the International Symposium on Networks-on-Chip (NOCS), 2008, pp. 193–202. 21

[37] N. Enright Jerger, L.-S. Peh, M. Lipasti, Virtual circuit tree multicasting: a case for on-chip hardware multicast support, in: Proceedings of the International Symposium on Computer Architecture (ISCA), 2008, pp. 229–240. 19

[38] H. Esmaeilzadeh, E. Blem, R. St. Amant, K. Sankaralingam, D. Burger, Dark silicon and the end of multicore scaling, in: Proceedings of the International Symposium on Computer Architecture (ISCA), 2011, pp. 365–376. 4, 6

[39] C. Fallin, C. Craik, O. Mutlu, CHIPPER: a low-complexity bufferless deflection router, in: Proceedings of the International Symposium on High-Performance Computer Architecture (HPCA), 2011, pp. 144–155. 20

[40] C. Fallin, G. Nazario, X. Yu, K. Chang, R. Ausavarungnirun, O. Mutlu, MinBD: minimally-buffered deflection routing for energy-efficient interconnect, in: Proceedings of the International Symposium on Networks on Chip (NOCS), 2012, pp. 1–10. 21

[41] K. Fatahalian, M. Houston, A closer look at GPUs, Commun. ACM 51 (10) (2008) 50–57. 5

[42] M. Flynn, P. Hung, Microprocessor design issues: thoughts on the road ahead, IEEE Micro 25 (3) (2005) 16–31. 4

[43] B. Fu, Y. Han, J. Ma, H. Li, X. Li, An abacus turn model for time/space-efficient reconfigurable routing, in: Proceedings of the International Symposium on Computer Architecture (ISCA), 2011, pp. 259–270. 11

[44] M. Galles, Spider: a high-speed network interconnect, IEEE Micro 17 (1) (1997) 34–39. 15

[45] D. Geer, Chip makers turn to multicore processors, Computer 38 (5) (2005) 11–13. 4

[46] C. Glass, L. Ni, The turn model for adaptive routing, in: Proceedings of the International Symposium on Computer Architecture (ISCA), 1992, pp. 278–287. 11

[47] P. Gratz, B. Grot, S. Keckler, Regional congestion awareness for load balance in networks-on-chip, in: Proceedings of the International Symposium on High-Performance Computer Architecture (HPCA), 2008, pp. 203–214. 19

[48] P. Gratz, C. Kim, K. Sankaralingam, H. Hanson, P. Shivakumar, S.W. Keckler, D. Burger, On-chip interconnection networks of the TRIPS chip, IEEE Micro 27 (5) (2007) 41–50. 8, 28

[49] P. Gratz, K. Sankaralingam, H. Hanson, P. Shivakumar, R. McDonald, S. Keckler, D. Burger, Implementation and evaluation of a dynamically routed processor operand network, in: Proceedings of the International Symposium on Networks-on-Chip (NOCS), 2007, pp. 7–17. 28, 29

[50] L. Hammond, B. Hubbert, M. Siu, M. Prabhu, M. Chen, K. Olukolun, The Stanford Hydra CMP, IEEE Micro 20 (2) (2000) 71–84. 4

[51] M. Hayenga, N. Enright Jerger, M. Lipasti, SCARAB: a single cycle adaptive routing and bufferless network, in: Proceedings of the International Symposium on Microarchitecture (MICRO), 2009, pp. 244–254. 20

[52] M. Hayenga, M. Lipasti, The NoX router, in: Proceedings of the International Symposium on Microarchitecture (MICRO), 2011, pp. 36–46. 22

[53] J. Henkel, W. Wolf, S. Chakradhar, On-chip networks: a scalable, communication-centric embedded system design paradigm, in: Proceedings of the International Conference on VLSI Design (ICVD), 2004, pp. 845–851. 7

[54] J.L. Hennessy, D.A. Patterson, Computer Architecture: A Quantitative Approach, fifth ed., Morgan Kaufmann Publishers Inc., San Francisco, CA, USA, 2012. 4

[55] R. Hesse, J. Nicholls, N. Enright Jerger, Fine-grained bandwidth adaptivity in networks-on-chip using bidirectional channels, in: Proceedings of the International Symposium on Networks-on-Chip (NOCS), 2012, pp. 132–141. 18

[56] G. Hinton, D. Sager, M. Upton, D. Boggs, D. Carmean, A. Kyker, P. Roussel, The microarchitecture of the Pentium® 4 Processor, Intel Technol. J. Q (1) (2001) 1–13. 4

[57] Y. Hoskote, S. Vangal, A. Singh, N. Borkar, S. Borkar, A 5-GHz mesh interconnect for a Teraflops Processor, IEEE Micro 27 (5) (2007) 51–61. 7, 30, 31, 37

[58] J. Howard, S. Dighe, Y. Hoskote, S. Vangal, D. Finan, G. Ruhl, D. Jenkins, H. Wilson, N. Borkar, G. Schrom, et al., A 48-core IA-32 message-passing processor with DVFS in 45nm CMOS, in: Proceedings of the International Solid-State Circuits Conference Digest of Technical Papers (ISSCC), 2012, pp. 108–109. 5

[59] J. Howard, S. Dighe, S. Vangal, G. Ruhl, N. Borkar, S. Jain, V. Erraguntla, M. Konow, M. Riepen, M. Gries, G. Droege, T. Lund-Larsen, S. Steibl, S. Borkar, V. De, R. Van Der Wijngaart, A 48-Core IA-32 Processor in 45 nm CMOS using on-die

message-passing and DVFS for performance and power scaling, IEEE J. Solid-State Circ. 46 (1) (2011) 173–183. 7, 30, 31, 34, 36

[60] M.S. Hrishikesh, D. Burger, N.P. Jouppi, S.W. Keckler, K.I. Farkas, P. Shivakumar, The optimal logic depth per pipeline stage is 6 to 8 FO4 Inverter Delays, in: Proceedings of the International Symposium on Computer Architecture (ISCA), 2002, pp. 14–24. 4

[61] J. Hu, R. Marculescu. DyAD—smart routing for networks-on-chip, in: Proceedings of the Design Automation Conference (DAC), 2004, pp. 260–263. 19

[62] L. Huang, Z. Wang, N. Xiao, Accelerating NoC-based MPI primitives via communication architecture customization, in: Proceedings of the International Conference on Application-Specific Systems, Architectures and Processors (ASAP), 2012, pp. 141–148. 20

[63] L. Huang, Z. Wang, N. Xiao, An optimized multicore cache coherence design for exploiting communication locality, in: Proceedings of the Great Lakes Symposium on VLSI (GLSVLSI), 2012, pp. 59–62. 18, 20

[64] L. Huang, Z. Wang, N. Xiao, VBON: toward efficient on-chip networks via hierarchical virtual bus, Microprocess. Microsyst. 37 (8, Part B) (2013) 915–928. 18

[65] Intel, Knights landing: Next Generation Intel Xeon Phi, in: Invited Presentation to SC 2013, 2013. 5

[66] Intel, Intel Xeon Phi Coprocessor—Datasheet, Technical report, Intel, June 2013. 5, 8

[67] Intel, Intel Xeon Phi Coprocessor System Software Developers Guide, Technical report, Intel, March 2014. 34, 35, 37

[68] S.A.R. Jafri, Y.-J. Hong, M. Thottethodi, T.N. Vijaykumar, Adaptive flow control for robust performance and energy, in: Proceedings of the International Symposium on Microarchitecture (MICRO), 2010, pp. 433–444. 20

[69] A. Joshi, M. Mutyam, Prevention flow-control for low latency torus networks-on-chip, in: Proceedings of the International Symposium on Networks-on-Chip (NOCS), 2011, pp. 41–48. 21

[70] N.P. Jouppi, Improving direct-mapped cache performance by the addition of a small fully-associative cache and prefetch buffers, in: Proceedings of the International Symposium on Computer Architecture (ISCA), 1990, pp. 364–373. 4

[71] J.A. Kahle, M.N. Day, H.P. Hofstee, C.R. Johns, T.R. Maeurer, D. Shippy, Introduction to the cell multiprocessor, IBM J. Res. Dev. 49 (4.5) (2005) 589–604. 8, 26

[72] P. Kermani, L. Kleinrock, Virtual cut-through: a new computer communication switching technique, Comput. Netw. 3 (4) (1979) 267–286. 11

[73] R.E. Kessler, The Alpha 21264 microprocessor, IEEE Micro 19 (2) (1999) 24–36. 4

[74] G. Kim, J. Kim, S. Yoo, FlexiBuffer: reducing leakage power in on-chip network routers, in: Proceedings of the Design Automation Conference (DAC), 2011, pp. 936–941. 22

[75] H. Kim, G. Kim, S. Maeng, H. Yeo, J. Kim, Transportation-network-inspired network-on-chip, in: Proceedings of the International Symposium on High Performance Computer Architecture (HPCA), 2014. 22

[76] J. Kim, Low-cost router microarchitecture for on-chip networks, in: Proceedings of the International Symposium on Microarchitecture (MICRO), 2009, pp. 255–266. 22

[77] J. Kim, J. Balfour, W. Dally, Flattened butterfly topology for on-chip networks, in: Proceedings of the International Symposium on Microarchitecture (MICRO), 2007, pp. 37–40. 17

[78] J. Kim, D. Park, T. Theocharides, N. Vijaykrishnan, C.R. Das, A low latency router supporting adaptivity for on-chip interconnects, in: Proceedings of the Design Automation Conference (DAC), 2005, pp. 559–564. 15, 19

[79] M. Kistler, M. Perrone, F. Petrini, Cell multiprocessor communication network: built for speed, IEEE Micro 26 (3) (2006) 10–23. 26, 27

[80] A. Kodi, A. Sarathy, A. Louri, iDEAL: inter-router dual-function energy and area-efficient links for network-on-chip (NoC) architectures, in: Proceedings of the International Symposium on Computer Architecture (ISCA), 2008, pp. 241–250. 22

[81] P. Kongetira, K. Aingaran, K. Olukotun, Niagara: a 32-way Multithreaded SPARC Processor, IEEE Micro 25 (2) (2005) 21–29. 7

[82] T. Krishna, L.-S. Peh, B.M. Beckmann, S.K. Reinhardt, Towards the ideal on-chip fabric for 1-to-many and many-to-1 communication, in: Proceedings of the International Symposium on Microarchitecture (MICRO), 2011, pp. 71–82. 20

[83] A. Kumar, The HP PA-8000 RISC CPU, IEEE Micro 17 (2) (1997) 27–32. 4

[84] A. Kumar, L.-S. Peh, N.K. Jha, Token flow control, in: Proceedings of the International Symposium on Microarchitecture (MICRO), 2008, pp. 342–353. 21

[85] A. Kumar, L.-S. Peh, P. Kundu, N.K. Jha, Express virtual channels: towards the ideal interconnection fabric, in: Proceedings of the International Symposium on Computer Architecture (ISCA), 2007, pp. 150–161. 21

[86] P. Kumar, Y. Pan, J. Kim, G. Memik, A. Choudhary, Exploring concentration and channel slicing in on-chip network router, in: Proceedings of the International Symposium on Networks-on-Chip (NOCS), 2009, pp. 276–285. 17

[87] M. Lai, L. Gao, S. Ma, X. Nong, Z. Wang, A practical low-latency router architecture with wing channel for on-chip network, Microprocess. Microsyst. 35 (2) (2011) 98–109. 22

[88] M. Lai, Z. Wang, L. Gao, H. Lu, K. Dai, A dynamically-allocated virtual channel architecture with congestion awareness for on-chip routers, in: Proceedings of the Design Automation Conference (DAC), 2008, pp. 630–633. 22

[89] M. Li, Q.-A. Zeng, W.-B. Jone, DyXY—a proximity congestion-aware deadlock-free dynamic routing method for network on chip, in: Proceedings of the Design Automation Conference (DAC), 2006, pp. 849–852. 19

[90] E. Lindholm, J. Nickolls, S. Oberman, J. Montrym, NVIDIA Tesla: a unified graphics and computing architecture, IEEE Micro 22 (2) (2008) 39–55. 5

[91] Z. Lu, A. Jantsch, Trends of terascale computing chips in the next ten years, in: Proceedings of the International Conference on ASIC (ASICON), 2009, pp. 62–66. 5

[92] Z. Lu, M. Liu, A. Jantsch, Layered switching for networks on chip, in: Proceedings of the Design Automation Conference (DAC), 2007, pp. 122–127. 21

[93] Z. Lu, B. Yin, A. Jantsch, Connection-oriented multicasting in wormhole-switched networks on chip, in: Proceedings of the International Symposium on Emerging VLSI Technologies and Architectures (ISVLSI), 2006, pp. 205–210. 19

[94] W. Luo, D. Xiang, An efficient adaptive deadlock-free routing algorithm for torus networks, IEEE Trans. Parallel Distributed Syst. 23 (5) (2012) 800–808. 21

[95] S. Ma, N. Enright Jerger, Z. Wang, DBAR: an efficient routing algorithm to support multiple concurrent applications in networks-on-chip, in: Proceedings of the International Symposium on Computer Architecture (ISCA), 2011, pp. 413–424. 16, 19

[96] S. Ma, N. Enright Jerger, Z. Wang, Supporting efficient collective communication in NoCs, in: Proceedings of the International Symposium on High-Performance Computer Architecture (HPCA), 2012, pp. 165–176. 20

[97] S. Ma, N. Enright Jerger, Z. Wang, Whole packet forwarding: efficient design of fully adaptive routing algorithms for networks-on-chip, in: Proceedings of the International Symposium on High-Performance Computer Architecture (HPCA), 2012, pp. 467–478. 11, 16, 18

[98] S. Ma, N. Enright Jerger, Z. Wang, L. Huang, M. Lai, Holistic routing algorithm design to support workload consolidation in NoCs, IEEE Trans. Comput. 63 (3) (2014) 529–542. 19

[99] S. Ma, Z. Wang, N. Enright Jerger, L. Shen, N. Xiao, Novel flow control for fully adaptive routing in cache-coherent NoCs. IEEE Trans. Parallel Distributed Syst. 25 (9) (2014) 2397-2407. 11, 18

[100] S. Ma, Z. Wang, Z. Liu, N. Enright Jerger, Leaving one slot empty: flit bubble flow control for torus cache-coherent NoCs. IEEE Trans. Comput. PrePrints (99) (2013). 9, 16, 21

[101] R. Manevich, I. Walter, I. Cidon, A. Kolodny, Best of both worlds: a bus enhanced NoC (BENoC), in: Proceedings of the International Symposium on Networks-on-Chip (NOCS), 2009, pp. 173–182. 18

[102] R. Marculescu, U.Y. Ogras, L.-S. Peh, N. Enright Jerger, Y. Hoskote, Outstanding research problems in NoC design: system, microarchitecture, and circuit perspectives, IEEE Trans. Comput. Aided Des. Integrated Circ. Syst. 28 (2009) 3–21. 6

[103] A. Marowka, Parallel computing on any desktop, Commun. ACM 50 (9) (2007) 75–78. 5

[104] H. Matsutani, M. Koibuchi, H. Amano, T. Yoshinaga, Prediction router: yet another low latency on-chip router architecture, in: Proceedings of the International Symposium on High-Performance Computer Architecture (HPCA), 2009, pp. 367–378. 22

[105] S. McIntosh-Smith, A next-generation many-core processor with reliability, fault tolerance and adaptive power management features optimized for embedded and high performance computing applications, in: Proceedings of the High Performance Embedded Computing Conference (HPEC), 2008, pp. 1–2. 5

[106] G. Michelogiannakis, J. Balfour, W. Dally, Elastic-buffer flow control for on-chip networks, in: Proceedings of the International Symposium on High-Performance Computer Architecture (HPCA), 2009, pp. 151–162. 22

[107] A.K. Mishra, N. Vijaykrishnan, C.R. Das, A case for heterogeneous on-chip interconnects for CMPs, in: Proceedings of the International Symposium on Computer Architecture (ISCA), 2011, pp. 389–400. 8, 17

[108] G. Moore, Cramming more components onto integrated circuits, Electronics 38 (8) (1965) 114–117. 3

[109] T. Moscibroda, O. Mutlu, A case for bufferless routing in on-chip networks, in: Proceedings of the International Symposium on Computer Architecture (ISCA), 2009, pp. 196–207. 20

[110] A. Moshovos, G.S. Sohi, Microarchitectural innovations: boosting microprocessor performance beyond semiconductor technology scaling, Proc. IEEE 89 (11) (2001) 1560–1575. 4

[111] T. Mudge, Power: a first-class architectural design constraint, Computer 34 (4) (2001) 52–58. 4

[112] R. Mullins, A. West, S. Moore, Low-latency virtual-channel routers for on-chip networks, in: Proceedings of the International Symposium on Computer Architecture (ISCA), 2004, pp. 188–197. 16, 22

[113] C.A. Nicopoulos, D. Park, J. Kim, N. Vijaykrishnan, M.S. Yousif, C.R. Das, ViChaR: a dynamic virtual channel regulator for network-on-chip routers, in: Proceedings of the International Symposium on Microarchitecture (MICRO), 2006, pp. 333–346. 22

[114] NVIDIA, Whitepaper: NVIDIAs Next Generation CUDATM Compute Architecture: Kepler TM GK110, Technical report, NVIDIA, 2011. 4, 5

[115] U.Y. Ogras, J. Hu, R. Marculescu, Key research problems in NoC design: a holistic perspective, in: Proceedings of the International Conference on Hardware/Software Codesign and System Synthesis (CODES+ISSS), 2005, pp. 69–74. 6

[116] C. Park, R. Badeau, L. Biro, J. Chang, T. Singh, J. Vash, B. Wang, T. Wang, A 1.2 TB/s on-chip ring interconnect for 45nm 8-core enterprise Xeon processor, in: Proceedings of the International Solid-State Circuits Conference Digest of Technical Papers (ISSCC), 2010, pp. 180–181. 8, 33, 34

[117] D. Patterson, J. Hennessy, Computer organization and design: the hardware/software interface, fourth ed., Morgan Kaufmann, San Francisco, CA, 2009. 4

[118] L. Peh, W. Dally, Flit-reservation flow control, in: Proceedings of the International Symposium on High-Performance Computer Architecture (HPCA), 2000, pp. 73–84. 21

[119] L.-S. Peh, W. Dally, A delay model and speculative architecture for pipelined routers, in: Proceedings of the International Symposium on High-Performance Computer Architecture (HPCA), 2001, pp. 255–266. 16, 22

[120] M.A. Postiff, D.A. Greene, G.S. Tyson, T.N. Mudge, The limits of instruction level parallelism in SPEC95 applications, ACM SIGARCH Comput. Archit. News 27 (1) (1999) 31–34. 4

[121] S. Przybylski, M. Horowitz, J. Hennessy, Characteristics of performance-optimal multi-level cache hierarchies, in: Proceedings of the International Symposium on Computer Architecture (ISCA), 1989, pp. 114–121. 4

[122] V. Puente, R. Beivide, J.A. Gregorio, J.M. Prellezo, J. Duato, C. Izu, Adaptive bubble router: a design to improve performance in Torus networks, in: Proceedings of the International Conference Parallel Processing (ICPP), 1999, pp. 58–67. 21

[123] V. Puente, C. Izu, R. Beivide, J.A. Gregorio, F. Vallejo, J. Prellezo, The adaptive bubble router, J. Parallel Distributed Comput. 61 (9) (2001) 1180–1208. 9, 21

[124] R.S. Ramanujam, V. Soteriou, B. Lin, L.-S. Peh, Design of a high-throughput distributed shared-buffer NoC router, in: Proceedings of the International Symposium on Networks-on-Chip (NOCS), 2010, pp. 69–78. 22

[125] C. Ramey, Tile-GX100 manycore processor: acceleration interfaces and architecture, in: Invited Presentation to Hot Chips 2011, 2011. 5

[126] S. Rodrigo, J. Flich, J. Duato, M. Hummel, Efficient unicast and multicast support for CMPs, in: Proceedings of the International Symposium on Microarchitecture (MICRO), 2008, pp. 364–375. 20

[127] F.A. Samman, T. Hollstein, M. Glesner, Wormhole cut-through switching: flit-level messages interleaving for virtual-channelless network-on-chip, Microprocess. Microsyst. 35 (3) (2011) 343–358. 21

[128] S. Sawant, U. Desai, G. Shamanna, L. Sharma, M. Ranade, A. Agarwal, S. Dakshinamurthy, R. Narayanan, A 32nm Westmere-EX Xeon® Enterprise processor, in: Proceedings of the International Solid-State Circuits Conference Digest of Technical Papers (ISSCC), 2011, pp. 74–75. 4

[129] S.L. Scott, G.M. Thorson, The Cray T3E Network: adaptive routing in a high performance 3D Torus, in: Proceedings of the the Symposium on High Performance Interconnects (HOTI), 1996, pp. 1–10. 9, 18

[130] L. Seiler, D. Carmean, E. Sprangle, T. Forsyth, P. Dubey, S. Junkins, A. Lake, R. Cavin, R. Espasa, E. Grochowski, T. Juan, M. Abrash, J. Sugerman, P. Hanrahan, Larrabee: a many-core x86 architecture for visual computing, IEEE Micro 29 (1) (2009) 10–21. 5, 32, 33, 34

[131] Semiconductor Industry Association, International Technology Roadmap for Semiconductors, 2010 Edition. http://www.itrs.net, 2010. 4

[132] D. Seo, A. Ali, W.-T. Lim, N. Rafique, M. Thottethodi, Near-optimal worst-case throughput routing for two-dimensional mesh networks, in: Proceedings of the International Symposium on Computer Architecture (ISCA), 2005, pp. 432–443. 10

[133] W. Shi, W. Xu, H. Ren, Q. Dou, Z. Wang, L. Shen, C. Liu, A novel shared-buffer router for network-on-chip based on hierarchical bit-line buffer, in: Proceedings of the International Conference on Computer Design (ICCD), 2011, pp. 267–272. 22

[134] V. Srinivasan, D. Brooks, M. Gschwind, P. Bose, V. Zyuban, P.N. Strenski, P.G. Emma, Optimizing pipelines for power and performance, in: Proceedings of the International Symposium on Microarchitecture (MICRO), 2002, pp. 333–344. 4

[135] T. Sterling, Multicore: the New Moores Law, in: Invited Presentation to ICS 2007, 2007. 5

[136] Y. Tamir, G. Frazier, High-performance multiqueue buffers for VLSI communication switches, in: Proceedings of the International Symposium on Computer Architecture (ISCA), 1988, pp. 343–354. 22

[137] M. Taylor, J. Kim, J. Miller, D. Wentzlaff, F. Ghodrat, B. Greenwald, H. Hoffman, P. Johnson, J.-W. Lee, W. Lee, A. Ma, A. Saraf, M. Seneski, N. Shnidman, V. Strumpen, M. Frank, S. Amarasinghe, A. Agarwal, The Raw microprocessor: a computational fabric for software circuits and general-purpose programs, IEEE Micro 22 (2) (2002) 25–35. 7, 23, 24

[138] TOP500 Supercomputer Site, TOP500 List at November 2013. http://www.top500.org/lists/2013/11/, Cited March 2014. 5

[139] L. Valiant, G. Brebner, Universal schemes for parallel communication, in: Proceedings of the ACM Symposium on Theory of Computing (STOC), 1981, pp. 263–277. 9

[140] S. Vangal, J. Howard, G. Ruhl, S. Dighe, H. Wilson, J. Tschanz, D. Finan, P. Iyer, A. Singh, T. Jacob, et al., An 80-tile 1.28 TFLOPS network-on-chip in 65nm CMOS, in: Proceedings of the International Solid-State Circuits Conference Digest of Technical Papers (ISSCC), 2007, pp. 98–589. 8, 29

[141] S. Vangal, J. Howard, G. Ruhl, S. Dighe, H. Wilson, J. Tschanz, D. Finan, A. Singh, T. Jacob, S. Jain, V. Erraguntla, C. Roberts, Y. Hoskote, N. Borkar, S. Borkar, An 80-Tile Sub-100-W TeraFLOPS Processor in 65-nm CMOS, IEEE J. Solid-State Circ. 43 (1) (2008) 29–41. 5, 29, 34

[142] F. Verbeek, J. Schmaltz, A decision procedure for deadlock-free routing in Wormhole networks, IEEE Trans. Parallel Distributed Syst. PrePrints (99) (2013). 18

[143] D.W. Wall, Limits of instruction-level parallelism, in: Proceedings of the International Conference on Architectural Support for Programming Languages and Operating Systems (ASPLOS), 1991, pp. 176–188. 4

[144] L. Wang, Y. Jin, H. Kim, E.J. Kim, Recursive partitioning multicast: a bandwidth-efficient routing for networks-on-chip, in: Proceedings of the International Symposium on Networks-on-Chip (NOCS), 2009, pp. 64–73. 20

[145] D. Wentzlaff, P. Griffin, H. Hoffmann, L. Bao, B. Edwards, C. Ramey, M. Mattina, C.-C. Miao, J.F. Brown III, A. Agarwa, On-chip interconnection architecture of the TILE Processor, IEEE Micro 27 (5) (2007) 15–31. 8, 24, 25

[146] C. Wittenbrink, E. Kilgariff, A. Prabhu, Fermi GF100 GPU architecture, IEEE Micro 31 (2) (2011) 50–59. 5

[147] Y. Xu, B. Zhao, Y. Zhang, J. Yang, Simple virtual channel allocation for high throughput and high frequency on-chip routers, in: Proceedings of the International Symposium on High-Performance Computer Architecture (HPCA), 2010, pp. 1–11. 22

[148] B. Zafar, J. Draper, T.M. Pinkston, Cubic Ring networks: a polymorphic topology for network-on-chip, in: Proceedings of the International Conference on Parallel Processing (ICPP), 2010, pp. 443–452. 18

Logic implementations

PART

II

The logic implementations of a network-on-chip (NoC) directly determine its overhead, performance, and efficiency. The network latency depends on router pipeline delays; a primary design goal of router architectures is to reduce pipeline delays in a low-cost way. Buffers consume a significant portion of NoC power and area budgets. Dynamically sharing buffers among virtual channels (VCs) or ports can reduce buffer amount requirements, while maintaining or improving performance. Also, avoiding network congestion is essential to optimize the throughput and latency. Efficiently supporting multicast or broadcast communications is important for the overall performance of many-core systems. While packet switching networks are appropriate for unicast communications, bus structures are more suitable for multicast communications. Combining bus structures and packet switching networks can satisfy the requirements of both types of communications. On the basis of these observations, this part describes the NoC logic implementations in three chapters.

With an increasing number of cores, the NoC communication latency becomes a dominant factor for system performance owing to complex operations per network node. In Chapter 2 we try to reduce the communication latency by proposing single-cycle router architectures with wing channels, which forward the incoming packets to free ports immediately with the inspection of switch allocation results. Also, the incoming packets granted wing channels can fill in the time slots of the crossbar switch and reduce the contentions with subsequent packets, thereby increasing throughput effectively. The proposed router supports different routing schemes and outperforms several existing single-cycle routers in terms of latency and throughput. Owing to fewer arbitration activities, it achieves power consumption gains.

Chapter 3 studies the design of efficient NoC buffers. We first introduce a dynamically allocated VC design with congestion awareness. All the buffers are shared among VCs, whose structure varies with the traffic condition. At a low rate, this structure extends the VC depth for continual transfers to reduce packet latencies. At a high rate, it dispenses many VCs and avoids congestion situations to improve the throughput. Then, this chapter presents a novel NoC router with a shared buffer based on a hierarchical bit-line buffer. The hierarchical bit-line buffer can be configured flexibly according to the traffic, and it has a low power overhead. Moreover, we propose two schemes to optimize the router further. First, a congestion-aware output port allocator is used to assign higher priorities to packets heading in lightly loaded directions to avoid network congestion. Second, an efficient run-time VC regulation scheme is proposed to configure the shared buffer, so that VCs are allocated according to the loads of the network.

Chapter 4 presents a virtual bus on-chip network (VBON), a new NoC topology of incorporating buses into NoCs in order to take advantage of both NOCs and buses in a hierarchical way. This design is proposed on the basis of the following observations. Compared with transaction-based bus structures, conventional packet-based NoCs can provide high efficiency, high throughput, and low latency for many-core systems. However, these superior features are only applied to unicast (one-to-one) latency noncritical traffic. Their multihop feature and inefficient multicast (one-to-many) or broadcast (one-to-all) support make conventional NoCs awkward for some kinds of communications, including cache coherence protocol, global timing, and control signals, and some latency critical communications. The proposed VBON topology dynamically uses point-to-point links of conventional NoC designs as bus transaction links for bus request. This can achieve low latency while sustaining high throughput for both unicast and multicast communications at low cost. To reduce the latency of the physical layout for the bus organization, the hierarchical redundant buses are used. The VBON can provide the ideal interconnect for a broad spectrum of unicast and multicast scenarios and achieve these benefits with inexpensive extensions to current NoC routers.

A single-cycle router with wing channels[†]

2

CHAPTER OUTLINE

2.1 INTRODUCTION

As semiconductor technology is continually advancing into the nanometer region, a single chip will soon be able to integrate thousands of cores. There is a wide consensus, from both industry and academia, that the many-core chip is the only efficient way to utilize the billions of transistors, and it represents the trend of future processor architectures. Recently, industry and academia have delivered several commercial or prototype many-core chips, such as the Teraflops [5], TILE64 [18], and Kilocore [10] processors. The traditional bus or crossbar interconnection structures encounter several challenges in the many-core era, including the sharply increasing

---

[†]Part of this research was first published at Microprocessors and Microsystems—Embedded Hardware Design [9].

Networks-on-Chip. http://dx.doi.org/10.1016/B978-0-12-800979-6.00002-0

wire delay and the poor scalability. The network-on-chip (NoC), as an effective way for on-chip communication, has introduced a packet-switched fabric to address the challenges of the increasing interconnection complexity [1].

Although the NoC provides a preferable solution to mitigate the long wire delay problem compared with the traditional structures, the communication latency is still a dominant challenge with increasing core counts. For example, the average communication latencies of the 80-core Teraflops and 64-core TILE64 processors are close to 41 and 31 cycles, since their packets being forwarded between cores must undergo complex operations at each hop through five-stage or four-stage routers. The mean minimal path of an $n \times n$ mesh is given by the formula $2n/3 - 1/3n$ [20]; the communication latency increases linearly with the expansion of the network size. In this way, the communication latency easily becomes the bottleneck of application performance for the many-core chips.

There has been significant research to reduce the NoC communication latency via several approaches, such as designing novel topologies and developing fast routers. Bourduas and Zilic [2] proposed a hybrid topology which combines the mesh and hierarchical ring to provide fewer transfer cycles. In theory, architects prefer to adopt high-radix topologies to further reduce average hop counts; however, for complex structures such as a flattened butterfly [6], finding the efficient wiring layout during the back-end design flows is a challenge in its own right.

Recently, many aggressive router architectures with single-cycle hop latencies have been developed. Kumar *et al.* [8] proposed the express virtual channel (EVC) to reduce the communication latency by bypassing intermediate routers in a completely nonspeculative fashion. This method efficiently closes the gap between speculative routers and ideal routers; however, it does not work well at some nonintermediate nodes and is suitable only for deterministic routing. Moreover, it sends a starvation token upstream every fixed n cycles to stop the EVC flits to prevent the normal flits of high-load nodes from being starved. This starvation prevention scheme results in many packets at the EVC source node having to be forwarded via a normal virtual channel (VC), which increases average latencies.

Another predictive switching scheme is proposed in Refs. [14, 16], where the incoming packets are transferred without waiting for the routing computation (RC) and switch allocation (SA) if the prediction hits. Matsutani *et al.* [11] analyzed the prediction rates of six algorithms, and found that the average hit rate of the best one was only 70% under different traffic patterns. This means that many packets still require at least three cycles to go through a router when the prediction misses or several packets conflict. Kumar *et al.* [7] presented a single-cycle router pipeline which uses advanced bundles to remove the control setup overhead. However, their proposed design works well only at a low traffic rate since it emphasizes that no flit exists in the input buffer when the advanced bundle arrives. Finally, the preferred path design [12] is also prespecified to offer the ideal latency, but it cannot adapt to the different network environments.

In addition to the single-cycle transfer property exhibited by some of the techniques mentioned above, we emphasize three other important properties for the design of an efficient low-latency router:

(1) A preferred technique that accelerates a specific traffic pattern should also work well for other patterns and it would be best to be suitable for different routing schemes, including both deterministic and adaptive ones.

(2) In addition to low latencies under light network loads, high throughput and low latencies under different loads are also important since the traffic rate is easily changed on an NoC.

(3) Some complex hardware mechanisms should be avoided to realize the cost-efficiency of our design, and these mechanisms include the prediction, speculation, retransmission, and abort detection logics.

To achieve these three desired properties, we propose a novel low-latency router architecture with wing channels in Section 2.2. Regardless of what the traffic rate is, the proposed router inspects the SA results, and then selects some new packets without port conflicts to enter into the wing channel and fill the time slots of crossbar ports, thereby bypassing the complex two-stage allocations and directly forwarding the incoming packets downstream in the next cycle. Here, no matter what the traffic pattern or routing scheme is, once there is no port conflict, the new packet at the current router can be delivered within one cycle, which is the optimal case in our opinion. Moreover, as the packets of the wing channel make full use of the crossbar time slots and reduce contentions with subsequent packets, the network throughput is also increased effectively.

We then modify a traditional router with few additional costs, and present the detailed microarchitecture and circuit schematics of our proposed router in Section 2.3. In Section 2.4 we estimate the timing and power consumption using commercial tools, and evaluate the network performance via a cycle-accurate simulator considering different routing schemes under various traffic rates or patterns. Our experimental results show that the proposed router outperforms the EVC router, the prediction, and Kumar's single-cycle router in terms of latency and throughput metrics. Compared with the state-of-the-art speculative router, our proposed router provides latency reduction of 45.7% and throughput improvement of 14.0% on average. The evaluation results for the proposed router also show that although the router area is increased by 8.1%, its average power consumption is reduced by 7.8% owing to fewer arbitration activities at low rates. Finally, Section 2.5 concludes this chapter.

2.2 THE ROUTER ARCHITECTURE

This section proposes the single-cycle router architecture which supports several different routing schemes. On the basis of the inspection of SA results, the proposed router selects the incoming packets forwarded to the free ports for immediate transfers at wing channels without waiting for their VC arbitration (VA) and SA operations. Hence, it can reduce the communication latency and improve the network throughput under various environments. On the basis of the analysis of a baseline NoC router, we first present the single-cycle router architecture and its pipeline structure in Section 2.2.1. Then, we explain the detailed design of the wing channel in Section 2.2.2.

2.2.1 THE OVERALL ARCHITECTURE

It is well known that wormhole flow control was first introduced to improve performance through fine-grained buffering at the flit level. Here, the router with a single channel, playing a crucial role in the implementation of cost-efficient on-chip interconnects, always supports low latency owing to its low hardware complexity. However, it is prone to head-of-line blocking, which is a significant performance-limiting factor.

To remedy this predicament, the VC provides an alternative to improve the performance, but it is not amenable to cost-efficient on-chip implementation. Some complex mechanisms, including the VA and the two-phase SA, increase the normal pipeline stage. By detailed analysis of a router with VCs, it can be found that the SA and switch traversal (ST) are necessary during the transfer of each packet, but the pipeline delay of the former always exceeds that of the latter [17]. Hence, we believe that the complex two-phase SA may be preferred because of the arbitration among multiple packets at high rates, but it would increase communication latencies at low rates, where the redundant arbitrations, which increase the pipeline delay, are unwanted since no contention happens.

Given the aforementioned analysis, we introduce another alternative to optimize the single-cycle router. When an input port receives a packet, it computes the state of the forwarding path on the basis of the output direction and SA results. If the forwarding path is free, this port grants the wing channel to the packet, and then bypasses the two-phase SA pipeline stage to forward the packet directly. For the purpose of illustration, we first introduce the original speculative router [17] and then describe the detailed modifications.

The main components of a speculative router include the input buffers and next routing computation (NRC), VA, SA, and crossbar units. When a header flit comes into the input channel, the router at the first pipeline stage parallelizes the NRC, VA, and SA using speculative allocation, which performs SA based on the prediction of the VA winner. When the VA operation fails owing to conflicts with other packets, the router cancels the SA operation regardless of its results. Then, at the second pipeline stage, the winning packets will be transferred through the crossbar to the output port.

Figure 2.1 illustrates our proposed low-latency router architecture modified on the basis of the speculative router mentioned above. First, each input port is configured with a single wing channel, which uses the same number of flit slots to replace any VC reference, thus keeping the buffering overhead the same. The function of the wing channel is to hold the packets without contentions with others and to assert the request signal to a fast arbiter immediately when packets arrive to implement the single-cycle

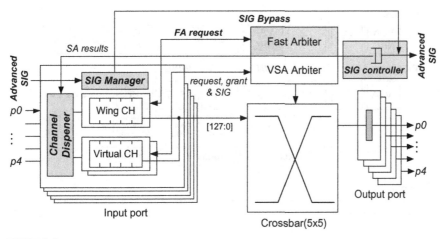

FIGURE 2.1

The low-latency router architecture.

transfer. Note that since the packet transfer at wing channels is established with the SA results from the previous cycle, it has lower priority than the normal transfers. Second, we add the fast arbiter, which is of low complexity, to handle the requests from wing channels of all inputs. Here, the winner of fast arbitration will now progress to crossbar traversal.

Third, each input introduces a channel dispenser unit. In contrast to the original router, the proposed router allocates the logical identifier of the free channel in the neighborhood. According to the logical identifier stored in the header flits, the dispenser unit grants the physical channel identifier to new packets. Besides, this unit is also responsible for selecting proper packets to use the wing channel, which is described in detail in Section 2.2.2. Finally, we use the advanced request signal to perform the RC in advance. In the original router, the RC or NRC must be completed before the ST, and it is difficult to combine them [13] in a single cycle without harming the frequency. Here, the advantage of the proposed advanced RC method is to decouple the routing and switch functions by precomputing the routing information before a packet arrives. As shown in Figure 2.1, because the signal manager has already received the request signal from the signal controller of the upstream neighborhood, the incoming packets can directly enter the ST stage without RCs being repeated.

Figure 2.2 depicts the original and proposed aggressive pipelines on which a header flit is transferred from router 0 to router 2. Here, the header flit in the original router requires at least two consecutive cycles to go through the VC and switch allocation (VSA)/NRC and ST pipeline stages. But in our proposed router, since the routing information is prepared prior to the arrival of the packet, the header flit can go through the crossbar switch within a single cycle once its input and output ports are both free. In the scenario shown in Figure 2.2, the flit arriving at router 0 inspects

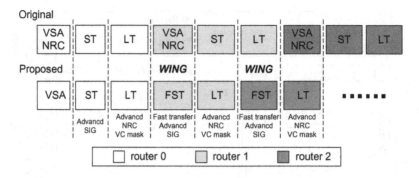

FIGURE 2.2

The aggressive pipeline.

the contentions with other packets and then it selects the normal VC to be forwarded to the right output through the original two-stage pipeline.

Here, once the header flit succeeds in winning the VC arbitration, it will send an advanced request signal (SIG) downstream in the next cycle for the routing precomputation at the downstream neighborhood. Then, the header flit is able to bypass the pipeline at router 1 by directly entering the wing channel on its arrival since its forwarding path is inspected to be free. In the previous link traversal (LT) pipeline stage, router 1 has generated a VC mask to determine the candidates for each output port among all the competing wing channels. In the fast switch transfer (FST) stage, router 1 chooses the single winning header flit to be transferred through the crossbar switch according to VC masks and propagates its advanced SIG downstream to router 2. At router 2, the header flit transfer is also completed in a single cycle, similarly to that at router 1.

For the purpose of comparison, Figure 2.3 illustrates the packet transfers of each input for a certain period when they are deployed with the original and proposed routers respectively. The x axis denotes the time, the y axis denotes the input port, the white boxes are the transfers at the wing channel, and the gray boxes are the transfers at normal VCs. At the beginning, while the packet addressed from input 4 to output 2 is blocked owing to the contention, other packets from inputs 1 and 3 are delivered toward outputs 2 and 4 respectively. In this scenario, we collect all the time slots of the crossbar switch, and it can be seen that, as compared with the original router (Figure 2.3a), the proposed router provides up to 45.1% reduction in latency and up to 31.7% improvement in throughput; these gains are mainly due to the single-cycle transfers and decreased contentions.

On the basis of a detailed analysis of the router behaviors, the proposed router obviously outperforms the original router, and the reasons for this are as follows. First, with the inspection of the SA results in the previous cycle, the proposed router allocates the free time slots of the crossbar switch for single-cycle transfers of new packets at wing channels. It can reduce the packet latencies and improve

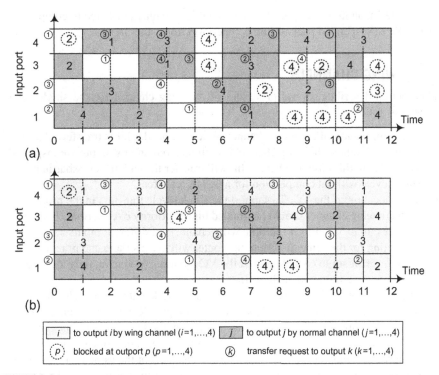

FIGURE 2.3

The packet transfer of the original and proposed routers.

the SA efficiency, thereby increasing throughput. Second, the single-cycle transfers of new packets at wing channels advance all the packets at the same input and significantly reduce the level of contentions with subsequent packets of other inputs, which translates directly to a higher throughput.

Next, we provide two case studies. The first case happens at cycle 4, where the packets from inputs 2, 3, and 4 rush to output 4 simultaneously. In the original router, the SA chooses only a single winner, and thus it leads to the idle cycles of inputs 3 and 4 as shown in Figure 2.3a. But wing channels are granted to these packets in our proposed router. At cycle 5, when the packet from the normal channel of input 4 is forwarded toward output 2, the new packet at input 2 undergoes the single-cycle transfer with wing channels immediately once it inspects the idle state of output 4, thereby improving the switch channel utilization. Next, let us consider the second case, where the new packet arrives at input 4 at cycle 2. Here, the packet in the original router is delivered to output 3 one cycle later than in our proposed router, and thus it directly suspends the transfers of subsequent packets forwarded to output 2 or output 4. As illustrated in Figure 2.3a, the contentions at output 2 or output 4 become higher after cycle 7 than in our proposed router, and this is mainly attributed to competition among these suspended packets and competition from other inputs.

2.2.2 WING CHANNELS

This section describes several principles of the wing channel for the single-cycle transfer as follows:

- The dispensation of the wing channel. When the header flit is transferred to the input, it is granted the wing channel if both its input and its output are inspected to be free. In this case, the header flit will transfer through the crossbar in a single cycle without competition for network resources.
- The allocation of the VC. To support the single-cycle transfers in the proposed router, a wing channel should be granted higher priority over normal channels to win the output VC. But if multiple wing channels of different inputs are competing for the same direction, the router will use a simple and practical VC mask mechanism to cancel most of the VC requests, producing only a single winner for one output at each cycle.
- The arbitration of the crossbar switch. The new packet at the wing channel sends its request to the fast arbiter (FSA) unit if its output VC has been granted or its VC mask is invalid. If it wins the fast arbitration, the header flit will be delivered to the right output and its body flits will come into heel in sequence. In this scenario, the transfer requests from normal channels of the same input will be cancelled until the wing channel becomes empty, or the tail of the packet leaves the router. However, if its fast arbitration fails, an idle cycle of the input port is inevitable and then the flits of other normal channels will go through the switch in the next cycle depending on the current SA results.
- The transmission of an advanced SIG. The SIG encodes the logical identifier of the channel, NRC results, and destination node. For an 8 × 8 mesh, this wiring overhead is 10 bits, around 7.8% of the forward wiring, assuming 16-byte-wide links. In the proposed router, if the VC mask of the new packet at the wing channel is invalid, its advanced SIG must be given the highest priority to be forwarded downstream immediately. However, for those packets with valid VC masks or at normal channels, their advanced SIGs are sent to the SIG FIFO unit one cycle later than the grant of output VCs. Obviously, the latter has lower priority than the former. With this principle, we ensure that the completion of the RC is prior to the arrival of the packet, because the advanced SIG always arrives at least one cycle earlier than header flits.

Next, Figure 2.4 shows a timing diagram for packet transfer from router 1 to router 5 with these principles. At cycle t_2, the header flit arriving at router 2 enters the wing channel by inspecting the SA results, and then it transfers through the router and forwards its advanced SIG in the next cycle. When the header reaches router 3, its RC is already completed. But at cycle t_4, this flit cannot be delivered immediately owing to its contentions with the packets of other wing channels. Here, the router

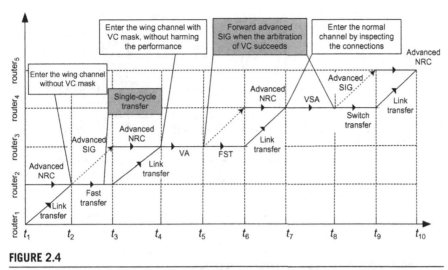

FIGURE 2.4

The packet transfer using wing channels.

sets up the VC mask and repeats the request of the output VC at [t_4, t_5]. Once the VA has been won at cycle t_5, the router immediately sends the advanced SIG to the FIFO unit and this flit's trip continues through the FST stage at the same time. Note that although the header flit stays at router 3 for more than one cycle, it allows the packets of other channels to be transferred in advance without affecting the network performance. Finally, owing to the contentions inspected at cycle t_7, this header flit selects the normal channel and transfers through the original pipeline stage, thereby leaving the wing channel to subsequent new packets.

On the basis of the above principles, the fast arbitration unit allocates the free time slots of the crossbar switch to new packets when they arrive to improve the network performance and reduce the power consumption. Figure 2.5 illustrates the effect of fast arbitration for the wing channel. At cycle 4, the transfer completion of the wing channel informs the normal channels at the same input to request their outputs. The packets that win the SA are then able to traverse the crossbar and leave no idle cycle on the input link of the switch, which translates directly to a higher throughput. Next, at cycle 8, the empty slot due to the SA failure of cycle 7 is filled by the single-cycle transfer of fast arbitration, which also leads to a significant improvement in throughput. However, the idle cycle of the input link still appears at cycle 10. Here, our proposed router restarts the output requests of the other channels at the same port when it inspects the empty wing channel FIFO unit owing to the transfer interrupt, and thus leaves one idle cycle. This scenario is unavoidable but there is no serious performance degradation because of its low frequency. On the other hand, our proposed router cancels the output requests of other channels during the transfer of the wing channel, which can decrease the arbitration power consumption in most cases.

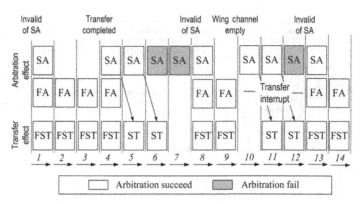

FIGURE 2.5

The fast arbitration for wing channels.

2.3 MICROARCHITECTURE DESIGNS

On the basis of an original speculative router, this section presents the main microarchitecture designs, including the channel dispenser, fast arbitration, SIG controller, and SIG manager units.

2.3.1 CHANNEL DISPENSERS

Figure 2.6 shows the microarchitecture of the channel dispenser, which is mainly composed of the VC assigner, VC tracker, and VC table. The VC table forms the core of the dispenser logic. It is a compact table indexed by the logical channel ID, and holds the physical channel ID and tail pointers at each entry. With the VC table, the VC tracker simply provides the next available channels by keeping track of all the normal ones.

When receiving the information from the VC tracker, the VC assigner decides whether to grant a wing channel or a normal channel to the new packet on the basis of the generated wing flag. Once any channel has been granted to the incoming packets, the VC assigner needs to provide the dispensed channel ID to the VC table to change the status of the available channels. Here, the VC assigner is also responsible for improving the throughput by fully utilizing all VCs at high rates. When all the normal VCs at the local input are exhausted, the VC assigner is forced to allocate the wing channel to the new packet. In this way, when the normal VCs are occupied and the generated wing flag is false, the wing channel will serve as a normal one, and the packet allocated the wing channel will apply for the channel and switch like those in the normal ones. Our experiments described in Section 2.4.3 validate the effectiveness of this technique under different routing schemes.

This channel dispenser also guarantees protocol- and network-level deadlock-free operations by adopting relatively low-cost methods. Toward that goal, we use different VC sets for request and response message types, respectively, to avoid

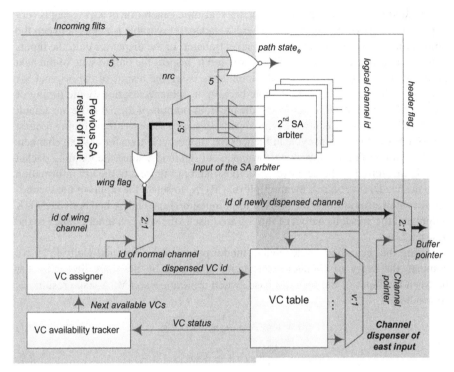

FIGURE 2.6

The channel dispenser.

protocol- and network-level deadlock situations. The proposed router includes two VC sets per port: the first VC set is composed of channel 0 (i.e., wing channel), channel 1, and channel 2, and is used by the request packets; the second VC set, comprising channels 0, 1, and 3, is used by the response packets.

Note that since channels 2 and 3 are used by request and response packets exclusively, we can provide deadlock-free operations in two aspects. First, in order to provide deadlock recovery in adaptive routing algorithms, channels 2 and 3 employ a deterministic routing algorithm to break the network deadlock. For any packet at the wing channel or channel 1, once it has been checked for possible deadlock situations, the channel dispenser unit of the neighborhood will grant channels 2 and 3 to this packet with higher priority, thereby using deterministic routing to break the network deadlock situation. Second, we introduce separate channels 2 and 3 for two different message types to break the dependencies between the request and response messages. When both the shared channel 0 and channel 1 at the neighborhood are holding other messages of different types, the current message can be transferred through the exclusive VC on the basis of its own type, thereby breaking the dependency cycle between messages of different classes.

To satisfy the tight timing constraint, the real-time generation of a wing flag is very important, as shown in Figure 2.6, where we adopt a favorable tradeoff between the limited timing overhead and the fast packet transfer. In the dispenser unit, the inputs of the second-phase switch arbitration are used to inspect the output state of the next cycle; however, the inspection of the input state using the results of the original SA pipeline would prove to be intolerable because it influences seriously the timing of the critical path. Instead, we consider the previous outputs of the SA to be the default input state, thereby reducing the critical path. In such a scenario, the winning request from the SA of the current cycle influences the single-cycle transfer of wing channels but does not harm the network performance at high loads, since the winning packet will be transferred in the next cycle. As illustrated in Figure 2.6, the *nrc* information from the header flit controls the multiplexer (MUX) to select the inputs of the second-phase switch arbitration on the basis of its output direction. Only when both the MUX output and the previous output of the SA are false is the wing channel granted to the new packets.

Using the technology-independent model proposed by Peh and Dally [17], we compare the delay of channel dispensation indicated by the thick line with that between the input of the second-phase switch arbitration and VSA stage results, as shown below:

$$t_1(n) = \log_4 2n + \frac{11}{15}n + 4, \tag{2.1}$$

$$t_2(n, v) = \frac{33}{10}\log_4 n + \frac{n}{3} + \log_4 v + 3\frac{5}{12}. \tag{2.2}$$

In the typical case of five ports and four channels, the former equation equals 9.3 fan-outs of 4 (FO4), which is close to the 9.9 FO4 of the latter one. In addition, it also introduces some timing overheads to generate the buffer pointer based on the LTs. With the same model, a delay of 3 FO4 is added when compared with the *v*-input MUX of the original router. Here, the delay of 1 mm wire after placement and routing in a 65 nm process is calculated be 8.2 FO4, which is far less than the 18 FO4 of the original critical path. Hence, this overhead would not harm the network frequency.

2.3.2 FAST ARBITER COMPONENTS

Taking the east port as an example, we show in Figure 2.7 the microarchitecture of a fast arbiter which includes the VC mask logic in the upper left corner and the transfer arbitration logic in lower right corner. First, the grant bit for the wing channel of a certain input is set in the case when the incoming header tends to undergo the single-cycle transfer from this input to output east as discussed in Section 2.2.1. In order to do so, the incoming header always needs to check the states of the wing channel and its forwarding path as shown in Figure 2.7. Then, on the basis of these grant bits for the wing channels, the VC mask logic shields multiple requests competing for the same output. Here, the state indicating that one input request has priority over another is stored in a matrix, which is updated each cycle in a round-robin fashion

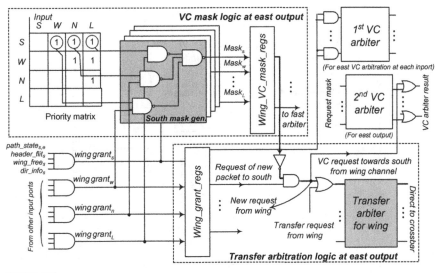

FIGURE 2.7

The fast arbiter and virtual channel arbitration components.

to reflect the new priorities. With the current matrix state, the mask bit of an input is set when any input request that has higher priority is asserted. Finally, this logic will assert the new request of a wing channel to compete for the output if the conditions of the grant bit and the VC mask bit are both satisfied. In our design, any valid bit of the wing grant register also generates a request mask signal to shield input requests of all the first-phase VC arbiters for a particular output, thereby reducing the energy overhead of VC arbitration.

The packets at the wing channels undergo single-cycle transfers based on the fast arbitration results, and thus the delay of the functions on the entire path through the VC mask register, fast arbiter, and crossbar traversal is calculated in Equation (2.3), where the result of 14.3 FO4 in the case of five ports and 16-byte width ($w = 128$) is lower than the 18 FO4 of the original critical path:

$$t_3(n, w) = 6\frac{7}{10}\log_4 n + 3\frac{1}{60} + \log_4 w. \tag{2.3}$$

In addition, with the same model, the timing overhead of the wing grant and VC mask based on the LT is calculated to be about 4.0 FO4, which also has no influence on the critical path.

2.3.3 SIG MANAGERS AND SIG CONTROLLERS

Figure 2.8 shows the SIG manager of the input port and the SIG controller of the output port. Here, the SIG manager completes the RC in the next cycle of advanced

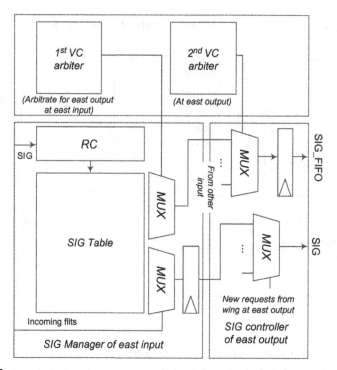

FIGURE 2.8

The SIG manager and controller.

SIG arrivals, and records the results in a compact SIG table, which holds the transfer destination and NRC information. For the incoming packet, the header flit uses its logical channel identifier to select and latch its SIG information on the basis of the SIG input table. By receiving SIGs from multiple inputs, the SIG controller decides on the winning SIG according to the new request from the wing channel in the next cycle. On the other hand, for the other packets at normal channels or blocked ones at wing channels, the two-phase VC arbitration logic provides their advanced SIG instead. For the purposes of low complexity, our proposed router adopts another two-phase VC arbitration [15], where the first-phase arbiters at each input gain winner requests from different directions and then the second-phase logic at each output generates a final winner at each cycle. With the winner of the first phase, the SIG selection, which is parallel with second-phase arbitration, is performed by the MUX. Then, according to the second-phase result, the SIG controller selects the SIG of final winners, thereby forwarding it to the SIG FIFO unit in the next cycle.

2.4 **EXPERIMENTAL RESULTS**

2.4.1 **SIMULATION INFRASTRUCTURES**

In this section, we present the simulation-based performance evaluation of the proposed router in terms of latency, throughput, pipeline delay, area, and power consumption, under different traffic patterns. To model the network performance, we develop a cycle-accurate simulator, which models all major router components at the clock granularity, including VCs, switch allocators, and flow controls. The simulator inputs are configurable, allowing us to specify parameters such as the network size, routing algorithm, channel width, and VC counts. And the simulator also supports various traffic patterns, such as random, hotspot, and nearest-neighbor patterns. Here, all simulations are performed in an 8×8 mesh network by simulating 8×10^5 cycles for different rates after a warm-up phase of 2×10^5 cycles. In more detail, each router is configured with five physical channels, including the local channel, and each port has a set of four VCs with a depth of four 128-bit flits. To take the estimation a step further for the original and proposed routers, we firstly synthesize their RTL designs with a 65 nm standard cell library by Synopsys Design Compiler, and then derive the pipeline delay with static timing analysis by Synopsys Timing Analyzer. For power consumption, we further finish the physical design of the above-described on-chip network and antimark the parasitic parameters derived from the circuit layout into netlist to evaluate the total power overhead. In this way, the switching activities of both routers and wires were captured through gate-level simulation using a core voltage of 1.2 V and a frequency of 500 MHz.

2.4.2 **PIPELINE DELAY ANALYSIS**

Figure 2.9 displays the distribution of the pipeline delay obtained by varying the numbers of ports and channels for two routers. It is evident that the wing channel incurs a modest timing overhead for the ST and LT stages, but this does not obviously affect the delay of the VSA stage, which is the critical path in our design. Here, with the number of ports and channels increasing to a certain extent, the difference of their critical path delay is quite small.

As shown in Figure 2.9, the critical path delay of the proposed router is increased by 8.9% in the case of five ports and two channels, but this difference is gradually reduced to 3.2% and 1.3% when the number of channels reaches four and eight respectively. This is because the delay between the input of the second-phase switch arbitration and the VSA stage results increases as the channel number (i.e., v) increases, while that of channel dispensation is constant. Then, in the case of four channels, it can be seen that the critical path delay of our router tracks that of the original one well, and this is because the change of $t_2(n, v)$ is close to that of $t_1(n)$ as n increases. As expected, the path through the fast arbiter and crossbar traversal has never been the critical one with increasing n in our timing analysis, and this proves that our proposed router is suitable for different port numbers without obvious

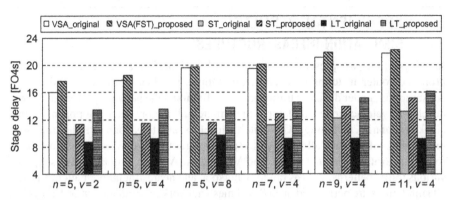

FIGURE 2.9

The effect of two parameters on the pipeline delay.

frequency degradation. As shown, the increase of the critical path delay is around 3.7% on average in different cases.

2.4.3 LATENCY AND THROUGHPUT

This section first compares three routers in terms of zero-load latency as shown in Figure 2.10. With increasing network size, the proposed router is clearly much more efficient at reducing the latency. The results show that the proposed router reduces the

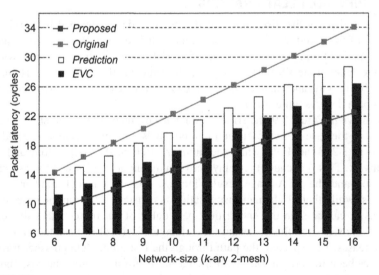

FIGURE 2.10

The comparison of zero-load latencies.

zero-load latency by 34.3% compared with the original one when the network size reaches 16-ary. Then, it also results in 22.6% and 16.1% reductions in average zero-load latency over prediction [11] and EVC [8] routers with different network sizes.

Next, we evaluate the network performance of the proposed router as a function of the injection rate and traffic pattern under different routing schemes, and compare it with those of the original router, prediction router [11], EVC router [8], and Kumar's single-cycle router [7] respectively. Here, we refer to Kumar's single-cycle router as SNR for simplicity, and all the prediction, EVC, and SNR structures are implemented and incorporated into our cycle-accurate simulator. To study the performance of EVCs, we configure one EVC within the entire set of four VCs and consider the dynamic EVC scheme with a maximum EVC length of three hops. For better performance of the prediction router, we use many 4-bit binary counters to construct the history table, which records the number of references toward all the outputs, and thus the router can achieve a higher hit rate by supporting the finite context prediction method. Finally, for a fair comparison, we use the statically managed buffers in SNR and assign the same number of buffers to each VC per port when modeling the behavior of major components.

Figure 2.11 displays the network performance of four routers under *XY* deterministic routing. For a random traffic pattern (Figure 2.11a), the proposed router latency of 13.1 cycles at a rate of 0.1 flits per node per cycle is closer to the ideal, e.g., 11.8 cycles, when compared with the 20.3, 18.6, 16.5, and 13.8 cycles of the others. Here, our proposed router performs single-cycle transfer at each node at a low rate; however, the prediction router always takes three cycles when the prediction misses and the EVC router still needs two cycles at some nonintermediate nodes along its express path. Then, with the increase of network loads, our proposed router uses the single-cycle transfer of the wing channel to fill in the free time slot of the crossbar switch and reduce the contentions with subsequent packets of other inputs, thereby leading to significant improvement of throughput. However, with increase of network traffic, the SNR's performance, which is almost identical to that of our proposed router at low rates, begins to degrade obviously. This is attributed to its strict precondition that no flit exists in the input buffer when the advanced bundle arrives. Under higher traffic rates, with SNR it is hard to forward most incoming packets within one cycle owing to there being many residual flits at router input ports. The results for our proposed router show 44.5%, 34.2%, 21.7%, and 36.1% reductions in average latency compared with the other four routers, while improving throughput by 17.1%, 9.3%, and 9.0% over the original, prediction, and SNR routers respectively.

For the hotspot pattern (Figure 2.11b), the wing channel is effective in reducing packet latency at many low-load nodes, while improving switch throughput by shortening service time and reducing residual packets at a few high-load nodes. Hence, the simulation results show that our proposed router reduces average latency by 46.8%, 38.8%, and 25.1% and improves throughput by 10.8%, 8.2%, and 4.1% when compared with the original, prediction, and SNR routers respectively. Then, we observe that the express channels at some low-load nodes are underutilized. Because the EVC router prevents EVC flits periodically to avoid the starvation of normal flits,

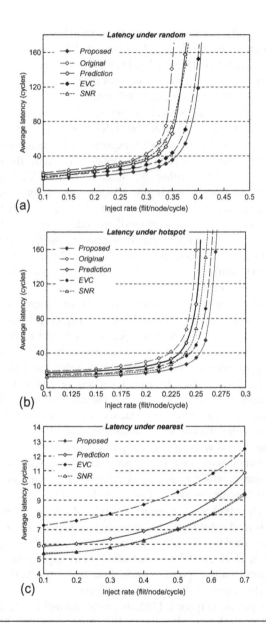

FIGURE 2.11

The network performance with deterministic routing. (a) Uniform random. (b) Hotspot. (c) Nearest neighbor.

many packets at low-load nodes cannot be forwarded through express channels that are occupied by others, thereby increasing packet latencies. As shown in Figure 2.11b, our proposed router reduces the average latency by 27.8% when compared with the EVC router.

For the nearest-neighbor pattern, the EVC router does not work well, and takes two cycles at each node for communications between neighborhoods. Then, the prediction router performs the single-cycle transfer at the neighbor but still needs three cycles at the source node when the prediction misses. In contrast, our proposed router and the SNR router perform better than the others, as shown in Figure 2.11c. This is because their packets can transfer through either the source or the destination in a single cycle. As shown in Figure 2.11c, the latency of our router is reduced by 27.0% and 10.2% when compared with the EVC and prediction routers respectively.

In addition, Figure 2.12 compares network performance under regional congestion awareness adaptive routing among all routers except the EVC router, which is restricted to the deterministic routing scheme. Here, we use the combinations of free VCs and crossbar demand to be the congestion metric [4]. First, the proposed router outperforms the original and SNR routers in terms of packet latency and saturated throughput in all cases. For the SNR router, the adaptive routing algorithm forwarding packets to the low traffic load regions reduces the possibility of contention, and thus helps in clearing the input buffers by delivering the incoming flits in time. As seen in Figure 2.12a and b, at low injection rates, the performance of the SNR router tracks that of the proposed router well. However, as network traffic increases, the increasing number of residual flits per port starts to cancel the single-cycle pipeline operations at the SNR router. Note that this issue can be completely avoided in our proposed design by guaranteeing fast transferring of the wing channel. It can be seen that our design outperforms the SNR router, with the latency reduction near saturation being nearly 44.2%. Second, for the prediction router, the latest port matching strategy to predict output is introduced to provide high hit rates in most cases, but only approximately 68% of predictions hit in our experiment. Once the prediction misses or the packet conflicts, it requires at least three cycles in contrast with two cycles for our proposed router to forward a header flit, thereby increasing the number of residual packets and saturating network throughput in advance. When compared with the prediction router, our results in Figure 2.12a and b show that average latency is reduced by 28.1% and 34.0% and network throughput is improved by 10.2% and 5.6% for random and hotspot patterns respectively.

Finally, we also investigate the performance of the proposed router under realistic network traces which are collected from the 64-processor multiprocessor simulator [3] using the `raytrace` and `barnes` applications from SPLASH2 benchmarks [19]. Figure 2.12c illustrates the simulation results for three different routers when compared with the original router under adaptive routing. For the `raytrace` application, the communication patterns among nodes correspond to low traffic load with large hop counts. In this case, the switch contention is low and the router ports are empty, and thus one-cycle delays through the SNR and proposed routers are realized, whereas the prediction router needs three cycles when the predictions miss according

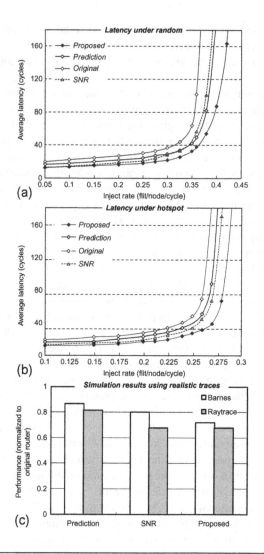

FIGURE 2.12

The network performance with adaptive routing. (a) Uniform random. (b) Hotspot. (c) Realistic traces.

to the traffic patterns or history tables. It can be seen that the SNR router and the proposed router provide up to 32.1% and 32.3% reductions in average latencies, in contrast with that for the prediction router, which is 19.4%. Then, for the `barnes` application, representing a moderate traffic rate with fewer hop counts, the multiple flits stored in the input buffer seriously influence the one-cycle transferring process through the SNR router. In contrast, our proposed router enables the single-cycle pipeline operations if only its delivering input and output ports are inspected to be free, regardless of how many flits are stored in the input buffers. Hence, the evaluation results in Figure 2.12c show that with the `barnes` trace, our proposed router, with a latency reduction of 28.6%, outperforms the SNR router, with the reduction in this case being around 19.3%.

2.4.4 AREA AND POWER CONSUMPTION

This section estimates the area of the original and proposed routers with a 65 nm standard cell library as shown in Figure 2.13a. Here, both of them are configured with 16 buffers per input port, and thus the wing channel will not incur any extra area overhead due to buffer resources. In our proposed router, we add the channel dispenser, the SIG manager, and the fast arbiter, and modify the VC arbitration component. However, the area overhead of these components contributes little to the

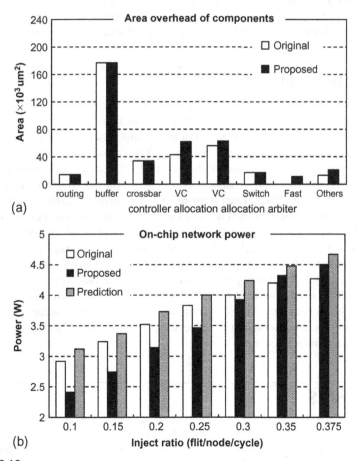

FIGURE 2.13

The router area and network power consumption. (a) Area overhead. (b) Power overhead.

total area, which is dominated by the buffers. Thus, the total area of our router is increased by nearly 8.1% compared with that of the original router.

In terms of power consumption, we investigate 8×8 mesh networks with adaptive routing at different rates in Figure 2.13b. Here, because the increasing number of wires incurs some extra link power as reported in Section 2.2.2, both routers and forward wires are considered in the power evaluation on the basis of the physical design of an on-chip network. Although more switching activities of the channel dispensation, SIG forwarding, and transmission lead to a modest increase of power consumption, the average power consumption of the proposed router when compared with the original one is still reduced by 7.8%. This is because most packets which undergo the single-cycle transfer at low rates avoid the complex two-stage switch and VC arbitration activities. For the prediction router, the high miss rate always results in the frequent switching activities of some complex logics, e.g., abort detection, copy retransmission, and flit killer, which are power consuming and completely avoided in our design. Hence, it can be seen that the power consumption of the proposed router is lower than that of the prediction router at all rates, with an average reduction of 15.1%.

2.5 CHAPTER SUMMARY

As continuing shrinkage of technology in the nanometer era is leading to an increase in the number of cores, the NoC as an effective communication fabric will face the tough challenges of low latency and high throughput in the future. In this chapter, we proposed a practical router architecture with a wing channel, which is granted to the incoming packets by inspecting SA results for the single-cycle transfers. With the wing channel, the packet transfers also improve the SA efficiency and reduce contentions with subsequent ones from other ports, thereby increasing throughput effectively. The proposed router is designed using a 65 nm CMOS process, and it only incurs little overhead of critical path delay while exhibiting excellent scalability with increasing numbers of channels and ports. Simulation results using a cycle-accurate simulator indicate that our proposed design outperforms the EVC, prediction, and SNR routers in terms of latency and throughput under various traffic patterns, and it provides 45.7% latency reduction and 14.0% throughput improvement as compared with the speculative router. Moreover, although the router area is increased by 8.1%, the average power consumption is reduced by 7.8%, which is attributed to fewer arbitration activities at low rates.

REFERENCES

[1] T. Bjerregaard, S. Mahadevan, A survey of research and practices of network-on-chip, ACM Computing Survey 38 (1) (2006) 1–51. 54
[2] S. Bourduas, Z. Zilic, A hybrid ring/mesh interconnect for network-on-chip using hierarchical rings for global routing, in: Proceedings of the International Symposium on Networks-on-Chip (NOCS), 2007, pp. 195–204. 54

[3] H. Chafi, J. Casper, B. Carlstrom, A. McDonald, C. Minh, W. Baek, C. Kozyrakis, K. Olukotun, A scalable, non-blocking approach to transactional memory, in: Proceedings of the International Symposium on High-Performance Computer Architecture (HPCA), 2007, pp. 97–108. 71

[4] P. Gratz, B. Grot, S. Keckler, Regional congestion awareness for load balance in networks-on-chip, in: Proceedings of the International Symposium on High-Performance Computer Architecture (HPCA), 2008, pp. 203–214. 71

[5] Y. Hoskote, S. Vangal, A. Singh, N. Borkar, S. Borkar, A 5-GHz mesh interconnect for a Teraflops Processor, IEEE Micro 27 (5) (2007) 51–61. 53

[6] J. Kim, J. Balfour, W. Dally, Flattened butterfly topology for on-chip networks, in: Proceedings of the International Symposium on Microarchitecture (MICRO), 2007, pp. 37–40. 54

[7] A. Kumar, P. Kundu, A. Singh, L.-S. Peh, N. Jha, A 4.6Tbits/s 3.6GHz single-cycle NoC router with a novel switch allocator in 65nm CMOS, in: Proceedings of the International Conference on Computer Design (ICCD), 2007, pp. 63–70. 54, 69

[8] A. Kumar, L.-S. Peh, P. Kundu, N.K. Jha, Express virtual channels: towards the ideal interconnection fabric, in: Proceedings of the International Symposium on Computer Architecture (ISCA), 2007, pp. 150–161. 54, 69

[9] M. Lai, L. Gao, S. Ma, X. Nong, Z. Wang, A practical low-latency router architecture with wing channel for on-chip network, Microprocessors and Microsystems 35 (2) (2011) 98–109. 53

[10] B. Levine, Kilocore: scalable, high performance and power efficient coarse grained reconfigurable fabrics, in: Invited Presentation at International Symposium on Advanced Reconfigurable Systems (ISARS), 2005. 53

[11] H. Matsutani, M. Koibuchi, H. Amano, T. Yoshinaga, Prediction router: yet another low latency on-chip router architecture, in: Proceedings of the International Symposium on High-Performance Computer Architecture (HPCA), 2009, pp. 367–378. 54, 69

[12] G. Michelogiannakis, D. Pnevmatikatos, M. Katevenis, Approaching ideal NoC latency with pre-configured routes, in: Proceedings of the International Symposium on Networks-on-Chip (NOCS), 2007, pp. 153–162. 54

[13] R. Mullins, A. West, S. Moore, Low-latency virtual-channel routers for on-chip networks, in: Proceedings of the International Symposium on Computer Architecture (ISCA), 2004, pp. 188–197. 57

[14] R. Mullins, A. West, S. Moore, The design and implementation of a low-latency on-chip network, in: Proceedings of the Asia and South Pacific Conference on Design Automation (ASP-DAC), 2006, pp. 164–169. 54

[15] C.A. Nicopoulos, D. Park, J. Kim, N. Vijaykrishnan, M.S. Yousif, C.R. Das, ViChaR: a dynamic virtual channel regulator for network-on-chip routers, in: Proceedings of the International Symposium on Microarchitecture (MICRO), 2006, pp. 333–346. 66

[16] D. Park, R. Das, C. Nicopoulos, J. Kim, N. Vijaykrishnan, R. Iyer, C. Das, Design of a dynamic priority-based fast path architecture for on-chip interconnects, in: Proceedings of the the Symposium on High Performance Interconnects (HOTI), 2007, pp. 15–20. 54

[17] L.S. Peh, Flow control and micro-architectural mechanisms for extending the performance of interconnection networks (Ph.D. thesis), Stanford, USA, 2001. 56, 64

[18] D. Wentzlaff, P. Griffin, H. Hoffmann, L. Bao, B. Edwards, C. Ramey, M. Mattina, C.-C. Miao, J.F. Brown III, A. Agarwa, On-chip interconnection architecture of the TILE processor, IEEE Micro 27 (5) (2007) 15–31. 53

[19] S. Woo, M. Ohara, E. Torrie, J. Singh, A. Gupta, The SPLASH-2 programs: characterization and methodological considerations, in: Proceedings of the International Symposium on Computer Architecture (ISCA), 1995, pp. 24–36. 71

[20] D. Yingfei, W. Dingxing, Z. Weimin, Exact computation of the mean minimal path length of n-mesh and n-torus, Chinese Journal of Computers 20 (4) (1997) 376–380. 54

Dynamic virtual channel routers with congestion awareness[†]

3

CHAPTER OUTLINE

3.1 INTRODUCTION

In on-chip networks, communication resources such as buffers and links are finite and often shared by many packets in the time-division multiplexing [24] or spatial-division multiplexing [20, 36] way. Under heavy traffic conditions, multiple packets

---

[†]Part of this research was first presented at the 45th Design Automation Conference (DAC-2008) [19] and the 29th International Conference of Computer Design (ICCD-2011) [33].

Networks-on-Chip. http://dx.doi.org/10.1016/B978-0-12-800979-6.00003-2
Copyright © 2015 China Machine Press/Beijing Huazhang Graphics & Information Co., Ltd.
Published by Elsevier Inc. All rights reserved.

contend for the precious on-chip resources, leading to congestion situations. When congestion occurs, many packets will be blocked in the network and overall performance falls dramatically. Congestion management was proposed to prevent networks from becoming saturated and improve the throughput in the network-on-chip (NoC). Congestion can be avoided or relieved using various methods from the link layer to the application layer. At the link layer, high performance or flexible link techniques [17] are often employed to enhance packet-transferring capability. At the network layer, various techniques such as adaptive routing [11], wormhole flow control [10], and virtual channels (VCs) [26] can greatly reduce the congestion. At the application layer, congestion-aware task mapping [6] and scheduling [7] are often considered as effective approaches to avoid congestion.

According to the method of tackling congestion situations, the techniques mentioned above can be categorized into two groups. The first group, which comprises congestion-aware task mapping/scheduling and adaptive routing, first predicates the possible congestion, then adjusts the commutation path to *avoid* the upcoming congestion. Congestion-aware task mapping and scheduling is not the topic of this book, while adaptive routing, which targets an evenly distributed traffic load over the network, will be investigated in Part III. However, congestion cannot be avoided absolutely and often occurs in any NoC. In this situation, use of the other techniques belonging to the second group will make sense. With the help of wormhole flow control and VC techniques, buffer utilization achieves significant enhancement and congestion may be *relieved*. Almost all the on-chip networks have already employed these two techniques.

How to further enhance network performance and reduce congestion is a hot research domain, one important direction of which is the microarchitecture optimization, including designing few pipeline stages [18], efficient allocators [3, 25], high-speed links [32], and efficiently managed buffers [4]. From previous work [31], we know that the key inhibitor of performance in conventional VCs is the fixed structure, which results in serious network blocking or a large number of packet cross-transfers. This chapter exploits dynamic VC (DVC) buffer designs to avoid or relieve congestion. We first focus on DVC structures and analyze how to improve performance. Then, corresponding congestion awareness schemes for different DVC structures are investigated. Some specific packets will be picked and given higher priorities for moving to downstream nodes. As a result, precious buffer resources in the congested node will be released quickly, and the congestion situation may be avoided or quickly solved.

In on-chip routers, buffers are commonly implemented using either register-based FIFO buffers [28] or SRAM-based FIFO buffers [37]. Aiming at the two types of FIFO buffers, we propose two kinds of DVC structures with congestion awareness. The congestion awareness schemes dynamically allocate buffer resources to improve buffer utilization. The structures of two DVC buffers will be described in Sections 3.2 and 3.3, respectively. On the basis of the proposed DVC structures, two NoC router architectures are designed, and they will be described in Sections 3.4 and 3.5.

In this chapter we make the following main contributions:

- We propose two kinds of DVC structures.
- We propose corresponding congestion awareness schemes for DVC buffers to further optimize buffer allocation.
- We propose microarchitectures of two on-chip routers based on dynamic VC structures.

3.2 DVC WITH CONGESTION AWARENESS
3.2.1 DVC SCHEME

In an on-chip network, head-of-line (HoL) blocking and low link utilization are critical issues, especially for communication-intensive applications. Adaptive routing, wormhole flow control, and VC techniques can overcome these issues. Wormhole flow control and VC techniques are often adopted by the generic on-chip router, while simple deterministic routing algorithms such as source-based routing [5, 12] and dimensional order routing [9, 13] are employed instead of adaptive routing in order to reduce the complexity of on-chip routers. Deterministic algorithms do not avoid traffic congestion, and thus the network performance is degraded when the traffic pattern has localities.

Although the VC flow control scheme providing an alternative to reduce HoL blocking may improve link utilization, it brings with it lots of area and power overheads. For instance, the VC buffers will take up to nearly 50% of the area and will account for nearly 64% of leakage power in a router implemented under 70 nm CMOS technology [8]. There have been significant works on VC organizations to save buffers and exploit performance. Huang *et al.* [14] customized VCs and achieved 40% buffer savings without any performance degradation. However, they used a static approach which was based on a detailed analysis of application-specific traffic patterns. Tamir and Frazier [35] proposed a dynamically allocated multiqueue structure for communication on a multiprocessor. This architecture did not adapt to the NoC owing to the complex controller, the limited channel number, and the three-cycle delay for each flit arrival/departure. Nicopoulos *et al.* [28] introduced a DVC regulator which allocated a large number of VCs in high traffic conditions. It could not be scaled well to apply it to different traffic conditions, and dispensed many VCs with much complexity, whose overhead especially for the VC control table would increase nonlinearly. For instance, in the case of 16 buffers per port and a nine-flit packet size, the table size supporting 10 VCs will reach 220 bytes, nearly 21.5% of the total buffers.

Ideally, allocating more VCs implies better utilization on a given link, but it decreases the VC depth because of the fixed buffer size. Rezazad and Sarbaziazad [31] revealed that the optimal number of VCs depends on the traffic patterns when the buffer size is fixed. At low rates, increasing VC depths resulted in better performance. At high rates, the optimal structure depended on the distributing patterns. It was advisable to increase the VC count under a uniform pattern but extend the VC depth under matrix transpose or hotspot patterns. The statically allocated VC structure lacks flexibility in various traffic conditions and corresponds to low buffer utilization. Supposing routers are configured with few deep VCs, many blocking due to HoL blocking or lack of VCs at high rates will lead to low throughput. Inversely, if many shallow VCs are arranged, the packets are distributed over a large number of routers. At low rates, the continual packet transfers will be interrupted by many contentions, thus increasing the latency.

On the basis of the analysis above, we introduce a DVC scheme to overcome the limitations of static VCs. Figure 3.1 shows the DVC structure in detail. It includes v linked lists and a buffer track unit. The linked lists share all the buffers and each VC corresponds to a list for the single packet without any buffer reservation. When the flit is incoming, the appropriate list is selected according to the VC identifier. And then the arrival flit is stored in the tail buffer, which is dynamically allocated by the track unit. For the departure flit, the track unit is responsible for releasing the head buffer of the winning VC, which is arbitrated by the switch allocation (SA) unit.

In our structure, the VCs may be dynamically regulated according to the traffic conditions. At low rates, only a few channels are used. Each input port has enough buffers to be allocated for arrival flits. This structure ensures the continual transfer of packets without flow control, and is helpful to form a few deep channels. The packets distributing among few routers will reduce the average latency. At high rates, lots of VCs are allocated for the packet propagation. The increasing number of packets will result in many shallow channels, which improves throughput in two aspects. First,

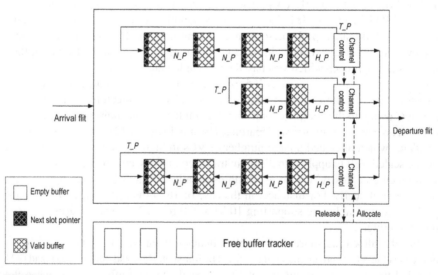

FIGURE 3.1

DVC structure.

the scheme by which each packet uses a separate list avoids the HoL blocking, which is frequent in static VCs. Dispensing more VCs for the local packets will increase the channel multiplexing of physical links. Second, more VCs granted to upstream packets will decrease the blocking due to lack of VCs. In generic routers, many VCs may be occupied by the packets whose tail flits are blocked at upstream routers. At this moment, lots of buffers may be unused but the packets at neighboring routers are blocked owing to lack of VCs. More dispensed VCs here will accommodate many packets from the neighbors, improving throughput with a higher buffer utilization.

3.2.2 CONGESTION AVOIDANCE SCHEME

Although a DVC scheme brings throughput improvement for on-chip networks, a congestion awareness scheme will further improve the performance at high traffic rates. From the point of link utilization, three key factors restraining throughput in generic routers involve the blocking without a VC, the blocking due to flow control, and the crossbar contention, where two individual packets from different ports competing for the same physical link will make some other link idle. Dispensing lots of VCs may reduce the first type of blocking, and is helpful for reducing the crossbar contention by increasing the channel multiplexing. However, allocating more VCs makes no contribution to performance when congestion happens, where all buffers at the port are exhausted and the flow control is generated to prevent the upstream packets. Especially under a hotspot pattern, a great number of packets transferring toward routers with few buffers will result in frequent congestion situations. Once the packets have been blocked at local routers, the congestion will degrade seriously the link throughput.

Figure 3.2a shows the effect of congestion. Packets p_1 and p_2 that transfer toward router C are blocked at router B owing to flow control or contention. A number of flits swarming into router B will lead to congestion at its west port. In the following, the upstream packets p_3 and p_4 will be blocked sequentially owing to flow control, incurring the idle state of links A→B, B→E, and B→D. If more buffers and VCs are allocated to packet p_3 or packet p_4 at router B, the link utilization may be increased linearly. But the buffers at the west port of router B have already been exhausted by packets p_1 and p_2 at this moment. Therefore, this scheme predicts the congestion of neighbors and inspects the traffic conditions around them. Once the local router senses a possible congestion in advance, it allocates buffers and grants high transmission priorities to packets traversing low traffic regions beyond the neighbor. Then, these packets will not be blocked by flow control at the neighboring router, and are good at avoiding the congestion by increasing channel multiplexing. As depicted in Figure 3.2b, supposing the west port of router B will not receive any flit in the next k cycles, the number of available buffers is aggregated k cycles ahead of time by predicting the minimum number of flits from the west port. If the aggregation value is no more than k, the continual transfer toward the west port of node B may cause possible congestion. In Figure 3.2b, the transfer request of packet p_2 is cancelled and the transmission priority is granted to packet p_3 or packet p_4,

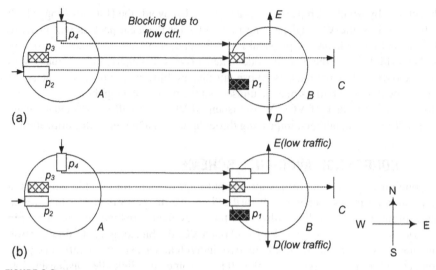

FIGURE 3.2

Congestion avoidance scheme.

which transfers toward the low-traffic routers beyond the neighbors. In the end, the congestion at the west port of node B is avoided and the throughputs of links A→B, B→E, and B→D are improved.

3.3 MULTIPLE-PORT SHARED BUFFER WITH CONGESTION AWARENESS

3.3.1 DVC SCHEME AMONG MULTIPLE PORTS

The dynamically allocated VC scheme is an efficient way to improve the NoC performance, and we introduced a dynamically allocated VC structure in Section 3.2.1. In that structure, buffers in each channel are shared by different VCs. There is another type of dynamically allocated VC scheme, i.e., buffers are shared by all the channels. Several multiple-port shared buffer structures have been introduced [21, 22, 27]. These structures employ multiple-port register files to implement centralized memories, which may occupy much area and account for much power dissipation. Ramanujam *et al.* [30] proposed a shared-buffer router architecture to emulate an output-buffered router, which results in a high throughout. In their design, several normal buffers are sandwiched in two crossbar stages. The buffer structures are simplified, but this may also bring some problems. A flit is usually stored in the input channel buffers and then moved to the middle memory, which may lead to extra flit traversals. Flits in the buffer were time-stamped in first come, first served order, and flits of the same packet cannot be transferred continuously, which may cause higher average packet latencies.

In this section, we propose a novel DVC structure based on a hierarchical bit-line buffer (HiBB) aiming at low cost and high performance. The VC buffers can be shared by several input channels according to the traffic status, and the input channel with heavier loads can win more VCs than the others. Furthermore, the proposed buffer structure brings potential advantages in power consumption and link utilization. When a VC is in the idle state, it can be power gated easily for the characteristic of the HiBB, which results in great power savings. In the generic router, multiple packets in input channels are arbitrated, and winning packets traverse the crossbar. The link utilization may be affected by the inefficient arbitration. In the HiBB, packets aiming at the same direction are switched to the corresponding output port inside the buffer, and a scheme named the output-port allocation is proposed to directly select a packet to serve. This scheme improves the link utilization, and it also reduces the router complexity.

In the HiBB structure, cells of each VC are connected to a sub bit-line, and the sub bit-line is connected to several bit-lines corresponding to input or output channels of the router via pass transistors. In order to reduce the area overhead due to the increased number of bit-lines, we further employ three schemes to optimize the HiBB structure: (1) single-ended 6T SRAM cells [34] are adopted to reduce area overheads and power consumption; (2) the higher metal layers are used for routing increased numbers of bit-lines to reduce the area; (3) each shared VC is shared by only a few input channels instead of all the input channels.

Figure 3.3a shows an example of the third scheme described above. There are five private VCs and eight shared VCs in the HiBB router. Each private VC is allocated to one input channel exclusively, and each shared VC can be shared between two neighboring input channels. For instance, the VC identified by "1" is shared between the west and north channels, while the VC identified by "3" is shared between the west and south channels. Figure 3.3b shows how buffers are customized under a specific traffic load. The west channel is allocated four shared VCs, and the south channel has two shared VCs. Two shared VCs are left unallocated, and they can be switched to be in a sleep state to achieve power reduction [1]. We use a set of registers to control the global word-lines to regulate VCs, and the process of buffer customization is accomplished by writing configuration information into the configuration registers.

The detailed structure of the proposed unified buffer is shown in Figure 3.4. For briefness, just one column of a VC is displayed in the figure. Each memory cell has only one write port and one read port, which connect to a sub write bit-line (sub wbl) and a sub read bit-line (sub rbl) respectively. Each input channel is connected to a write bit-line (wbl) and each output channel is connected to a read bit-line (rbl). Sub bit-lines and bit-lines are linked via pass transistors, and the global word-lines are responsible for turning on and off pass transistors. When memory cells are assigned to an input channel, the sub wbl is switched to a wbl. When a packet enters the VC, the routing computation (RC) results are used to switch the sub rbl to a specific rbl. The structure of the VC remains unchanged during the process, so the flits of a packet traverse from the input channel to an output channel, which results in power optimization owing to reduced switch activities.

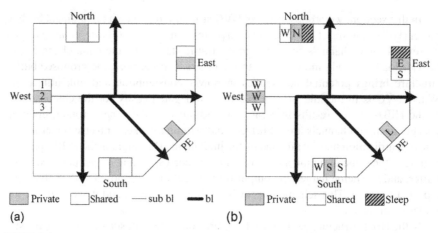

FIGURE 3.3

Router with nonuniform VC configuration.

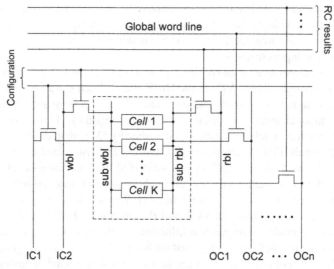

FIGURE 3.4

The detailed structure of unified buffers. IC, input channel; OC, output channel.

3.3.2 CONGESTION AVOIDANCE SCHEME

For the DVC structure, buffers are dispensed or adjusted under two conditions. First, when one input channel is congested, more VCs will be established to avoid HoL blocking. Second, the buffer is dispensed with higher priorities for those packets aiming at lowly loaded directions in order to avoid exhausting precious buffer resources. Similarly to the DVC scheme, a congestion awareness scheme was also

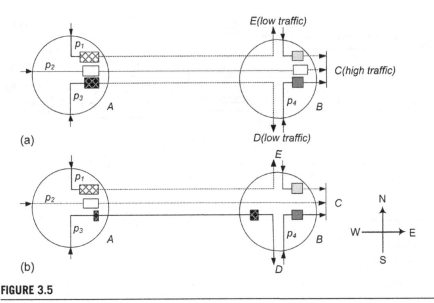

FIGURE 3.5

Congestion avoidance scheme for the HiBB.

proposed for the HiBB to further optimize the performance. The packets heading to an output channel are first transferred to a neighboring input channel of the next hop, and are then transferred in one direction according to the RC information. As shown in Figure 3.5, if packet p_2 heading in a congested direction is selected and served, it may be blocked in the next hop, B. As a result, the limited buffer resources are exhausted and packets (p_1 and p_3) heading in lightly loaded directions may also be blocked for the lack of buffers. If packet p_1 or packet p_3 is served, the congestion of the total network will be greatly relieved.

For the HiBB, the VC is shared by several input channels, and VCs can be dispensed to those inputs with heavier traffic loads. This feature is similar to that of the DVC buffer. The DVC router dispenses buffers in an individual channel with flit granularity, while the HiBB router dispenses buffers among multiple channels with VC granularity. The main goal of the congestion avoidance scheme for the DVC buffer is to avoid exhausting the buffer rapidly, and the goal of the congestion avoidance scheme for the HiBB is to release the VC as soon as possible. The essence of the two congestion avoidance schemes is to improve buffer utilization. As a result, more buffer resources are dispensed to improve packet transmission.

3.4 DVC ROUTER MICROARCHITECTURE

The generic VC router was presented by Peh and Dally [29]. It was an input-queued router with four direction ports and a local port. The key components included the buffer storages, VC controller, routing module, VC allocation module,

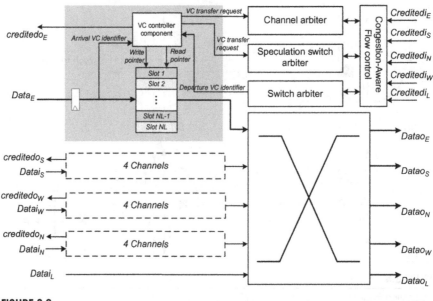

FIGURE 3.6

The DVC router architecture.

SA module, and switch component. The routers used VCs to avoid deadlock and improve throughput. On the basis of the generic router and DVC buffers, we design a DVC router whose architecture is shown in Figure 3.6. The main difference between the DVC router and the generic router is the VC organization. When a flit arrives at an input channel of the DVC router, it can be written into any slot of the buffer according to the command of the VC controller. When a flit in the buffer wants to transfer to the next hop, the congestion-aware flow control module will pick the packet aiming at directions with low loads. In this section, we describe the microarchitecture of the VC controller and VC allocation modules, as well as the design of the simple congestion avoidance logic.

3.4.1 VC CONTROL MODULE

The generic routers are deployed with several fixed VCs to share physical links. Each arrival/departure flit selects its individual FIFO controller unit according to the VC identifier. Instead, each input port in our modified routers adopts the uniform VC control logic, which generates the buffer indexes by VC identifiers in one-cycle delay. The port includes v linked lists, and the relative information about each VC involves the valid tag, the link head pointer (HP), the tail pointer (TP), the requested channel identifier, and the transfer directions at the local hop (DIR) as well as the transfer directions at the next hop (NDIR). The main information about each list item includes the buffer storage and next flit pointer (NP) to the following flit.

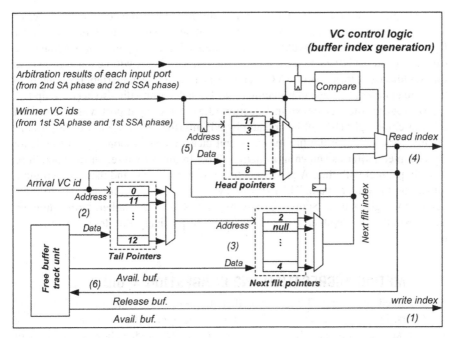

FIGURE 3.7

VC control logic for flit arrival/departure.

Figure 3.7 shows the VC control logic for flit arrival/departure in detail. The key elements include the HP register file, the TP register file, the next slot pointer register file, and a free buffer track unit. For each arrival flit, the write index is generated when three operations have been completed as shown in Equations (3.1)–(3.3):

$$data\,[tracker()] \; <= flit\_incoming, \tag{3.1}$$

$$VC\,[i]\,.TP \; <= tracker(), \tag{3.2}$$

$$NP\,[VC\,[i]\,.TP] \; <= tracker(). \tag{3.3}$$

At the beginning of each cycle, the track unit returns an available buffer index to accommodate the arrival flit. Also, with use of the VC identifier, the TP information of the right link and the NP information of the right tail are updated by the available buffer index at the next clock rising edge.

On the other hand, each departure flit corresponds to another three operations as shown in Equations (3.4)–(3.6):

$$flit\_departure \; <= data\,[VC\,[i]\,.HP], \tag{3.4}$$

$$VC\,[i]\,.HP \; <= NP\,[VC\,[i]\,.HP], \tag{3.5}$$

$$tracker\,[VC\,[i]\,.HP] \; <= 1. \tag{3.6}$$

As depicted in Figure 3.7, the compare unit first judges the continual transfer of a certain packet by comparing VC identifiers of successive requestors. If the latter flit follows the former one within the same channel, the compare unit generates the true result and the NP content selected by the last read index will be the current read index. Otherwise, the departure VC identifier is used to select the right HP content to be the read index as shown in Equation (3.4). Each departure flit also needs to update the HP register file. As shown in Equation (3.5), if the input port wins the second SA phase, the right HP will be updated by NP information of the last departure flit. Finally, for each departure flit, the track unit clears the corresponding bit to release the occupied buffer as shown in Equation (3.6). In this structure, all the operations work in parallel with the SA process within one-cycle delay, and there is negligible impact on the critical path. The extra hardware cost is also limited owing to there being few register files. Supposing the flit width bits, VC number, buffer number, and maximum VC depth are 128, 10, 16, and 9 respectively, the storage overhead is nearly 7.0%.

3.4.2 METRIC AGGREGATION AND CONGESTION AVOIDANCE

As discussed in Section 3.2.2, we predict the congestion situation at immediate neighbors. Figure 3.8 shows the congestion metric aggregation module. Supposing the current time stamp is t, then $nc_d(t,k)$ $(d = E, \ldots, N)$ predicts how many flits will leave from the immediate neighbor in direction d over the next k cycles. Further, $b_d(t,k)$ predicts the availability of its input buffers in k cycles ahead of time. The right box in Figure 3.8 shows the input port of the neighbor. Its key components include the first SA unit and a prediction unit. The SA unit arbitrates among all the requests from the same port and grants transfer priority to the winner. The prediction unit calculates the number of departure flits from its input port over the next k cycles. The neighbor has already aggregated the availability information $b_q(t,k)$ in the winner's output direction. In the wormhole switching scheme, the winning packets are served continuously until the tail flit. When any condition happens, the first allocation phase grants the transfer priority in a round-robin order to other packets. Thus, the prediction unit of the neighbor calculates the number of departure flits as shown in Equation (3.7):

$$nc_d(t + 1, k) = \min\{b_q(t, k), k, l_w\}(d, q \in \{E, S, W, N\}), \qquad (3.7)$$

where k denotes the stride and l_w denotes the list length of the winner. The left box in Figure 3.8 represents the local router. The canonical router records the available buffers of its immediate neighbors for flow control, and we name them as $x_d(t)$. The modified module aggregates the available buffers at its neighbor input port k cycles ahead of time by the sum of $x_d(t)$ and $nc_d(t,k)$, as depicted in Figure 3.8. If the prediction value is no more than k, the congestion metric $cong_d$ will be true. This means that the packet transfers to direction d may cause congestion.

Then, the transmission priority is granted to packets which transfer toward low-traffic regions beyond neighbor routers to avoid congestion. As shown in Figure 3.8,

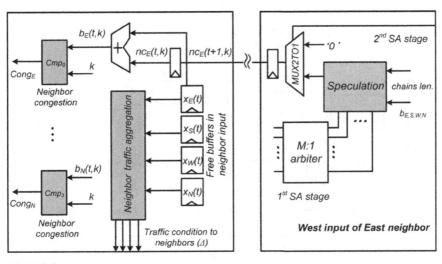

FIGURE 3.8

Contention metric aggregation module.

each router collects the available buffers at neighbor inputs, and aggregates the traffic metrics, Δ where E, S, W, and N denote metrics in different directions. We performed a detailed empirical evaluation to determine the threshold for metric, and found that a value which was a little more than stride k was reasonable for packet bypass. Since the routers can sense the possible congestion situations at neighbors, they may use traffic metrics to advance the packets which transfer toward low-traffic regions beyond neighbors.

As shown in Figure 3.9, the modified flow control module consists of two parts. First, similarly to the generic router, each channel compares its remote VC length with the FIFO depth to determine the flow control. Our modified module adopts the FIFO depth, which varies with the number of VCs. At a high rate, the VC depth at the neighbor is reduced reasonably for more packet propagations due to many VCs. Second, the transfer requests toward high-traffic regions beyond neighbors are masked when predicting congestion. Each input port is deployed with several avoidance units which correspond to other output directions. Utilizing DIR information, each channel exports its NDIR information to the right avoidance unit by the demultiplexer unit. The avoidance unit collects all NDIR information within the same port and specifies the mask bit of each VC. It uses multiplexers to select the traffic metric of the next output direction. The relative mask bit will be set to be true when the packet transfers toward the high-traffic region beyond the neighbor. However, if all the packets are transferring toward high-traffic regions beyond the same neighbor, the avoidance unit will not mask any request and will set all bits to be false.

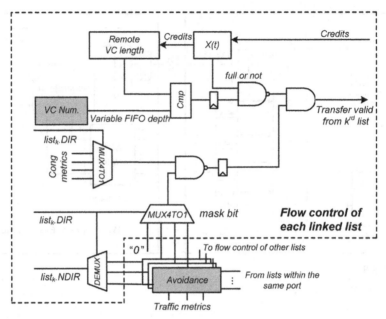

FIGURE 3.9

Flow control module.

In the modified module, each list uses its DIR information to select the congestion metric and the mask bit. When both conditions are true, the transfer valid signal will be canceled. With use of this approach, the packets heading toward low-traffic regions beyond neighbors are transferred in advance when predicting congestions. Note that for the packets from different input ports, the transfer priority is decided by the second SA phase according to the traffic metrics. If the prediction metric $nc_d(t,k)$ is summarized into 3 bits, plus 3 bits for the traffic metrics, this scheme requires an extra 6 bits per link, and the wire overhead is just 4.1%. This scheme has no influence on the critical path, and the area overhead from the prediction and avoidance units is confirmed to be low in the following synthesis.

3.4.3 VC ALLOCATION MODULE

The VC allocation in the generic router was performed in two stages, which arranged $v{:}1$ and $5v{:}1$ arbiters respectively. With increasing number of VCs, the VC allocation module, which relies on the critical path, will influence the performance. The modified VC allocation module adopts another two-phase arbitration. The first phase arranges five $v{:}1$ arbiters at each input port and every arbiter gains a winner request in the corresponding direction. The second-phase logic distributes at each output port, which arranges a 5:1 arbiter to generate the final winner to occupy dispensed VCs. Although the proposed module allocates a single VC in each direction, it has negligible impact on the performance. Even if more VCs are dispensed, the flits still transfer toward the output port one by one. The main advantage of the modified structure is that the area overhead and critical path length are reduced owing to there being few 5:1 arbiters instead of many $5v$:1 ones. In addition, each output port arranges a VC manage unit for VC dispensation. It dispenses lots of VCs for request packets until the VCs are exhausted, and releases VCs according to the credits from neighbors. For the allocation scheme, the second phase arbitrates for the winners according to the traffic metrics from neighbors, granting the VC allocation priority to the packets which transfer toward the low-traffic regions beyond neighbors.

3.5 HiBB ROUTER MICROARCHITECTURE

On the basis of the analysis presented in Section 3.3, we introduce a new shared-buffer router which employs an HiBB to overcome the limitations of existing buffer structures. Figure 3.10 shows the proposed HiBB router architecture. In the new router, VCs are shared among different input channels. These VCs will be allocated to each channel dynamically according to the traffic situation of the channel, i.e., heavily loaded channels will get more VCs, and lightly loaded channels will get fewer. Note also that the VC allocation (VA) module, the SA module, and the crossbar switch in our HiBB router are highly simplified compared with those in the generic router.

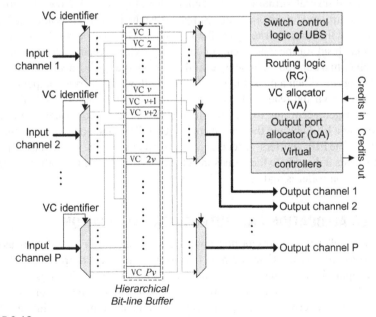

FIGURE 3.10

The HiBB router architecture.

When a flit arrives at an input channel of the HiBB router, it is written into a specific VC. If the flit is a header flit, the output port direction is computed by the RC module. At the same time, the read port of the VC switches to an output channel according to the result of the RC. As a result, all the VCs containing packets with the same direction are connected to a single output channel. After the header flit enters a VC, it requests a VC in the next hop and the VA module is responsible for the VC allocation. Then, the packet which is assigned to a VC can transfer to the output channel. As there may be several packets applying for traversal, instead of using the traditional switch allocator, we use an output port allocator (OA) module to select a packet for traversal. The structure and principle of the OA will be explained in detail in Section 3.5.2.

If the buffers are implemented using register-based FIFO buffers, there will be many multiplexer gates to select registers for specific input or output channels. If the buffers are implemented using SRAM-based FIFO buffers, each cell will have many read ports and write ports. Both of these implementations increase the design complexity, leading to more area and power overhead. The HiBB has an inherent characteristic of low power [15], and it avoids the problem suffered by multiple-port buffers.

3.5.1 VC CONTROL MODULE

To reduce the hardware overhead of the VC control logic, we introduce virtual controllers to control VCs. For each FIFO body, there is a write controller, a read controller, and several virtual write/read controllers. Figure 3.11 shows the structure of the VC control module in our router. Suppose that the FIFO body is allocated to the west port, and the virtual write controller is omitted. Because packets from the west channel may head for the other four directions, four virtual read controllers are used here. Each virtual controller contains several items: a valid tag (v), a VC identifier, and an empty tag (e). When a packet in the FIFO body flows to a certain output channel, the tag v in the virtual read controller of this direction is set to be valid. The FIFO body is managed according to the information in the virtual controller. If a channel needs to read a FIFO buffer, the virtual controller will send a read request to the read controller of the FIFO buffers. After receiving a read request, the read controller reads a flit and sends it directly to the requesting channel. At the same time, the tag e in the virtual read controller is updated.

3.5.2 VC ALLOCATION AND OUTPUT PORT ALLOCATION

Before the packet in the present hop flows downstream, the VA module is usually employed to allocate a VC in the next hop to it. In the generic router, packets in different input channels may request the same VC, and two allocation stages are often used [29]. In our router, the VA module is much simpler than the corresponding module in the generic router. Specifically, an allocation module similar to the technique used in Ref. [28] is presented, as shown in Figure 3.12. All the VCs requesting the same output channel have a valid tag ($v = 1$) in the virtual FIFO

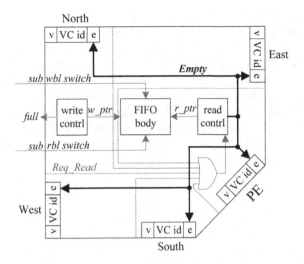

FIGURE 3.11

VC control module.

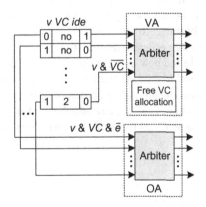

FIGURE 3.12

Simple VA and OA of the HiBB router.

controllers. This information can be used in the process of VA. All the virtual controllers in an output channel send requests to an arbiter to determine which one will gain a VC, and then the free VC allocation module allocates an available VC recorded in the free VC table to the winner.

In the generic router, the crossbar switch transfers packets or flits from input channels to output channels. The SA unit arbitrates among all VCs requesting access to the crossbar and grants permissions to the winners. Then, the winning flits are able to traverse the crossbar. In order to fully utilize the link bandwidth, highly efficient scheduling algorithms have been exploited [2]. However, it is very difficult to achieve a maximum size matching between input channels and output channels with high-performance implementation because of their complexity [23].

On the basis of the proposed HiBB, the SA and crossbar can be simplified. When a new packet enters a VC, the sub bit-line of the VC is switched to the bit-line of the corresponding output channel. As a result, all the VCs containing packets heading in the same direction are connected to a single output channel. In each clock cycle, the output channel reads a flit out of the VCs connected to it directly and then sends it to the output link. The OA module, which is also shown in Figure 3.12, can fulfill the requirements of the packet traversal in the HiBB router. Owing to the simplified structure, several other benefits are obtained:

(1) a maximum size matching can be achieved with simple hardware overhead;
(2) a VC maintains its read port connection until the tail flit leaves the VC, which may result in lower power consumption compared with the more frequent switching in the crossbar in a generic router.

The packets heading to an output channel are firstly transferred to a neighboring input channel of the next hop, and are then transferred in one direction according to the RC information. If the packet heading in a congested direction is selected and served, it may be blocked in the next hop. As a result, the limited buffer resources are exhausted and packets heading in lightly loaded directions may also be blocked because of the lack of buffers. To further improve the performance of the total network, we introduce a more efficient arbiter structure with congestion awareness, as shown in Figure 3.13. In the new arbiter structure, we classify the requests into *n* types according to the possible packet directions in the next hop and arbitrate them separately. At the same time, the possible directions of the next hop are also arbitrated using a dynamic priority arbiter [16]. As a result, the packet heading in the direction with the highest priority is selected to serve. The priorities of the dynamic priority

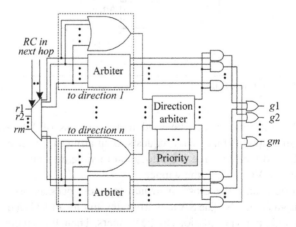

FIGURE 3.13

Improved arbiter with congestion awareness.

arbiter are determined by the congestion situation of each direction in the next hop, i.e., the congested directions will be assigned with lower priorities while others will be assigned with higher priorities.

3.5.3 VC REGULATION

Since each VC may be used by several input channels, configuration information is needed for the VC regulation. We use a set of configuration registers to store the configuration information. To utilize the flexibility of the HiBB further, we introduce a run-time configuration scheme. During the process of communications in an NoC, we predicate the importance of a VC to each input channel. If an input channel needs a new VC aggressively, the idle VC will be assigned to this channel with higher priority. Otherwise, an idle VC may be reassigned to other channels or set to be in a sleep state.

Figure 3.14 shows the run-time regulation scheme of the HiBB router. Each output channel predicts the possibility of a new VC request in the next k cycles and sends it to the neighboring input channel. The precondition of requesting a new VC is that the following three conditions must be satisfied. The first condition is that the remaining length of the packet being served is less than k. In the wormhole switching scheme, the winning packets are served continuously until the tail flit, flow control, or empty channel. If this condition is satisfied, the packet may not be served continuously and a new VC may be needed. The second condition is that the result of the direction arbiter in the OA and the direction of the winning packet in VA are not equal. If the two values are equal, there is already a packet heading in the same direction as the winner of VA which can be served in the following cycles. The third condition is that the output of the VC arbiter is not equal to zero. If the output of the VC arbiter is equal to zero, this means that there is no packet in the output channel requesting a new VC. After the output channel has predicted a possible VC request, the predicted information

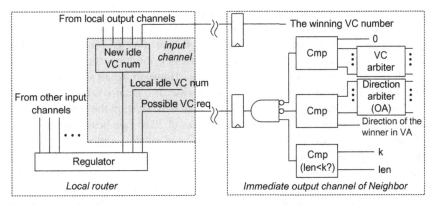

FIGURE 3.14

Run-time regulation scheme of the HiBB router.

will be used by its neighboring input channel to calculate whether a new VC should be assigned. Besides, the existing available idle VC number (N_{idle}) and potential idle VC number ($N_{potential}$) are also used. $N_{potential}$ indicates how many VCs containing packets will be released in the next k cycles. The information on existing VCs and the VCs being served from local output channels is taken to estimate whether the tail of the packet will leave the current VC. If the sum of N_{idle} and $N_{potential}$ is smaller than the predicted VC request number, the corresponding input channel requires a new VC allocation imminently. If the sum is much bigger than the predicted VC request number, the idle VCs in the input channel are redundant and they can be reclaimed for power reduction. The regulation module collects the information described above to do the VC switching. In this book, we set k as the number of cycles for a VC allocation.

3.6 EVALUATION

3.6.1 DVC ROUTER EVALUATION

We use a cycle-accurate simulator to evaluate the performance. The generic router employs a simple dimensional order routing algorithm and adopts a lookahead scheme which corresponds to a three-stage pipeline scheme. In the experiment, each input of the routers is deployed with 32 buffers. There are two kinds of generic routers, denoted as T-2 and T-4 respectively, according to the VC number, in contrast with 10 VCs at most in our DVC routers. We arrange different 8×8 mesh networks using three routers and perform simulations under uniform and hotspot patterns with a packet size of nine flits (one flit for the head plus eights flits for the body, 16 bytes per flit). The results were obtained by simulating 1×10^6 cycles for different rates or patterns after a warm-up phase of 2×10^5 cycles. First, the effect of prediction stride k on average latency is investigated. The latency is shown as a function of stride k for different traffic rates in Figure 3.15. When k increases from 0 to 8, the latencies remain steady at a low rate but decrease rapidly at a high rate. A stride value of 0 means that the congestion awareness scheme is not adopted. In this case, many allocated VCs may reduce the blocking caused by lack of VCs, but are of no use for the flow control, which will leave many links idle. With the increase of k, the congestion situations will be gradually reduced to improve link utilization. However, when stride k increases from 8 to 16, the average latency begins to increase. If a greater value is used for k, it may always predict congestion situations which will interrupt the continual packet transfers. Thus, we set k to be 8 with a buffer size of 32 to gain better performance.

In the following, the performances of DVC and generic routers are evaluated. Figure 3.16a illustrates the results obtained under a random pattern. At a low rate, T-2 and DVC routers may provide deep channels, which correspond to low latency. But with an increase of the rate, the performance of the T-2 router tends to be saturated rapidly. The T-4 router uses more VCs to improve the throughput, but its performance is still limited by serious HoL blocking or blocking caused by lack of VCs. Here, we observe the advantages of the DVC router in detail. It avoids HoL blockings

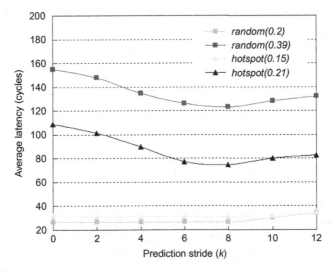

FIGURE 3.15

Effect of prediction stride on average latency.

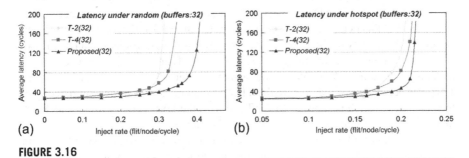

FIGURE 3.16

Average latency under different traffic condition (32 buffers).

and dispenses many VCs at a high rate. Also, it avoids the congestions to improve throughput. As a result, the saturated throughput of the DVC router outperforms the T-2 and T-4 routers by nearly 33.9% and 22.1% respectively. Figure 3.16b illustrates the results obtained under a hotspot pattern. At a low rate, the DVC router also has low latency owing to deep channels. With increase of the rate, a few routers become the high-traffic nodes and the others are still in the low-traffic state. DVC routers allocate VCs according to local conditions. At low-traffic nodes, they extend the VC depth to decrease average latency. Around the hotspot, they dispense many VCs to increase channel multiplexing. In particular, it grants transfer priority to packets heading toward low-traffic regions, and then many packets heading toward low-traffic destinations will not be blocked around the hotspot node. In Figure 3.16b we select

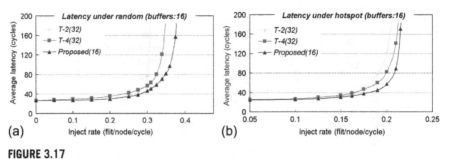

FIGURE 3.17

Average latency under different traffic conditions (16 buffers).

typical injection rates from 0.10 flits per cycle to saturated points, and the average latency drops by nearly 30.2% and 25.2% respectively.

In the generic routers, the buffer utilization is low owing to the limited number of VCs. We also evaluate the performance of the DVC router when the buffer size is reduced to 16. Using the same quantified method, we set stride k to be 4 for better performance in all cases. Figure 3.17 shows the average latency as a function of the injection rate. Under a random pattern, the saturated rates of the three routers are about 0.38, 0.34, and 0.31 flits per cycle respectively. Compared with the T-2 router, the average latency of the DVC router is decreased by 16.5% when we select 0.1, 0.2, 0.275, and 0.3 flits per cycle as typical rates. With rates of 0.1, 0.2, 0.3, and 0.325 flits per cycle, the latency of the DVC router outperforms the T-4 router by about 21.6%. Then, the saturated rates are measured to be 0.20, 0.21, and 0.22 flits per cycle under the hotspot pattern, and the DVC router achieves reductions in latency of 20.3% and 17.6% respectively, compared with the other two routers. On average, compared with the T-4 router, which is superior to the T-2 router, the DVC router provides an 8.3% increase in throughput and a 19.6% decrease in latency.

Finally, we complete the RTL-level description of the DVC router using 16 buffers and 10 VCs. The VLSI design results for the 90 nm process show that the router can operate at 600 MHz. The area information of each module is illustrated in Figure 3.18. The main area overhead comes from the VC control module owing to it accessing index generation and congestion avoidance logic. But halving the number of buffers and a simplified VC allocation module result in greater area reduction, and the total area saving is nearly 27.4%. Using PrimePower, we also estimated the power consumption of the routers, excluding the wire power consumption. The power consumption of the DVC router decreases by 28.6% compared with that of the T-4 router.

3.6.2 HiBB ROUTER EVALUATION

In this section, we implement the HiBB under 90 nm CMOS technology. The memory array of the HiBB is implemented by full custom design flow and its control logic is implemented by standard cell-based design flow. Figure 3.19 shows the area and

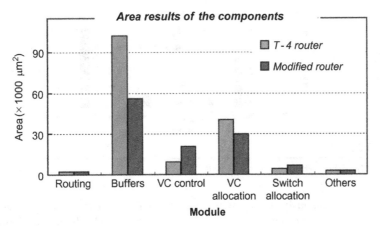

FIGURE 3.18

Area results of the DVC router.

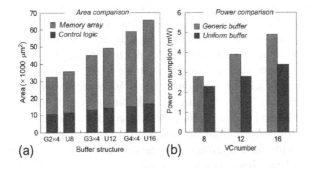

FIGURE 3.19

Area and power comparison of generic and unified buffers.

power comparison results for the HiBB and the generic buffer with different VC numbers. The generic buffer consists of four independent small buffer structures, and each small buffer structure contains several VCs of 128-bit width and eight-flit depth. In the HiBB, equivalent VCs constitute a unified buffer. From Figure 3.19a, we can conclude that the area of the HiBB is a little bigger than that of the corresponding generic buffer under the same capacity. However, the area does not increase much owing to the use of virtual VC controllers and the three schemes proposed in Section 3.3.1. The increased area of the HiBB is mainly because of the introduction of pass transistors connecting two levels of bit-lines and the sleep transistors.

The powers needed to access the traditional buffer and the HiBB are compared for a frequency of 500 MHz. The power result shown in Figure 3.19b is just the power consumed by the activity of one port. The total power is related to the read-write number, so it may increase linearly when there are several ports

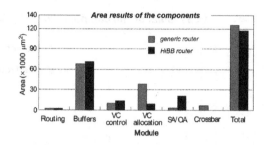

FIGURE 3.20

Area comparison of generic and HiBB routers.

working simultaneously. It is obvious that the HiBB is much more power efficient than the generic buffer under the same capacity. The hierarchical bit-line reduces the load capacitance of the bit-line, which may be the reason for the power reduction.

For 90 nm technology, the proposed HiBB router can operate at 500 MHz. The area overheads of the generic router and the HiBB router are shown in Figure 3.20. Each input channel of the generic router has four VCs, while the HiBB router has a unified buffer with the same number of VCs. It can be calculated that the HiBB router incurs a little more area overhead for the buffers and VC control logic. The area of the OA is greatly raised owing to the use of the improved arbiter with the congestion awareness scheme. The complexity of VA is greatly reduced, leading to an area reduction of about $29.5 \times 10^3 \mu m^2$. In addition, the crossbar switch is eliminated in the HiBB router, and plenty of silicon area is saved. As a result, the HiBB router can achieve a 6.9% total area saving compared with the generic router. Besides, if the new arbiter with the congestion awareness scheme is not adopted in the HiBB router, the area of the simple OA logic is just $2.1 \times 10^3 \mu m^2$, and the HiBB router can achieve an area saving of about 22.1%.

To evaluate the performance of the proposed HiBB router architecture, we built two 8×8 mesh cycle-accurate NoC simulators using *XY* deterministic routing. The first simulator employs the generic router, while the second one adopts the HiBB router. During simulation, we assume the nodes inject packets at regular intervals. In each input channel of the generic router there can be one, two, and four VCs respectively. The unified buffer in the HiBB router can be configured with 4, 8, and 16 VCs. To evaluate the power consumption of the HiBB router, two 8×8 mesh NoC gate-level implementations are also completed. Then, different traffic patterns are inputted into the gate-level netlist of the NoC to obtain value change dump (VCD) files. Finally, simulated gate-level power consumptions are collected using PrimePower.

The performance and the power consumption of the HiBB router are first evaluated under uniform traffic patterns. As illustrated in Figure 3.21a, the performance of the HiBB network is a little better than that of the generic network. Because the traffic is uniform, each input channel of the HiBB router is configured with the same VCs. The performance improvement is due to the OA scheme. In each cycle, one output

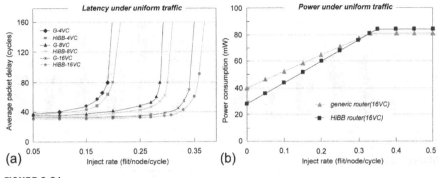

FIGURE 3.21

Performance and power comparisons under uniform traffic.

port selects a requesting packet connected to it and the conflicts between multiple output ports are eliminated. As a result, the link efficiencies are enhanced. The power consumptions of the generic and HiBB routers are shown in Figure 3.21b. Under light traffic, the power consumption of the HiBB router is lower than that of the generic one. This is because some idle VCs in the HiBB router can be set to the sleep state to save power under light traffic. When the whole network is idle, the HiBB router can achieve power savings of about 30%. However, the HiBB router consumes a little more power than the generic router under heavy traffic. The reason is that all the VCs in the HiBB are utilized and little power is saved. At the same time, more flits are transferred owing to the performance improvement of the HiBB router, causing a power consumption increase. The power consumption of the HiBB router may be up to 3.7% higher than that of the generic router.

Then, we configure the generic router with four uniformly allocated VCs across all links and configure the HiBB router with an HiBB employing the same VCs. Three nonuniform traffic patterns are chosen, which are hotspot, matrix transpose, and random. In the random traffic pattern, we ensure that just several nodes are communicating with each other and the others are idle. Each traffic pattern can employ different traffic loads. For instance, the hotspot patterns with a light load and a heavy load are labeled as hotspot-L and hotspot-H respectively in Figure 3.22. The performance and power consumption for the six traffic patterns are shown in Figure 3.22.

For hotspot and transpose patterns, the average packet delay of the generic router increases rapidly as the traffic load becomes heavy, while the average delay of the HiBB router increases less sharply. For random traffic, the average packet delay of the two networks is much smaller than the delay under the hotspot and transpose patterns. This is because just several neighboring routers are picked for communication, and the communication links are often short.

Under light loads, the average power consumption of the HiBB router is lower than that of the generic router as there are abundant VCs in the sleep state. As the traffic load increases, the power consumption of the HiBB router becomes 5.7% and

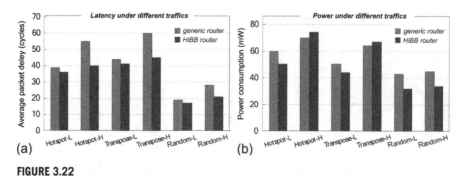

FIGURE 3.22

Performance and power comparison under nonuniform traffic.

4.6% higher than that of the generic router for the hotspot and transpose patterns respectively. The power saving from the HiBB is surpassed by the power consumption increase from additional transferred flits. However, the power consumption of the HiBB router is still less than that of the generic router under random traffic. In random traffic, there are a lot of idle routers and the idle VCs can be set to the sleep state, resulting in great power savings. When the traffic load becomes heavy for just several links, the average power consumption of the HiBB router, which is about 75% of that of the generic router, remains almost the same.

3.7 CHAPTER SUMMARY

In on-chip networks, many packets contend for finite communication resources. The VC organization structure has a great effect on network throughput. In this chapter we proposed two DVC structures, and also corresponding congestion awareness schemes to enhance further the resource utilization. In the first type of DVC, buffers in each channel are shared by different VCs. For low-rate traffic, it extends deep VCs to reduce packet latency, while it increases the VC number and avoids congestion situations to improve throughput under high-rate traffic. In the second type of DVC, buffers can be shared by several input channels according to the traffic information, and the input channel with the heaviest load can win more available VCs than the input channels. We proposed a shared buffer based on an HiBB to implement the second type of DVC. Finally, two DVC routers were designed and evaluated.

REFERENCES

[1] A. Agarwal, H. Li, K. Roy, A single-Vt low-leakage gated-ground cache for deep submicron, IEEE J. Solid-State Circ. 38 (2) (2003) 319–328. 83
[2] M. Ali, H. Nguyen, A neural network implementation of an input access scheme in a high-speed packet switch, in: Proceedings of Global Telecommunications Conference (GLOBECOM), 1989, pp. 1192–1196. 93

[3] D.U. Becker, W.J. Dally, Allocator implementations for network-on-chip routers, in: Proceedings of the Conference on High Performance Computing Networking, Storage and Analysis (SC), 2009, pp. 1–12. 78

[4] D.U. Becker, N. Jiang, G. Michelogiannakis, W.J. Dally, Adaptive backpressure: efficient buffer management for on-chip networks, in: Proceedings of the International Conference on Computer Design (ICCD), 2012, pp. 419–426. 78

[5] T. Bjerregaard, M.B. Stensgaard, J. Sparsoe, A scalable, timing-safe, network-on-chip architecture with an integrated clock distribution method, in: Proceedings of the Design, Automation & Test in Europe Conference & Exhibition (DATE), 2007, pp. 1–6. 79

[6] E. Carvalho, N. Calazans, F. Moraes, Congestion-aware task mapping in NoC-based MPSoCs with dynamic workload, in: Proceedings of the Symposium on VLSI (ISVLSI), 2007, pp. 459–460. 78

[7] H.-L. Chao, Y.-R. Chen, S.-Y. Tung, P.-A. Hsiung, S.-J. Chen, Congestion-aware scheduling for NoC-based reconfigurable systems, in: Proceedings of the Design, Automation & Test in Europe Conference & Exhibition (DATE), 2012, pp. 1561–1566. 78

[8] X. Chen, L.-S. Peh, Leakage power modeling and optimization in interconnection networks, in: Proceedings of the International Symposium on Low Power Electronics and Design (ISLPED), 2002, pp. 90–95. 79

[9] W. Dally, B. Towles, Route packets, not wires: on-chip interconnection networks, in: Procceddings of the Design Automation Conference (DAC), 2001, pp. 684–689. 79

[10] W. Dally, B. Towles, Principles and Practices of Interconnection Networks, first ed., Morgan Kaufmann Publishers Inc., San Francisco, CA, USA, 2003. 78

[11] W.J. Dally, H. Aoki, Deadlock-free adaptive routing in multicomputer networks using virtual channels, IEEE Trans. Parallel Distributed Syst. 4 (1993) 466–475. 78

[12] R. Dobkin, R. Ginosar, A. Kolodny, QNoC asynchronous router, Integration VLSI J. 42 (2) (2009) 103–115. 79

[13] P. Guerrier, A. Greiner, A generic architecture for on-chip packet-switched interconnections, in: Proceedings of the Design, Automation & Test in Europe Conference & Exhibition (DATE), 2000, pp. 250–256. 79

[14] T.-C. Huang, U. Ogras, R. Marculescu, Virtual channels planning for networks-on-chip, in: Proceedings of the International Symposium on Quality Electronic Design (ISQED), 2007, pp. 879–884. 79

[15] A. Karandikar, K. Parhi, Low power SRAM design using hierarchical divided bit-line approach, in: Proceedings of the International Conference on Computer Design (ICCD), 1998, pp. 82–88. 92

[16] D. Kinniment, Synchronization and Arbitration in Digital Systems, first ed., Wiley Publishing, West Sussex, England, 2007. 94

[17] A. Kodi, A. Sarathy, A. Louri, iDEAL: inter-router dual-function energy and area-efficient links for network-on-chip (NoC) architectures, in: Proceedings of the International Symposium on Computer Architecture (ISCA), 2008, pp. 241–250. 78

[18] A. Kumar, P. Kundu, A. Singh, L.-S. Peh, N. Jha, A 4.6Tbits/s 3.6GHz single-cycle NoC router with a novel switch allocator in 65nm CMOS, in: Proceedings of the International Conference on Computer Design (ICCD), 2007, pp. 63–70. 78

[19] M. Lai, Z. Wang, L. Gao, H. Lu, K. Dai, A dynamically-allocated virtual channel architecture with congestion awareness for on-chip routers, in: Proceedings of the Design Automation Conference (DAC), 2008, pp. 630–633. 77

[20] A. Leroy, P. Marchal, A. Shickova, F. Catthoor, F. Robert, D. Verkest, Spatial division multiplexing: a novel approach for guaranteed throughput on NoCs, in: Proceedings

of the International Conference on Hardware/Software Codesign and System Synthesis (CODES+ISSS), 2005, pp. 81–86. 77

[21] L.-F. Leung, C.-Y. Tsui, Optimal link scheduling on improving best-effort and guaranteed services performance in network-on-chip system, in: Proceedings of the Design Automation Conference (DAC), 2006, pp. 833–838. 82

[22] J. Liu, J. Delgado-Frias, A shared self-compacting buffer for network-on-chip systems, in: Proceedings of the International Midwest Symposium on Circuits and Systems (MWSCAS), 2006, pp. 26–30. 82

[23] N. Mackeown, The iSLIP scheduling algorithm for input-queued switches, IEEE/ACM Trans. Netw. 7 (2) (1999) 188–201. 94

[24] A. Mello, L. Tedesco, N. Calazans, F. Moraes, Virtual channels in networks on chip: implementation and evaluation on Hermes NoC, in: Proceedings of the Symposium on Integrated Circuits and System Design (SBCCI), 2005, pp. 178–183. 77

[25] G. Michelogiannakis, N. Jiang, D. Becker, W. Dally, Packet chaining: efficient single-cycle allocation for on-chip networks, in: Proceedings of the International Symposium on Microarchitecture (MICRO), 2011, pp. 33–36. 78

[26] R. Mullins, A. West, S. Moore, Low-latency virtual-channel routers for on-chip networks, in: Proceedings of the International Symposium on Computer Architecture (ISCA), 2004, pp. 188–197. 78

[27] M. Neishaburi, Z. Zilic, Reliability aware NoC router architecture using input channel buffer sharing, in: Proceedings of the Great Lakes Symposium on VLSI (GLSVLSI), 2009, pp. 511–516. 82

[28] C.A. Nicopoulos, D. Park, J. Kim, N. Vijaykrishnan, M.S. Yousif, C.R. Das, ViChaR: a dynamic virtual channel regulator for network-on-chip routers, in: Proceedings of the International Symposium on Microarchitecture (MICRO), 2006, pp. 333–346. 78, 79, 92

[29] L.-S. Peh, W. Dally, A delay model and speculative architecture for pipelined routers, in: Proceedings of the International Symposium on High-Performance Computer Architecture (HPCA), 2001, pp. 255–266. 85, 92

[30] R.S. Ramanujam, V. Soteriou, B. Lin, L.-S. Peh, Design of a high-throughput distributed shared-buffer NoC router, in: Proceedings of the International Symposium on Networks-on-Chip (NOCS), 2010, pp. 69–78. 82

[31] M. Rezazad, H. Sarbaziazad, The effect of virtual channel organization on the performance of interconnection networks, in: Proceedings of the International Parallel and Distributed Processing Symposium (IPDPS), 2005, pp. 1–8. 78, 79

[32] D. Schinkel, E. Mensink, E.A.M. Klumperink, E. Tuijl, B. Nauta, Low-power, high-speed transceivers for network-on-chip communication, IEEE Trans. Very Large Scale Integration (VLSI) Syst. 17 (1) (2009) 12–21. 78

[33] W. Shi, W. Xu, H. Ren, Q. Dou, Z. Wang, L. Shen, C. Liu, A novel shared-buffer router for network-on-chip based on hierarchical bit-line buffer, in: Proceedings of the International Conference on Computer Design (ICCD), 2011, pp. 267–272. 77

[34] J. Singh, D. Pradhan, S. Hollis, S. Mohanty, J. Mathew, Single ended 6T SRAM with isolated read-port for low-power embedded systems, in: Proceedings of the Design, Automation & Test in Europe Conference & Exhibition (DATE), 2009, pp. 917–922. 83

[35] Y. Tamir, G. Frazier, High-performance multiqueue buffers for VLSI communication switches, in: Proceedings of the International Symposium on Computer Architecture (ISCA), 1988, pp. 343–354. 79

[36] P. Wolkotte, G. Smit, G. Rauwerda, L. Smit, An energy-efficient reconfigurable circuit-switched network-on-chip, in: Proceedings of the International Parallel and Distributed Processing Symposium (IPDPS), 2005, pp. 155–162. 77

[37] Y. Xu, B. Zhao, Y. Zhang, J. Yang, Simple virtual channel allocation for high throughput and high frequency on-chip routers, in: Proceedings of the International Symposium on High-Performance Computer Architecture (HPCA), 2010, pp. 1–11. 78

Virtual bus structure-based network-on-chip topologies[†]

4

CHAPTER OUTLINE

[†]Part of this research was first presented at the work-in-progress (WIP) session of the 48th ACM/IEEE Design Automation Conference (DAC-2011) [12]. The journal extension edition of the DAC-2011 WIP paper was published in *Microprocessors and Microsystems* [13].

Networks-on-Chip. http://dx.doi.org/10.1016/B978-0-12-800979-6.00004-4

4.1 INTRODUCTION

The need for scalable and efficient on-chip communication in future many-core systems has resulted in the network-on-chip (NoC) design emerging as a popular solution [11, 31]. NoCs are conceived to be more cost-effective than the bus in terms of traffic scalability, area, and power in large-scale systems. Such networks are ideal for component reuse, design modularity, plug-and-play, and scalability, and they avoid the uncertainty of global wire delay perspectives. Recent proposals, including the 64-core TILE64 processor from Tilera [31] and Intel's 80-core Teraflops chip [11], have successfully demonstrated the potential effectiveness of NoC designs. However, most existing NoCs are based on the assumption that the vast majority of traffic is of a one-to-one (unicast) nature [6]. Their multihop feature and inefficient multicast (one-to-many) or broadcast (one-to-all) support have made it awkward to perform some kinds of latency-critical and collective communications. Examples include cache coherence protocols, global timing and control signals, and some latency-critical communications [20].

The packet-based NoC designs often use wormhole switching as the flow control mechanism; this delivers a packet in a pipelined fashion [5]. For long messages, the latency is insensitive to the distance traversed. However, it would be a problem for some short and urgent messages [18]. Another problem of current NoC design is that the performance of multicast is much lower than that of unicast in general. Multicast communication in NoCs is achieved by extending unicast, which sends each packet to all destination nodes or a subset of destinations, each of which in turn forwards the message to one or more other destinations in a multicast tree or path [19]. However, owing to inherent multicycle packet delivery latency and network contention, the multicast-packet overhead in NoCs is very high. A common approach is wormhole switching with multicast support [6, 19, 26]. But this method also needs routing time to the destinations before sending data and multihop time for sending data. So, the multicast overhead is still high.

It is interesting that conventional buses usually show lower packet latency than NoCs. The broadcast or multicast can be efficiently supported by a shared bus within one transaction. So, many recent NoC designs have taken advantage of transaction-based buses by integrating buses physically [20, 23, 25, 29]. Though on-chip wires are relatively abundant, use of a large number of dedicated point-to-point links leads to a large area footprint and low channel utilization. Some researchers have proposed the idea of express channels in NoCs, namely using dedicated NoC connections as buses that are dynamically selected such as express virtual channels (EVCs) [17] and NoC hybrid interconnect (NOCHI) EVCs [15]. These approaches can reduce the effective network diameters of NoC designs, but they do not consider multicast or broadcast problems.

In this chapter, we present the virtual bus (VB) on-chip network (VBON), a new architecture for incorporating buses into NoCs in order to take advantage of both NOCs and buses. It uses the VB as the bus transaction link dynamically for bus requests, which is built upon the point-to-point links of conventional NoC designs. The VBON topology has the advantage that it does not use extra physical links. This achieves low latencies while supporting high throughput for both unicast and multicast communications at low costs. To reduce the latency of the physical layout for the bus organization, two methods are used. The first one is the hierarchical clustered bus structure, which divides the chip into several clusters with bus-connecting routers in a cluster. Each of these clusters is connected via intermediate routers. The second one is called the redundant bus, which uses several buses instead of one bus to constrain the longest distant of the network to a bearable bound.

4.2 BACKGROUND

In this section, we discuss related work on improving the performance metrics of packet-switched NoCs.

Improving the performance metrics of packet-switched NoCs by integrating a second switching mechanism has been addressed in several works, such as integrating a physical bus [20, 23, 25, 29]. These buses typically serve as a local mechanism in the NoC interconnect hierarchy [23, 25]. Then the delivery of data over short distances does not involve the multihop network. In Ref. [29], the NoC network was reported to be created totally on the basis of a bus in a hierarchical way. Lower energy consumption and a simple network/protocol design are claimed. The hybrid bus-enhanced NoC design adds a global bus as a low-latency broadcast/multicast/unicast medium [20]. The reconfigurable NoC proposed in Ref. [28] reduces the hop count by physical bypass paths. Though our works also have some similarities with this reconfigurable NoC, we reduce the hop count by virtually bypassing the intermediate routers and further consider the multicast problem.

Another approach is the express topology, which employs long links between nonlocal routers to reduce the effective network diameter. The long links can be created physically (adding extra router ports, larger crossbars, and extra physical channels [10, 14]) or virtually (opportunistically bypassing router pipelines at intermediate network hops [15–17, 21, 22]). The EVC approach [17] improves the energy-delay throughput in the network that lets packets bypass intermediate routers. Further improvements are done by NOCHI EVCs [15], enabling more bypassing of nodes and reducing the traversal latency and dynamic power with global interconnect circuits.

Though the express topology can reduce the unicast latency of NoC communication, it does not consider the multicast or broadcast communication pattern, which is the main problem this chapter tries to solve.

The support for multicast communications in NoCs may be implemented in software or hardware. The software-based approaches [3] rely on unicast-based message passing mechanisms to provide multicast communication. Implementing the required functionality partially or fully in hardware has proved to improve the performance of multicast operations such as the connection-oriented multicast scheme in wormhole-switched NoCs [19], the XHiNoC multicast router [26], and virtual circuit tree multicasting [6].

This chapter proposes the use of a VB as an attractive solution for packet-based NoCs in latency-critical and multicast communications. Though the concept of a VB has been explored for off-chip networks [4], as power, area, and latency constraints for off-chip routers versus on-chip routers differ substantially, prior off-chip routers are not directly suitable for on-chip usage. The proposed VBON design with its microarchitecture and protocol are discussed in detail to overcome the wire delay limitation of the conventional on-chip bus.

4.3 MOTIVATION

Before describing the VBON architecture, we first introduce the baseline on-chip interconnection networks and their problems, and then briefly discuss the advantages of transaction-based bus communication.

4.3.1 BASELINE ON-CHIP COMMUNICATION NETWORKS

The huge communication demands of applications have made the on-chip communication network the most important part of chip design. The interconnect structure has been changed from bus-based communication to an NoC approach as will be discussed in detail in the following subsections.

4.3.1.1 Transaction-based bus

Traditionally, shared-bus-based communication architectures are popular choices for on-chip communication. Figure 4.1a shows a bus example architecture which is regarded as a transaction-based communication architecture. The bus is a simple set of wires interconnected to all the master nodes, slave nodes, and an arbiter in a single way. A bus transaction generally consists of first establishing a connection between communicating pairs before sending data (a role endorsed by the bus arbiter). So the bus can be a centric architecture in which all the components are tightly coupled. In one transaction, the master node can reach all the slave nodes directly and immediately. This is very efficient when connecting only a few tens of intellectual property cores. However, these transactions are necessarily serialized, causing heavy contention and poor performance when more cores are integrated. Thus, the bus structure fails to satisfy the requirements of future parallel applications. Furthermore,

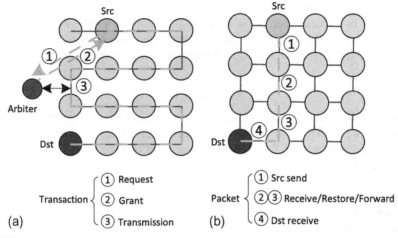

FIGURE 4.1

The baseline interconnect structure. (a) Transaction-based bus. (b) Packet-based mesh NoC.

the global wires introduced by the bus also cause scalability, energy consumption, and performance problems.

4.3.1.2 Packet-based NoC

NoC communication architectures connect the processing and storage resources via a network. Communication among various cores is realized by generating and forwarding packets through the network infrastructure via routers. By eliminating the global wires used in the bus, the NoC approach can provide high bandwidth and scalability. The 2D mesh is by far the most exploited topology since it is particularly easy to implement on current planar CMOS technology. Figure 4.1b shows an example of the mesh network. It consists of 16 tiles arranged in a 4 × 4 grid, where each tile is connected to its four neighbors (with the exception of the edge tiles). Because only neighboring nodes are connected, packets that need to travel long distances suffer from large hop counts. Current packet-based on-chip networks have made each packet need to go through router pipelines, typically four to five stages at each hop [5], and involve multiple hops according to the distance. This widens the gap between the ideal interconnect and the NoC design. A more serious problem may occur when multicast or broadcast communications are activated in that their latency and throughput features usually do not satisfy the demand of applications.

4.3.2 ANALYSIS OF NoC PROBLEMS

We will introduce a model to evaluate the latency of NoC communication with both unicast and multicast/broadcast traffic. The model considers conventional pipeline routers [9] and targets wormhole flow control under deterministic routing algorithms.

4.3.2.1 Multihop problem

The latency of a unicast message is regarded as the time from the generation of the unicast message at the source node until the time when the last flit of the message is absorbed by the destination node. So the service time of a unicast packet, T_{NoC}, is given by

$$T_{\text{NoC}} = T_s + T_b, \tag{4.1}$$

where T_s is the router service time for the header flit and T_b is the number of cycles required by a packet to cross the channel. Since the remaining flits follow the header flit in a pipelined fashion, T_b is simply the quotient of the packet size, S, and the channel width, W:

$$T_b = S/W. \tag{4.2}$$

We note that T_s is a function of the router design and its hop number, H, including the time to traverse the router (t_R) and the link (t_L):

$$T_s = \sum_1^H t_{\text{hop}} = \sum_1^H (t_R + t_L)$$

$$= \sum_1^H (t_{\text{crossbar}} + t_{\text{BW}} + t_{\text{VA}} + t_{\text{SA}} + t_{\text{contention}} + t_L). \tag{4.3}$$

Here, t_{hop} is the latency of each router hop, t_{BW} is the time the flit spends in the buffers, t_{VA} and t_{SA} are the times the flit spends in arbitrating the buffer and switching resources, and t_{crossbar} is the time to actually traverse the router.

In the ideal case, the unicast delay for NoC networks, T_{ideal}, should be the transmission delay of the physical link. For each hop, the ideal network latency, indicating the intrinsic network delay, can be written as

$$t_{\text{ideal}} = t_L + t_{\text{crossbar}}. \tag{4.4}$$

So the total ideal delay, T_{ideal}, can be written as

$$T_{\text{ideal}} = \sum_1^H (t_{\text{ideal}}) + T_b = \sum_1^H (t_L + t_{\text{crossbar}}) + T_b. \tag{4.5}$$

Compared with the ideal network latency, T_{gap}, indicating the extra router pipeline and resource contention latency in the NoC network design, can be written as

$$T_{\text{gap}} = T_{\text{NoC}} - T_{\text{ideal}} = \sum_1^H (t_{\text{BW}} + t_{\text{VA}} + t_{\text{SA}} + t_{\text{contention}}). \tag{4.6}$$

From Equation (4.6), it can be seen that if we want to reduce the unicast latency of the NoC, we should reduce the number of transmission hops, the processing time of router pipelines, and the contention time caused by multiple packets waiting for transmission.

4.3.2.2 Multicast problem

The multicast latency can be defined as the time from the generation of the message at the source node until the time when the last flit of the multicast message is absorbed by the last destination of the multicast message among messages leaving injection ports. We use unicast-based message passing mechanisms to provide multicast communication. These multicast packets are broken down into multiple unicasts by the network interface controllers as the packet-switched routers are not designed to handle multiple destinations for one packet. Then the service time of a multicast packet, T_{NoC_mc}, is given by

$$T_{NoC\_mc} = \max\{t_{inject} + t_{NoC\_mc}\}$$

$$= \max\left\{ t_{inject} + (S/W) + \sum_{1}^{H}(t_{crossbar} + t_{BW} \right.$$

$$\left. + t_{VA} + t_{SA} + t_{contention} + t_L) \right\}, \tag{4.7}$$

where t_{inject} is the injection time of the multicast flit. Since a multicast packet has multiple destinations, this causes competition at the network interface for injection into the network. This adds significant delays to the packet communications. From Equation (4.7), it can be seen that we can reduce the multicast delay in two ways. One is to reduce the unicast traffic caused by multicast packets so that the competition for network resources can be reduced. The other is to choose a good routing algorithm so that the communication path to each destination is the shortest.

4.3.3 ADVANTAGES OF A TRANSACTION-BASED BUS

When the traffic is transmitted by a conventional bus, it allows only one master to transfer data over the bus at a time. Since each packet has many flits to transmit, the bus generally supports burst transfer. The latency of the burst transfer in a bus can be represented as

$$T_{bus} = T_s + T_b$$

$$= t_{arbiter} + \sum_{1}^{H}(t_L) + (S/W). \tag{4.8}$$

Compared with the conventional NoC, the decreased latency can be written as

$$\Delta_{bus\_decreased} = T_{NoC} - T_{bus}$$

$$= \sum_{1}^{H}(t_{BW} + t_{VA} + t_{SA} + t_{contention} + t_{crossbar}) - t_{arbiter}$$

$$= \sum_{1}^{H}(t_{gap} + t_{crossbar}) - t_{arbiter}$$

$$= T_{gap} + \sum_{1}^{H}(t_{crossbar}) - t_{arbiter}. \tag{4.9}$$

Equation (4.9) shows that the bus mechanism eliminates the delay caused by the NoC multihop feature, and its delay is close to that of the ideal communication mechanism. However, the competition for transmission resources in the shared bus is more serious than in the distributed NoC, which increases the time of bus arbitration, $t_{arbiter}$.

Another advantage of the bus is that it transmits all the data in a broadcast way, so the communication amount generated by a multicast or broadcast packet is the same as with a unicast packet as shown in Equation (4.8). This means that the bus has a natural advantage for multicast communication.

Though the benefits of the bus are appealing, the bus is poorly scalable. When the number of nodes that connect to the bus is larger than a certain number (typically more than 10), the performance and power metric of the bus will decline rapidly. Therefore, how to incorporate the bus mechanism in the context of an on-chip network to maximize its advantages requires further study.

4.4 **THE VBON**

4.4.1 **INTERCONNECT STRUCTURES**

The VBON architecture is introduced as a simple and efficient NoC design to offer low latency for both unicast and multicast/broadcast communication. The key idea of the VBON is to integrate the VB into the network, and to use the VB to bypass intermediate routers by skipping the router pipeline or transmit to multiple destinations in a broadcast way. Differently from one set of proposals which employ additional physical links to construct the local buses [20, 23, 25], the bus transaction link in a VBON is constructed dynamically for bus requests from the existing point-to-point links of conventional NoCs.

The concept architecture for a VBON with a baseline 8×8 mesh is presented in Figure 4.2. To improve the characteristics of the bus, we organize the VBs in a hierarchical way, which can reduce the physical layout effort for the bus organization. The chip is divided into several clusters of cores, with VBs connecting routers in each cluster. These VBs between clusters are then connected via intermediate cores. As illustrated in Figure 4.2, four VB clusters (4×4, 5×4, 5×5, 4×5) are configured in an 8×8 NoC to ensure that the intermediate routers exist between each cluster pair.

In each cluster, if all the routers are connected through one global bus, then the length is of $O(N)$ complexity. To reduce this physical length of the VB, the *redundant bus* method is introduced, which uses several buses instead of one to control the VB length within a bearable bound. In the example shown in Figure 4.2, the 2D mesh is configured with row and column VBs in each cluster. Thus, each VBON router is connected to one row VB and one column VB. Instead of one transaction in the global bus [29], two transactions are needed for communications between different row and column VBs. But the length increases at a much lower rate of $O(\sqrt{N})$. This is important for the physical layout of VB organization. Furthermore, since every row

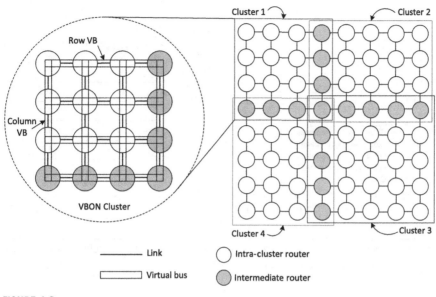

Cluster 1   Cluster 2
Row VB
Column VB
VBON Cluster
Cluster 4   Cluster 3

——— Link   ◯ Intra-cluster router
▭ Virtual bus   ● Intermediate router

FIGURE 4.2

The VBON interconnect structure.

or column bus has its own access control, multiple independent bus transactions can coexist at the same time.

Since the VB shares network links with the underlying NoCs and it applies a simplified flow control, it is a light-weight bus, and is much less expensive than conventional system buses in terms of area, power, and system complexity. The VB does not utilize the segmentation, spatial reuse, split transactions, and other costly throughput boosting mechanisms. The VB provides a high-efficiency and low-latency communication structure that outperforms the conventional NoC in terms of power and latency for short unicast, broadcast, and multicast transactions.

When the bus transaction needs to traverse several VB segments, we retain the same arbitration scheme for different VB segments. This organization essentially allows each transmission across different VBs to be pipelined, which not only increases throughput, but also reduces the contention for the buses, bringing performance improvement.

4.4.1.1 Wire delay consideration

Technology scaling has resulted in a steady increase in transistor speed. However, unlike transistors, global wires that span the chip show a reverse trend of getting slower with the shrinkage process. Modern processors are severely constrained by the wire delay. However, wire delays are somewhat tolerable when repeaters are judiciously employed [2]. Figure 4.3 shows the interconnect delay scaling from the prediction of the International Technology Roadmap for Semiconductors 2008

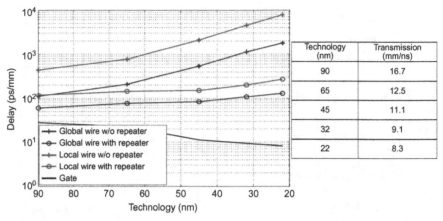

FIGURE 4.3

The interconnect delay scaling from the prediction of the International Technology Roadmap for Semiconductors 2008 report [27].

report [27]. This wire delay prediction can help us to decide the optimal length of VBs in many-core processor chips. The table in Figure 4.3 shows the wire transmission length per nanosecond under different technology nodes. We can see that with repeaters, the delay of transmitting signals on global wire can be reduced from 2000 to 120 ps/mm with 22 nm technology, i.e., the wire data can be transmitted at a rate of 8 mm/ns.

If we assume the chip die area is $10 \times 10 \, mm^2$ and the frequency is 1 GHz, the data can be transmitted in a relatively long range across the chip in one cycle. Therefore, in this chapter, we consider only the case that a VB can transmit in one cycle. If a higher frequency is required, the maximal length of the VB may be decreased or one VB segment can be transmitted in multiple cycles.

4.4.2 THE VB MECHANISM

4.4.2.1 The VB construction

Figure 4.4 illustrates the mechanism for constructing the VB. It is established by bypassing the existing datapaths of the conventional router. Though only the concept structure of the row/column VB implementation is presented, this bypassing method can be used in any other type of VB. Each node is simply composed of some switches that can establish internal connections among its ports. In addition, each node has some simple controller logic. The controller can read control signals from bus arbiters and configure the internal switches. Four possible node configurations for the internal switches are displayed in Figure 4.4a, i.e., (N → S), (S → N), (W → E), and (E → W). The connections among the controller and the ports are not shown in Figure 4.4 for simplicity.

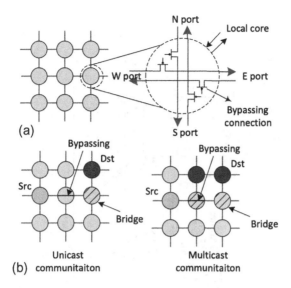

FIGURE 4.4

The row/column VB implementation concept. (a) Concept structure of row/column VB. (b) Example VB configuration.

Figure 4.4b gives two examples for how to construct the row/column VB. There are bypassing nodes for transmission in each row or column VB, and bridge nodes for transmission between the row and column VBs. So the real transmission in the VB link can be completed by bypassing existing router pipelines. This is different from the multihop transmission in the conventional NoC design. Through a little logic overhead in the basic router and dynamic link sharing between NoCs and buses, the VBON improves the conventional NoC while preserving all its essential features, including high efficiency, high throughput, and high bandwidth for unicast traffic, as well as scalability and low power consumption.

The transmission to multiple destinations on a single row or a single column is enough with only one VB. However, in several cases, the destinations are scattered so as to be distributed on multiple rows or columns. For multiple destinations on different rows or different columns, a single-row VB and multiple-column VBs have to be established. Their arbitrations and data transmissions are done in a dimensional order fashion.

4.4.2.2 VB arbitration

Figure 4.5 shows the signal lines for one row VB in an 8 × 8 hierarchical VBON system. Since signal lines for column VBs are the same as for the row VBs, they are not shown in the figure. The row VB presented is segmented into two row VBs, and each has an arbiter router in the middle of the VB. The *Request* and *Grant* signals are used for the bus arbitration. To arbitrate for the VB, a router must send the *Request* signal to a central arbiter structure, which then sends back the *Grant* signal to one of

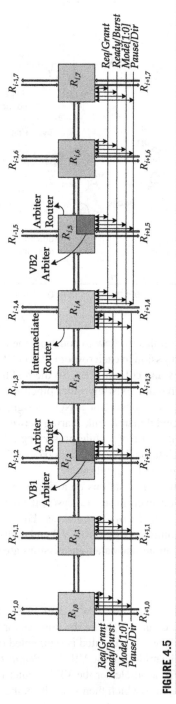

FIGURE 4.5

Signal lines for one row VB in an 8 × 8 hierarchical VBON system.

```
1:  alogrithm VBArbitration()
2:    if Ready then
3:      Sᵢ= Select_CircularPriority({i : Rquest[i]==1})
4:      Grant[Sᵢ]=1;
5:      Set Burst and Mode;
6:      for i=1; i<M; i++
7:        if i < Sᵢ then Dir = 0; Pause =1;
8:        else if i > Sᵢ then Dir = 1; Pause =1;
9:        endif
10:     end for
11:   endif
12: end algorithm
```

FIGURE 4.6

The VB arbitration algorithm.

the requesters. To keep the arbiter design simple, we assume that each node has only one outstanding bus request, the request signal is activated until the *Grant* signal is received, and there is no buffering of requests at the arbiter. The *Burst* signal is used to specify whether the bus communication is a burst mode (transmitting more than two flits consecutively) or not.

The bus mode signal (*Mode[1:0]*) indicates the bus transaction types, such as unicast, multicast, and broadcast. The bus ready signal (*Ready*) serves two purposes. One is related to the VB start-up procedure. When a router is trying to get the bus grant through the bus request signal *Request*, the other routers on the same row/column VB assert the *Ready* signal. The other purpose is for the end-to-end flow control of the VB transaction. When the buffer in the destination is filled with flits, the destination deasserts a *Ready* signal, then the arbiter will be able to pause the bus transaction.

The *Pause* and *Dir* signal pair is used to control the pausing logic for bypassing the intermediate routers. The *Pause* signal specifies whether the pausing logic is activated, and the *Dir* signal specifies the direction of the VB transmission link.

The VB arbitration algorithm is illustrated in Figure 4.6. It assumes that the number of VBs is *M*. It first checks whether the VB node is ready for transmission by reading the *Ready* signal. If it is ready, the arbitration is activated. To fairly access the VB, we use the circular priority to select the winning node for further processing. Lines 6-10 in the algorithm generate the pausing signals for the bypassing nodes. Since the two ends of the VB and the data sender are obviously not the bypassing nodes, pause signals should not be generated for them. The time complexity of this arbitration algorithm is low because *M* is small ($M = 4$ in our VB configuration). It is not the timing-critical part in the proposed router.

4.4.2.3 Packet format

In order to support multicasting, we expand the packet format into that shown in Figure 4.7. We explain the packet fields as follows. The 38-bit packet format for unicast is shown in Figure 4.7a. The packet consists of the header flit followed by

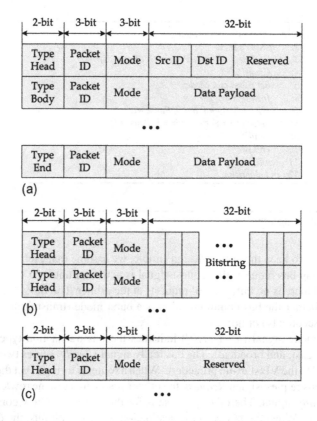

FIGURE 4.7

The VBON packet format. (a) Unicast packet format. (b) Multicast packet format for header flit. (c) Broadcast packet format for header flit.

payload flits. Three additional 3-bit heads are the Type, ID (identity), and Mode bits. The Type can be header, data body, and the end of data body (last flit). The Mode can be the unicast, multicast, and broadcast communication pattern for different packet formats. The source and target addresses of the packet are asserted in the header flit. Passing a communication segment of the NoC, each packet has the same local identity number (ID tag) to differentiate it from another packet. The local ID tag of the data flits of one packet will vary over different communication segments in order to provide a scalable concept. Figure 4.7b shows the packet format for multicast services. The target addresses are specified in the *bitstring* field. Each bit in the *bitstring* represents a node, the hop distance of which from the source node corresponds to the position of the bit in the *bitstring*. The status of each bit indicates whether the visited node is a target of the multicast or not. For broadcast communication as shown in Figure 4.7c, all nodes in the NoC are receiver nodes, so they can be specified implicitly.

4.4.2.4 VB operation

We apply the procedure used in the connection-oriented multicasting NoC [19] in the VB operation. The VBs can be dynamically established, and all conflicting ongoing messages can be paused during the VB transmission. This consists of three phases, VB setup, communication, and VB release:

(1) VB setup. First the source node sends a request to the arbiter to access the VB. In order to resolve VB requests from many sources, the row ID of a node stands for its priority. The row ID of highest priority is able to continue its VB transaction. As a result, others withdraw their VB requests, which will be retried afterwards. When the bypassing circuits for the VB are established, the existing routing data are frozen intact in the flit buffers.

(2) Communication. The VB data is being transmitted to the destination nodes. Except for the broadcast packet, the destination information is needed in multicast or urgent unicast packets. At the beginning of data transmission, the source transfers encoded destination vectors through point-to-point links, and then transfers the following raw data. Once the vectors have been received, the nodes can determine which nodes are destinations or bridges connected to the destinations in the other dimension.

(3) VB release. After the VB transmission, the datapath seized by the VB is released. The paused routing data can continue to proceed to its destination.

The transition state of VB operation is illustrated in Figure 4.8. If the source and destination nodes belong to the same bus segment (the same row or column), the packet can be transferred by using one VB operation ($IDLE \rightarrow SETUP \rightarrow BUSY \rightarrow IDLE$). If they are in different bus segments (different rows or columns), multiple VB operations are used.

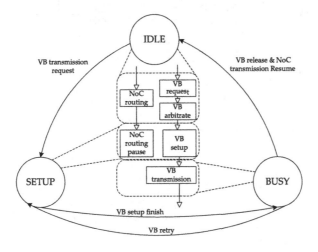

FIGURE 4.8

The VB operation.

The bridge node is responsible for transition from one row VB to one column VB. When this happens, the flits from the row VB can be stored in the bridge node and then they request another VB operation. Since a row VB has only one packet at a time, there is no contention for the buffer in the same bridge node.

A simple and illustrative timing diagram of one VB operation is shown in Figure 4.9. The VB transmits four flits in the burst mode, where three different roles of nodes are involved: the source node *Src*, the intermediate node *Inter*, and the destination node *Dst*. The signal names begin with the type of node that connects to the signal. When the *Src* node requests the VB transmission by asserting the *Request* signal, it also asserts the *Burst* signal. Then the VB arbiter grants the VB access according to the *Ready* signals from other nodes and the priority information. At the same time, it generates the *Pause* and *Dir* signals for bypassing *Inter* nodes. Then the flit data can be sent to the *Dst* node. After the VB transmission, the *Src* node withdraws the *Request* signal, and the arbiter will release the VB in the next cycle.

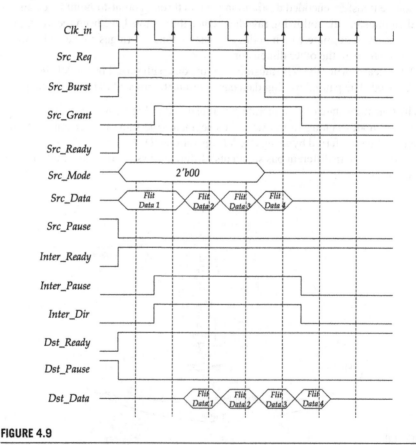

FIGURE 4.9

The timing diagram of VB operation with the burst mode.

4.4.2.5 A simple example for VB communication

Figure 4.10 shows a simple example for VB communication, which includes three packets. We assume that the transmission times of three packets overlap and the packet ID indicates the message generation order. *Packet1* and *Packet2* are unicast messages, and these two packets can be transfered concurrently since their required channels are disjoint. In this time, node N_{13} generates a multicast packet *Packet3* which goes to destination nodes: N_{01}, N_{10}, N_{12}, and N_{31}. Two VBs are involved in the communication process for *Packet3*. The row VB (VB_{R1}, from node N_{10} to node N_{13}) is first established after the arbitration. Then the column VB (VB_{C1}, from node N_{01} to node N_{31}) is established. Node N_{11} is the bridge between two VBs. When the flits of *Packet3* are sent to node N_{11}, they are stored in VB buffers and node N_{11} requests the column VB transmission. The ongoing flits of *P-U1* and *P-U2*, which share the same datapath with the two VBs, are just frozen, because VBs have the highest priority to transmit.

4.4.3 STARVATION AND DEADLOCK AVOIDANCE

There are two cases causing starvation scenarios in the VBON. The first case is when a node along the VB path always has incoming VB flits to serve, and flits buffered locally at the node may never get a chance to use the physical channel. This is because higher priority is given to VB flits. A simple solution to avoid such a scenario is to maintain a count of the number of consecutive VB operations for which it has served. After serving for n consecutive VB operations, the VB node sends a *starvation on* token to the arbiter if it has conventional NoC flits which are getting starved.

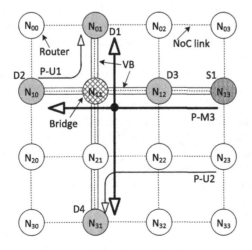

FIGURE 4.10

A communication example in the VBON.

Upon receiving this token, the VB arbiter stops granting VB requests for the next p consecutive cycles. Hence, the starved locally buffered NoC flits can now be serviced. The n and p are design parameters which can be set empirically.

The second case for starvation scenarios is caused by unfair priority policies between VB operations. We suggest that the row VB independently increases the bus priority of the respective nodes by way of a bus priority shifter. Thus, all the nodes on the same row have equal opportunities in the row VB. The bus priority of the column VB rotates like that of the row VB, but in contrast the column VB alters simultaneously the priority of all the columns. In this way, all the nodes on the same column have equal opportunities.

Another critical issue when designing NoCs is to guarantee deadlock-free operation. Routing-dependent deadlocks occur when there is a cyclic dependency of resources created by the packets on the various paths in the network. In the case of a VBON communication using VBs, the deadlock conditions are not generated, since VBs are used in order of the row VB and the column VB in the dimensional order fashion.

4.4.4 THE VBON ROUTER MICROARCHITECTURE

Figure 4.11 shows the microarchitecture of the proposed VBON router. It shows the implementation details of one input port and one output port of the router. The differences from a generic router are shaded. In order to implement the VB functionality, three components are added or modified: the VB channel, pause logic, and VB allocator. For each port, there is a VB channel (VC0) devoted to establishing VB connections. There are two datapaths in the VB channel: the virtual channel (VC) and the VB bypassing datapath. When the router is used as a bridge (between row and column VBs, or between different bus segments) or the injection node, the VC is used. When the router is used as a bypassing node, the VB bypassing datapath is used. To implement the bypass path, a two-to-one multiplexer is required at the output of the buffer, and this can isolate the buffer and connect the input to the output through the crossbar switch.

The pause logic is responsible for the VB protocols. It receives the control signals from the VB arbiter and generates inner control signals for VB channels and the VB allocator. If the pause signal is not activated, then the VC is selected on the basis of the outcome of the routing function, VC allocator, and switch allocator, as in traditional packet-switched networks. Otherwise, the VB data is directed to the crossbar input. Since VB data have higher priority over NoC packets, the NoC packets overlapping with the VBs, when forming the VBs, should be paused by the pause logic. However, these ongoing routing packets interrupted by the VB should be resumed after VB operation, so a register for each link is required in order to keep track of the routing information of the paused packets.

Each output port contains a VB allocator which takes the signals from the pause logic as well as the output of the switch allocator as inputs, and allocates the output

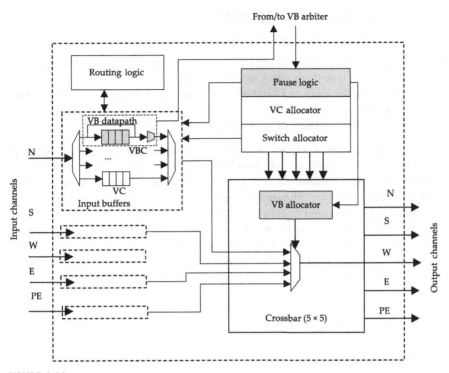

FIGURE 4.11

The VBON router architecture.

port to the input port of the VB when the pause signals activate. Otherwise, the allocation is done according to the output of the switch allocator, as in traditional packet-switched NoCs.

Supporting VBs in a baseline packet-switched router involves adding a bypassing datapath in each input port, the pause logic for the VB protocol, a simple VB allocator at each output port, and some wires to propagate the control signals. There is minimal impact on the baseline router architecture and a negligible area overhead is imposed.

4.5 EVALUATION

In this section, we present a detailed evaluation of the VBON with a baseline 2D mesh topology. We will describe the evaluation method, followed by results obtained using both synthetic and application traffic patterns. Afterwards, the effect of the proposed method on the NoC hardware cost is evaluated.

4.5.1 SIMULATION INFRASTRUCTURES

The purposes of our experiments are

(1) to compare VBON multicasting with state-of-the-art router designs unicasting multiple packets with each in a separate packet,
(2) to investigate the impact of the multicast traffic on the unicast traffic in a mixed unicast-multicast network, and
(3) to evaluate the scalability of the multicast scheme.

To evaluate the proposed VBON approach, we implemented its architecture using a SystemC-based cycle-level NoC simulator, which is modified from a NIRGAM simulator [8]. The simulator models a detailed five-stage router pipeline. We can change various network configurations, such as the network size, topology, buffer size, routing algorithm, and traffic pattern. In the following text, some specific issues related to the simulation, including the reference designs, network configuration, and traffic generation, are introduced. We integrated Orion [30], an architecture-level network energy model, in our simulator to calculate the power consumption of the networks. The power results reported by Orion are based on an NoC in 90 nm technology.

4.5.1.1 Router choices for comparison

We compare various characteristics of the proposed VBON schemes against two existing baseline packet-switched networks: the basic router [9] and the NOCHI EVC router [15].

The basic router is representative of conventional NoC designs. It originally had a four-stage router pipeline. The first stage is the buffer write; the routing computation (RC) occurs in the second stage. In the third stage, VC allocation (VA) and switch allocation (SA) are performed. In the fourth stage, the flit traverses the switch. Each pipeline stage takes one cycle, followed by one cycle to do the link traversal to the next router. To shorten the pipeline, the basic router uses lookahead routing [7] and the speculative method [24]. Lookahead routing determines the output port of a packet one hop in advance. That is to say while the flit is traversing the switch, a lookahead signal is traveling to the next router to perform the RC. In the next cycle, when the flit arrives, it will proceed directly to SA. The speculative method can eliminate the VA and SA stage by causing flits to directly enter the switch traversal pipeline.

The NOCHI EVC router was chosen because its concept is similar to our work. It is an optimized design based on the EVC router design [17]. A conventional EVC router sets up virtual express paths in the network that let packets bypass intermediate routers, thus improving the energy-delay throughput of NoC interconnects. We can regard this virtual express path as a multiple-stage VB. Though it uses the low-latency benefit of the bus mechanism for unicast communication, it does not use the broadcast feature of the bus, leading to some limitations. First, buffers at an EVC's end point must be managed conservatively to ensure that the destination router (the EVC end point) can accept the traffic. This leads to buffer overprovision and underutilization. The second deficiency is that VCs must be partitioned between different express paths

Table 4.1 Comparison of Router Designs

Parameter	Basic Router	NOCHI EVC Router	VBON Router
Overhead	–	G-lines, EVC controller	VB control lines, VBON controller
Bus link	–	Virtual	Virtual
No. of hops	Multiple hops	Multiple hops	Single hop for each VB
No. of pipelines	Multiple stages	Single stage with intermediate router	No stage with intermediate router
Routing	Lookahead routing	Bypassing routing	Bypassing routing
Unicast	Pipelining in each router	EVC in one dimension	Hierarchical VB besides pipelining
Multicast	Supported by unicast	Supported by unicast	Support with hierarchical VB

statically, and the control latency limits the number. To overcome these limitations, the NOCHI EVC router is proposed to enable single-cycle control communication across all nodes in a row or column of a mesh network through G-line. Using timely information reduces the demand on router buffers as well as the number of buffers needed to sustain a specific bandwidth. In a sentence, the NOCHI EVC approach is a very effective method for unicast communication, but it does not support multicast communication.

Table 4.1 lists the comparison results for the three different NoC designs. The comparison parameters are

(1) the hardware overhead of the basic router,
(2) the setup mechanism of the bus link,
(3) the number of hops for each unicast packet,
(4) the number of pipeline stages for each router,
(5) the RC method,
(6) the communication mechanism for the unicast traffic, and
(7) the communication mechanism for multicast traffic.

We extended the network model to capture these aspects of the VB protocol and G-line EVC protocol.

4.5.1.2 Network configuration
Table 4.2 lists the network configurations across all the studies. The mesh network is simulated with two configurations: one is that only cluster 1 with a 4 × 4 mesh is enabled and the other is that all 8 × 8 hierarchical clusters are enabled. It should be noted that this is a fair comparison between the proposed VBON architecture and other architectures since these different network configurations are just caused by their different architectures.

Each simulation experiment is run until the network reaches its steady state. The destinations of unicast messages at each node are selected randomly. For unicast-only scenarios, once the number of hops of the unicast message is not less than three (the threshold in a 4×4 mesh) or five (the threshold in an 8×8 mesh), then the messages are transmitted by VBs. For multicast scenarios, all the multicast messages are transmitted through VBs. To decrease the contention caused by the VB transmission, the threshold for VB transmission is decreased to four (4×4 mesh) or eight (8×8 mesh). The multicast destinations and their numbers are generated randomly at the beginning of the simulation.

4.5.1.3 Traffic generation

For the sake of comprehensive study, numerous validation experiments have been performed for several combinations of workload types and network sizes. In what follows, the capability of the VBON will be assessed for both synthetic and realistic traffic. The synthetic traffic patterns used in this research are the uniform random, transpose, complement, and Cauchy patterns to have a more specific evaluation for different traffic patterns [5]. Each simulation runs for 1×10^6 cycles. To obtain

Table 4.2 The Network Configurations

Design	Parameter	Measure
Basic design	Topology	2D mesh with single 4×4, hierarchical 8×8
	Basic routing	Dimensional order XY routing
	Router ports	5
	VCs per port	4
	Buffers per VC	16
	Channel width/flit size	128 bits
	Packet size	8 flits
NOCHI EVC design	Structure	Row and column G-lines with 8 nodes
	l_{max}	7 (maximum length of EVC)
	Buffers per port	15
VBON design	Structure	Row and column VBs
	Hierarchy	Level 1: 4×4, 5×4, 4×5, 5×5; Level 2: 8×8
	VB usage	Multicasting with 4 or 8 hops, unicasting with more than 2 or 4 hops
	Threshold: n	4 (maximum number of consecutive VB operations)
	Threshold: p	8 (number of cycles to stop granting the VB request)

stable performance results, the initial 1×10^5 cycles are used for simulation warm-up and the following 9×10^5 cycles are used for analysis. When destinations are chosen randomly, we repeat the simulation run five times and get the average of the values obtained in each run. We also studied the VBON architecture using real application network traffic. We ran Splash-2 benchmarks [32] on a chip multiprocessor simulator, M5[1], with shared second-level caches distributed in a tiled manner. Each core consists of a two-issue in-order SPARC processor with two-way 16 kB first-level ICache and two-way 32 kB DCache. Each tile also includes a 1 MB second-level cache. We used the MESI-based directory protocol for cache coherence.

4.5.2 SYNTHETIC TRAFFIC EVALUATIONS
4.5.2.1 Single-level 4 × 4 VBON
We first evaluate the performance of a VBON compared with other NoC designs in the environment of a single-level 4 × 4 mesh. Figure 4.12 shows the average latencies of the networks when all the packets are unicast ones. It can be seen that the conventional NoC routers perform very well when the traffic load is low. The latencies of the VBON, NOCHI EVC, and NoC designs are almost the same. But as the traffic load increases, the NOCHI EVC and VBON designs gradually reduce the latency of unicast communications. This is because unicast communication with long latency can be performed by global channels in the NOCHI EVC design or by VBs (when the number of hops exceeds two). This feature is very useful when some urgent messages need to be transmitted. The VBON design also outperforms

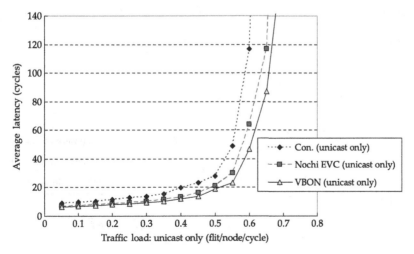

FIGURE 4.12

Performance results for unicast communication with uniform traffic for a 4 × 4 mesh.

the NOCHI EVC design when the traffic load is heavy. This is because the real VB transmission can be fulfilled by one cycle, while the EVC transmission is completed in a pipelined fashion, which takes several cycles.

Figure 4.13 shows the performance results when 10% of packets are multicast ones. In conventional NoCs, the multicast transmission path is defined by the source node. The performance significantly degrades and it saturates at a rate less than 0.3 flits per node per cycle. This is because a few multicast packets can cause a lot of the NoC workload, leading to significant throughput degradation. It can be seen that although the optimized NoC design, NOCHI EVC, alleviates this situation by decreasing the pipelining cycles of the routing path, it does not solve the multicast problem. The VBON design can greatly outperform the NOCHI EVC design for multicast transmission. The support for multicast in the VBON design leads to only a small increase in the average latency. This is because the VB can transmit the message to multiple nodes in a broadcast way. And for each multicast transmission, only a few VB transactions are involved.

4.5.2.2 Hierarchical 8 × 8 VBON

For an 8 × 8 mesh structure, the comparison results for average latencies for unicast communication are shown in Figure 4.14. These results are given with different traffic patterns: uniform random, transpose, complement, and Cauchy patterns. The packet communication in this hierarchical VBON may involve four VB clusters. Compared with a 4 × 4 mesh, the average latency of unicast communications in an 8 × 8 conventional NoC design increases significantly owing to the increased number of transmission hops. In this case, the performance improvement caused by the NOCHI EVC and VBON designs is more obvious. For all traffic patterns, the VBON design

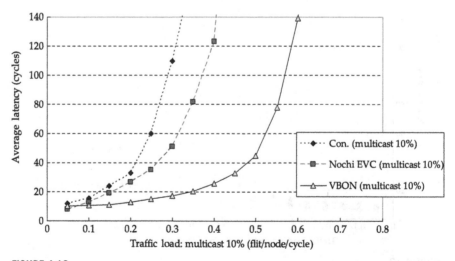

FIGURE 4.13

Performance results for multicast communication with random traffic for a 4 × 4 mesh.

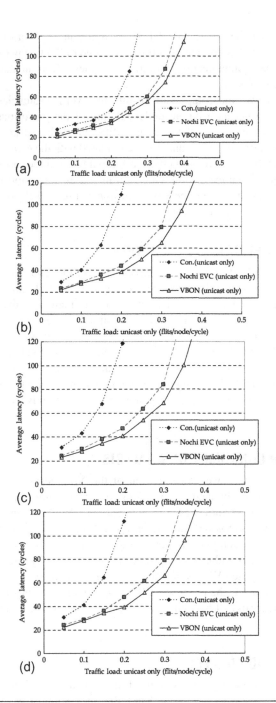

FIGURE 4.14

Performance results for unicast communication with various traffic patterns for an 8 × 8 mesh. (a) Uniform random traffic pattern. (b) Transpose traffic pattern. (c) Complement traffic pattern. (d) Cauchy traffic pattern.

also performs better than the NOCHI EVC design, ranging from 8% for uniform random traffic to 19% for complement traffic at a rate of 0.3 flits per node per cycle. This is because the communication of multiple-segment VBs can take less time than that of pipelined EVCs.

Figure 4.15 shows the comparison results for average latencies for multicast communication. To get an accurate understanding of the performance impact for multicast communications, we provide three different multicast percentages in the whole communication: 5%, 10%, and 20%. Compared with the NOCHI EVC design, the proposed VBON design shows better throughput for all three cases. Furthermore, as the percentage of multicast communication increases, this improvement is more obvious. This demonstrates that the VBON design can scale well for large multicast communication in a hierarchical way.

4.5.3 REAL APPLICATION EVALUATIONS

We also ran the application traffic to evaluate the VBON and use memory access latency as the performance metric for various NoC designs. Figures 4.16 and 4.17 illustrate the comparison results for memory latency normalized to that of the baseline NoC design for 4×4 mesh and 8×8 mesh structures respectively. We see that the VBON has the best overall performance for memory access. It outperforms the baseline and NOCHI EVC designs by 4.3% and 9.3% respectively in an 8×8 mesh structure. This performance improvement is mainly because the benchmarks have many long-latency and multicasting packets for maintaining the cache coherency. The results of application executions have demonstrated the effectiveness of the VBON design.

4.5.4 POWER CONSUMPTION ANALYSIS

Figure 4.18 shows the power consumption of the NoC designs. The power consumption results were obtained by executing application traffic in an 8×8 mesh topology. As the figure indicates, the conventional NoC design has the largest power consumption, and is used as the baseline design. The NOCHI EVC design reduces the power consumption by an average of about 14%. This is mainly due to a reduction in buffer and crossbar power consumption in bypassing routers. In a similar way, the proposed VBON design can also reduce the power consumption by about 20%. In summary, the proposed VBON design clearly outperforms the other two NoC designs across all benchmarks not only in performance but also in power consumption.

4.5.5 OVERHEAD ANALYSIS

In this subsection, we analyze the overhead introduced by the VBON design from performance and area aspects. On one hand, the integration of the VB mechanism into the NoC design will pause the original point-to-point transmissions when their

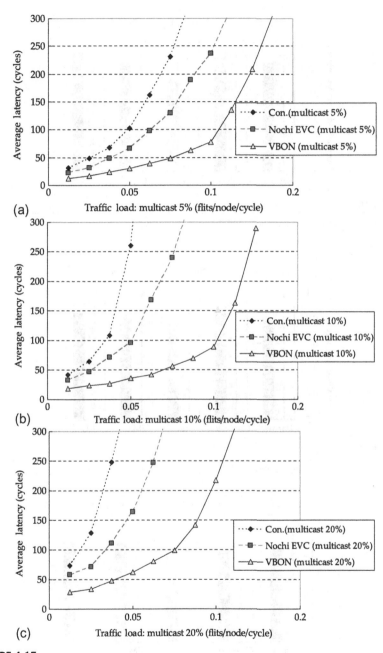

FIGURE 4.15

Performance results for various multicast communications with uniform traffic for an 8×8 mesh. Multicast communication: (a) 5%, (b) 10%, (c) 20%.

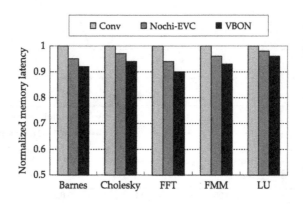

FIGURE 4.16

Performance results for application traffic in a 4 × 4 mesh.

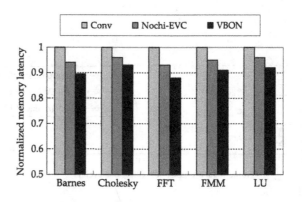

FIGURE 4.17

Performance results for application traffic in an 8 × 8 mesh.

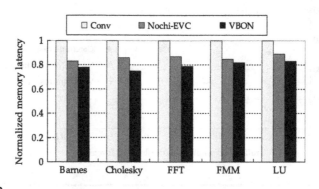

FIGURE 4.18

Power consumption results for application traffic in an 8 × 8 mesh.

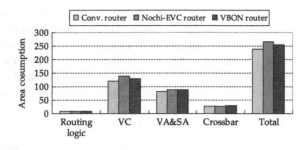

FIGURE 4.19

Area overhead results for router designs.

links are right on the VB path. This may introduce latency overheads for these paused packet transmissions. On the other hand, the VB mechanism can reduce the latency of critical transmissions, including multicast and long-latency packet transmissions. This will result in overall performance improvement as shown in the previous evaluation.

The modifications of conventional routers in the VBON will also introduce some area overheads. In order to estimate the hardware cost, we implemented the conventional NoC router, the NOCHI EVC router, and the VBON router in Verilog and performed the logic synthesis using Synopsys Design Compiler to get the area information. We used a Taiwan Semiconductor Manufacturing Company 90 nm CMOS generic process technology in logic synthesis. The area of the control network (with the bus arbiter and control wires) can be ignored compared with the area of the 128-bit-wide routers and 128-bit bidirectional links of the NoC network. In addition to the control network, the impacts of the proposed router microarchitecture modifications are illustrated in Figure 4.19. From the figure, about 6% area overhead for VBON is observed. This is due to the added logic, such as the pause logic, VC control logic for the VB, and some buffers to store the information on existing router states. The NOCHI EVC router has a higher area overhead, about 10% compared with the conventional router. This is mainly because the number of buffers added in VC logic is large.

To get resource consumption information that is relevant to the CMOS library, we also provide the gate number results. The total number of gate equivalents for the conventional router is about 28,000, while for the VBON router it is about 31,000. The NOCHI EVC router has the highest gate number, about 32,800.

4.6 CHAPTER SUMMARY

In this chapter, the VBON, a hybrid architecture with a packet-based NoC and a transaction-based VB, has been proposed. The point-to-point links of conventional NoC designs can be used as bus transaction links dynamically for VB requests.

To reduce the latency of the physical layout for the bus organization, we used several buses instead of one bus to limit the longest distance of the network to a bearable bound. Furthermore, the hierarchical clustered bus structure was configured, dividing the chip into several clusters of cores with redundant bus structures connecting routers in each cluster. Detailed experimental results confirmed that the VBON can achieve low latency while sustaining high throughput for both unicast and multicast communications at low cost.

REFERENCES

[1] N.L. Binkert, R.G. Dreslinski, L.R. Hsu, K.T. Lim, A.G. Saidi, S.K. Reinhardt, The M5 simulator: modeling networked systems, IEEE Micro 26 (4) (2006) 52–60. 129

[2] S. Borkar, Networks for multi-core chips—a contrarian view, in: Invited Presentation to ISLPED *2007*, 2007. 115

[3] E.A. Carara, F.G. Moraes, Deadlock-free multicast routing algorithm for wormhole-switched mesh networks-on-chip, in: Proceedings of the Symposium on VLSI (ISVLSI), 2008, pp. 341–346. 110

[4] J.H. Choi, B.W. Kim, K.H. Park, K.-I. Park, A bandwidth-efficient implementation of mesh with multiple broadcasting, in: Proceedings of the International Conference on Parallel Processing (ICPP), 1999, pp. 434–443. 110

[5] W. Dally, B. Towles, Principles and Practices of Interconnection Networks, first ed., Morgan Kaufmann Publishers Inc., San Francisco, CA, USA, 2003. 108, 111, 128

[6] N. Enright Jerger, L.-S. Peh, M. Lipasti, Virtual circuit tree multicasting: a case for on-chip hardware multicast support, in: Proceedings of the International Symposium on Computer Architecture (ISCA), 2008, pp. 229–240. 108, 110

[7] M. Galles, Spider: a high-speed network interconnect, IEEE Micro 17 (1) (1997) 34–39. 126

[8] M.S. Gaur, B.M. Al-Hashimi, V. Laxmi, R. Navaneeth, N. Choudhary, L. Jain, M. Ahmed, K.K. Paliwal, Varsha, Rekha, Vineetha, NIRGAM: a simulator for NoC interconnect routing and application modeling, in: Proceedings of the Design, Automation & Test in Europe Conference & Exhibition (DATE), 2007. 126

[9] P. Gratz, C. Kim, R. McDonald, S.W. Keckler, D. Burger, Implementation and evaluation of on-chip network architectures, in: Proceedings of the International Conference on Computer Design (ICCD), 2006, pp. 477–484. 111, 126

[10] B. Grot, J. Hestness, S.W. Keckler, O. Mutlu, Express cube topologies for on-chip interconnects, in: Proceedings of the International Symposium on High-Performance Computer Architecture (HPCA), 2009, pp. 163–174. 109

[11] Y. Hoskote, S. Vangal, A. Singh, N. Borkar, S. Borkar, A 5-GHz mesh interconnect for a Teraflops Processor, IEEE Micro 27 (5) (2007) 51–61. 108

[12] L. Huang, Z. Wang, N. Xiao, VBON: toward efficient on-chip networks via hierarchical virtual bus, in: WIP Session of the the Design Automation Conference (DAC), 2011. 107

[13] L. Huang, Z. Wang, N. Xiao, VBON: toward efficient on-chip networks via hierarchical virtual bus, Microprocess. Microsyst. 37 (8, Part B) (2013) 915–928. 107

[14] J. Kim, J. Balfour, W. Dally, Flattened butterfly topology for on-chip networks, in: Proceedings of the International Symposium on Microarchitecture (MICRO), 2007, pp. 37–40. 109

[15] T. Krishna, A. Kumar, L.-S. Peh, J. Postman, P. Chiang, M. Erez, Express virtual channels with capacitively driven global links, IEEE Micro 29 (2009) 48–61. 109, 126

[16] A. Kumar, L.-S. Peh, N.K. Jha, Token flow control, in: Proceedings of the International Symposium on Microarchitecture (MICRO), 2008, pp. 342–353.

[17] A. Kumar, L.-S. Peh, P. Kundu, N.K. Jha, Express virtual channels: towards the ideal interconnection fabric, in: Proceedings of the International Symposium on Computer Architecture (ISCA), 2007, pp. 150–161. 109, 126

[18] Z. Li, J. Wu, L. Shang, R.P. Dick, Y. Sun, Latency criticality aware on-chip communication, in: Proceedings of the Design, Automation & Test in Europe Conference & Exhibition (DATE), 2009, pp. 1052–1057. 108

[19] Z. Lu, B. Yin, A. Jantsch, Connection-oriented multicasting in wormhole-switched networks on chip, in: Proceedings of the International Symposium on Emerging VLSI Technologies and Architectures (ISVLSI), 2006, pp. 205–210. 108, 110, 121

[20] R. Manevich, I. Walter, I. Cidon, A. Kolodny, Best of both worlds: a bus enhanced NoC (BENoC), in: Proceedings of the International Symposium on Networks-on-Chip (NOCS), 2009, pp. 173–182. 108, 109, 114

[21] H. Matsutani, M. Koibuchi, H. Amano, T. Yoshinaga, Prediction router: yet another low latency on-chip router architecture, in: Proceedings of the International Symposium on High-Performance Computer Architecture (HPCA), 2009, pp. 367–378. 109

[22] S. Murali, D. Atienza, P. Meloni, S. Carta, L. Benini, G. De Micheli, L. Raffo, Synthesis of predictable networks-on-chip-based interconnect architectures for chip multiprocessors, IEEE Trans. Very Large Scale Integration (VLSI) Syst. 15 (2007) 869–880. 109

[23] N. Muralimanohar, R. Balasubramonian, Interconnect design considerations for large NUCA caches, in: Proceedings of the International Symposium on Computer Architecture (ISCA), 2007, pp. 369–380. 108, 109, 114

[24] L.-S. Peh, W. Dally, A delay model and speculative architecture for pipelined routers, in: Proceedings of the International Symposium on High-Performance Computer Architecture (HPCA), 2001, pp. 255–266. 126

[25] T.D. Richardson, C. Nicopoulos, D. Park, V. Narayanan, Y. Xie, C. Das, V. Degalahal, A hybrid SoC interconnect with dynamic TDMA-based transaction-less buses and on-chip networks, in: Proceedings of the International Conference on VLSI Design (VLSID), 2006, pp. 657–664. 108, 109, 114

[26] F.A. Samman, T. Hollstein, M. Glesner, Multicast parallel pipeline router architecture for network-on-chip, in: Proceedings of the Design, Automation & Test in Europe Conference & Exhibition (DATE), 2008, pp. 1396–1401. 108, 110

[27] Semiconductor Industry Association, International Technology Roadmap for Semiconductors, 2008 Edition. http://www.itrs.net, 2008. 116

[28] M.B. Stensgaard, J. Spars, ReNoC: a network-on-chip architecture with reconfigurable topology, in: Proceedings of the International Symposium on Networks-on-Chip (NOCS), 2008, pp. 55–64. 109

[29] A.N. Udipi, N. Muralimanohar, R. Balasubramonian, Towards scalable, energy-efficient, bus-based on-chip networks, in: Proceedings of the International Symposium on High-Performance Computer Architecture (HPCA), 2010, pp. 247–258. 108, 109, 114

[30] H.-S. Wang, X. Zhu, L.-S. Peh, S. Malik, Orion: a power-performance simulator for interconnection networks, in: Proceedings of the International Symposium on Microarchitecture (MICRO), 2002, pp. 294–305. 126

[31] D. Wentzlaff, P. Griffin, H. Hoffmann, L. Bao, B. Edwards, C. Ramey, M. Mattina, C.-C. Miao, J.F. Brown III, A. Agarwa, On-Chip Interconnection Architecture of the TILE Processor. IEEE Micro 27 (5) (2007) 15–31. 108

[32] S. Woo, M. Ohara, E. Torrie, J. Singh, A. Gupta, The SPLASH-2 programs: characterization and methodological considerations, in: Proceedings of the International Symposium on Computer Architecture (ISCA), 1995, pp. 24–36. 129.

Routing and flow control

The routing algorithm and flow control allocate network resources, including buffers and channels, to packets. They directly determine the network-on-chip (NoC) performance. Owing to the limited parallelism of a single application, multiple applications will run concurrently on a many-core platform. This workload consolidation scenario requires routing algorithms not only to provide high performance, but also to maintain the application isolation. In addition, the resource allocation must avoid network deadlock. There are several deadlock avoidance theories for routing and flow control designs. Yet, most theories were originally presented for off-chip networks, whose traffic characteristics are quite different from those of on-chip networks. These theories have been used for about one or two decades, and it is now the right time to re-examine them, and propose more efficient deadlock-free routing and flow control for NoCs. On the basis of these insights, this part delves into NoC routing and flow control designs in three chapters.

In Chapter 5 a holistic approach is applied to design efficient routing for workload consolidation scenarios. Existing locally adaptive algorithms do not consider enough status information to avoid network congestion. Globally adaptive routing algorithms attack this issue by utilizing network status beyond neighboring nodes. However, they may suffer from interference, coupling the otherwise independent applications. To address these issues, we propose a novel selection strategy, the destination-based selection strategy (DBSS), for many-core systems running consolidation workloads. The key aspects of routing design include adaptivity, path selection strategy, virtual channel (VC) allocation, isolation, and hardware implementation cost; these aspects are not independent. We holistically consider all aspects to ensure an efficient design. DBSS leverages both local and nonlocal network status to provide more effective adaptivity. More importantly, by integrating the destination into the selection procedure, DBSS mitigates interference and offers dynamic isolation among applications.

Fully adaptive routing algorithms provide high routing flexibility. Yet, their performance may be limited by flow control mechanisms. Chapter 6 focuses mainly on efficient flow control designs for deadlock-free fully adaptive routing. Existing fully adaptive routing algorithms apply conservative VC reallocation: only empty VCs can be reallocated, which limits performance. In this chapter we propose two novel flow control designs. First, whole packet forwarding reallocates a nonempty VC if the VC has enough free buffers for an entire packet. Whole packet forwarding does not induce deadlock if the routing algorithm is deadlock-free using conservative VC reallocation. It is an important extension to several deadlock avoidance theories. Second, we extend Duato's theory to apply aggressive VC reallocation on escape VCs without deadlock. Finally, we propose a design which maintains maximal routing flexibility with low hardware cost.

The torus NoC needs additional effort to avoid deadlock. Chapter 7 delves into the design of high-performance, low-cost, and deadlock-free flow control for torus NoCs. Existing designs for torus networks cannot efficiently handle variable-size packets existing in cache-coherent NoCs. For deadlock-free operations, one design uses two VCs, which increases the router complexity and negatively affects the router frequency. Some optimizations use one VC. Yet, they must regard all packets as maximum-length packets, inefficiently utilizing the precious NoC buffers. We propose the flit bubble flow control theory, which maintains one free flit-size buffer slot to avoid deadlock. Flit bubble flow control uses one VC, and does not treat short packets as long ones. On the basis of this theory, we present two implementations. Both implementations achieve high frequencies and efficient buffer utilization. They are suitable for the torus NoCs.

Routing algorithms for workload consolidation†

5

CHAPTER OUTLINE

†Part of this research was first presented at the 38th Annual International Symposium on Computer Architecture (ISCA-2011) [37]. The ISCA-2011 paper was also published in *ACM SIGARCH Computer Architecture News* [38]. The journal extension edition of the ISCA-2011 paper was published in *IEEE Transactions on Computers* [40].

Networks-on-Chip. http://dx.doi.org/10.1016/B978-0-12-800979-6.00005-6
Copyright © 2015 China Machine Press/Beijing Huazhang Graphics & Information Co., Ltd.
Published by Elsevier Inc. All rights reserved.

5.1 INTRODUCTION

Given the difficulty of extracting parallelism, it is quite likely that multiple applications will run concurrently on a many-core system [3, 25, 36], often referred to as workload consolidation. Significant research exists on maintaining isolation and effectively sharing on-chip resources such as caches [53] and memory controllers [46]. The network-on-chip (NoC) [11] is another, less-explored example of a shared resource where one application's traffic may degrade the performance of another. This work focuses on improving performance and providing isolation for workload consolidation via the routing algorithm.

Several requirements are placed on the routing algorithm for high performance. First, the routing algorithm should leverage available path diversity to provide sufficient adaptivity to avoid network congestion. Closely related to the issue of adaptivity is virtual channel (VC) [9] allocation (VA) and the need to provide deadlock freedom. Second, it should not leverage superfluous information leading to inaccurate estimates of network status. Finally, the routing algorithm should be implemented with low hardware cost. Workload consolidation has an additional requirement: dynamic isolation among applications. Existing routing algorithms are unable to meet all these needs. Oblivious routing, such as dimensional order routing (DOR), ignores network status, resulting in poor load balancing. Adaptive routing offers the ability to avoid congestion by supporting multiple paths between a source and the destination; a selection strategy is applied to choose between multiple outputs. Most existing selection strategies do not offer both adaptivity and isolation.

The selection strategy should choose the output port that will route a packet along the least congested path. A local selection strategy (LOCAL) leverages only the status of neighboring nodes, which tends to violate the global balance intrinsic to traffic [22]. Globally adaptive selection strategies, such as regional congestion awareness (RCA) [22], utilize a dedicated network to gather global information; it introduces excess information when selecting the output port and offers no isolation among applications, leading to performance degradation for consolidated workloads.

This excess information can be classified as intra-application and interapplication interference. Interference makes the performance less predictable.

Considering the future prevalence of server consolidation and the need for performance isolation, an efficient routing algorithm should combine high adaptivity with dynamic workload isolation. Therefore, we believe utilizing precise information is preferable; redundant or insufficient information easily leads to inferior performance. On the basis of this insight, we introduce the destination-based selection strategy (DBSS), which is well suited to workload consolidation.

We design a low-cost congestion information propagation network to leverage both local and nonlocal network status, giving DBSS high adaptivity. Furthermore, DBSS chooses the output port by considering only the nodes that a packet may traverse, while ignoring nodes located outside the minimum quadrant defined by the current location and the destination. Thus, it eliminates redundant information and dynamically isolates applications.

Integral to the evaluation of a new selection strategy is the underlying routing function. To provide a thorough exploration, we analyze the offered path diversity of several fully and partially adaptive routing functions. We also consider the VC reallocation scheme [12] that is required for deadlock freedom. On the basis of an appropriate routing function, we evaluate DBSS against other selection strategies for both a regular mesh and a concentrated mesh (CMesh). Our experimental results show that DBSS outperforms the other strategies for all evaluated network configurations.

In this chapter we make the following contributions:

- We analyze the limitations of other selection strategies and propose DBSS, which affords sufficient adaptivity for congestion avoidance and dynamic isolation among applications.
- We explore the effect of interference and demonstrate that the amount of congestion information considered impacts performance, especially for consolidated workloads.
- We design a low-cost congestion status propagation network with only 3.125% wiring overhead to leverage both local and nonlocal information.
- We holistically consider path diversity and VC reallocation to provide further insight for routing algorithm design.

5.2 BACKGROUND

In this section, we discuss related work in application mapping and adaptive routing algorithm design.

Since the arrival order and execution time of consolidated workloads cannot be known at the design time, run-time application mapping techniques are needed [7, 8, 27, 34]. Mapping each application to a near convex region is optimal for workload consolidation [7, 8]. Most application mapping techniques consider the Manhattan distance between the source and the destination but not the routing paths [8, 34]; routing algorithm design is complementary to these techniques.

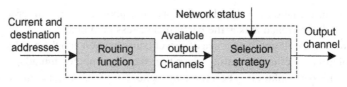

FIGURE 5.1

The structure of an adaptive routing algorithm.

As shown in Figure 5.1, an adaptive routing algorithm consists of two parts: the routing function and the selection strategy [1]. The routing function computes the set of possible output channels, while the selection strategy chooses one of them on the basis of the network status. A routing function must achieve deadlock freedom. This can be achieved by removing cycles from the channel dependency graph [10]. Duato [15, 16] proved that cycles are allowed in the channel dependency graph once there is a deadlock-free routing subfunction. Duato's theory is powerful for designing fully adaptive routing functions which allow all minimal paths for packet routing. Turn model functions avoid deadlock by offering only a subset of all minimal paths [6, 19, 21]. In addition to the offered path diversity, partially and fully adaptive functions differ in the VC reallocation. Partially adaptive routing can utilize an aggressive VC reallocation scheme, while fully adaptive routing can only leverage a conservative scheme. VC reallocation strongly impacts performance and cannot be overlooked in the design of routing algorithms [19, 39].

Off-chip networks are constrained by pin bandwidth, but the abundant wiring resources in NoCs allow easier implementation of congestion propagation mechanisms. Therefore, the NoC paradigm has sparked renewed interest in adaptive routing algorithms. DyAD combines the advantages of both deterministic and adaptive routing schemes [26]. DyXY uses dedicated wires to investigate the status of neighboring routers [35]. A low-latency adaptive routing algorithm performs lookahead routing and preselects the optimal output port [31]. The selection strategies of these designs [26, 31, 35] all leverage the status of the neighboring nodes. Many oblivious selection strategies have also been evaluated, including zigzag, XY, and no turn [13, 18, 21, 42, 52]. Neighbors-on-path (NoP) makes a selection based on the condition of the nodes adjacent to neighbors [1].

RCA is the first strategy utilizing global information to improve load balancing in NoCs [22]; however, it introduces interference. Redundant information may degrade the quality of the congestion estimates; to combat this, packet destination information can be leveraged to eliminate excess information [49]. Per-destination delay estimates steer the output selection. Despite leveraging a similar observation, our design is quite different. Moreover, we focus on the performance with workload consolidation. The bLBDR design [50] provides strict isolation among applications by statically configuring connectivity bits. In contrast, we offer more dynamic isolation.

5.3 MOTIVATION

The need for a novel selection strategy was motivated from two directions. First, the selection strategy should have enough information about network conditions to offer effective congestion avoidance. Both local and NoP [1] selection strategies lack enough information, leading to suboptimal performance. Second, intra-application and interapplication interference should be minimized. The RCA strategy [22] utilizes global network information but does not consider interference. DBSS offers a middle ground between these extremes.

5.3.1 INSUFFICIENT INFORMATION

The local selection strategy (LOCAL) leverages the conditions of neighboring nodes. These conditions may be free buffers [1, 22, 26, 31, 35], free VCs [13, 22], crossbar demands [22], or a combination of these [22]. Figure 5.2a shows a packet at router (0,2) that is routed to (2,0). A selection strategy is needed to choose between the west and south ports; LOCAL uses only the information about the nearest nodes ((0,1) and (1,2)). Without any information about the condition beyond neighbors, it cannot avoid congestion more than one hop away from the current node.

NoP uses the status of nodes adjacent to neighboring nodes as shown in Figure 5.2b. The limitation is that NoP ignores the status of neighbors ((0,1) and (1,2)); it makes decisions only on the basis of the conditions of nodes two hops away. In the example, for west output evaluation, it considers nodes (0,0) and (1,1). For the south port, it considers nodes (2,2) and (1,1). This strategy works well with an odd-even turn model [1]. However, with a fully adaptive routing function, its performance degrades owing to limited knowledge.

5.3.2 INTRAREGION INTERFERENCE

Three RCA variants have been proposed: RCA-1D, RCA-Fanin, and RCA-Quadrant [22]. No single RCA variant provides the best performance across all traffic patterns; therefore, we use RCA-1D as a baseline. Figure 5.2c shows an intraregion[1] scenario for RCA-1D. All 16 nodes run the same application. When evaluating the west output, RCA-1D considers nodes (0,1) and (0,0). Similarly, it considers nodes (1,2), (2,2), and (3,2) when evaluating the south port. For destination node (2,0), the information from node (3,2) causes interference as it lies outside the minimum quadrant defined by (0,2) and (2,0); the packet will not traverse this node. This interference may result in poor output port selection, especially considering traffic locality.

We show the average hop count (AHP) to measure locality for several synthetic traffic patterns [12] in Figure 5.3. Most traffic has an AHP of less than 5.6 (average 5.58) and 3 (average 2.63) for 8×8 and 4×4 meshes respectively. These patterns exhibit locality as most packets travel a short distance. Thus, we need strategies to mitigate intra-application interference.

---

[1] We use "region" and "application" interchangeably.

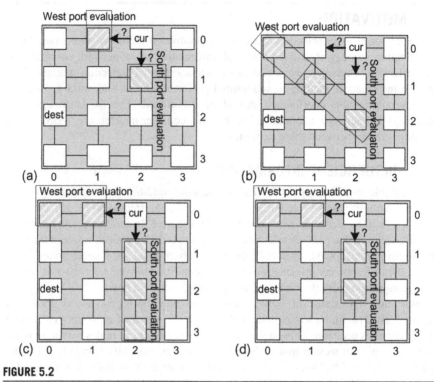

FIGURE 5.2

Packet routing examples. The current and destination routers are (0,2) and (2,0). (a) A scenario for LOCAL, (b) a scenario for NoP, (c) an intraregion interference scenario for RCA-1D, and (d) a scenario for DBSS.

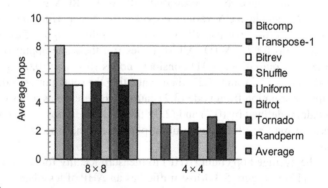

FIGURE 5.3

The average hop count for synthetic traffic.

5.3.3 INTER-REGION INTERFERENCE

Figure 5.4 illustrates a workload consolidation example for an 8×8 mesh; similar scenarios will be prevalent in many-core systems. Here, there are four concurrent applications and each one is mapped to a 4×4 region. Region R0 is defined by nodes (0,0) and (3,3), region R1 is defined by nodes (0,4) and (3,7), region R2 is defined by nodes (4,0) and (7,3), and region R3 is defined by nodes (4,4) and (7,7). Figure 5.4 shows a packet in R0 whose current location and destination are (0,2) and (2,0) respectively. Even though traffic in R0 is isolated from traffic in other regions, RCA-1D considers the congestion status of nodes in R2 when selecting output ports for traffic belonging to R0. Obviously, this method introduces interference and reduces performance isolation.

To evaluate the effect of inter-region interference, we assign transpose-1 traffic [6, 12] to R0 and uniform random traffic to R1, R2, and R3.[2] Figure 5.5 shows the performance of R0. The *RCA-uni_region* curve represents a single region (R0) in a 4×4 network; this latency reflects perfect isolation and no inter-region interference. The saturation throughput of RCA-1D is approximately 65%. However, without isolation, the saturation throughput drops dramatically to approximately 50%, as shown by the *RCA-multi_regions(4%)* curve, where R1, R2, and R3 all have an injection rate of 4%. For the *RCA-multi_regions(64%)* curve, R1 has an injection rate of 64%, while the injection rate remains 4% for R2 and R3; in this unbalanced scenario, the saturation throughput of R0 further decreases to 47%. Clearly, the

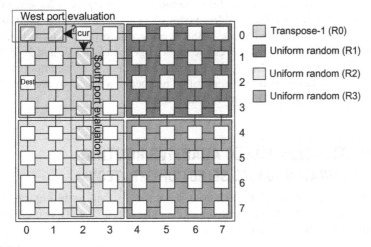

FIGURE 5.4

An inter-region interference scenario for RCA-1D. The current router is (0,2) and the destination is (2,0).

---

[2]See Section 8.5 for the full experimental method.

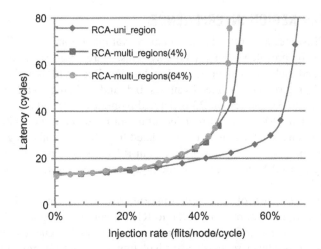

FIGURE 5.5

The load-latency graph of region R0.

congestion information of R1, R2, and R3 greatly affects the routing selection in R0. RCA-1D couples the activities of otherwise independent applications, which is not desirable for workload consolidation.

DBSS aims to reduce the interference by considering only nodes in the minimum quadrant defined by the current and destination nodes. In Figure 5.2d, when evaluating the west output, DBSS considers nodes (0,1) and (0,0); for the south output, it considers nodes (1,2) and (2,2). DBSS leverages information from both local and nonlocal nodes; it has more accurate knowledge than LOCAL and NoP. DBSS does not consider node (3,2), which eliminates interference. More importantly, for workload consolidation with each application mapped to a near convex region [7, 8], DBSS dynamically isolates routing for each region.

5.4 DESTINATION-BASED ADAPTIVE ROUTING
5.4.1 DESTINATION-BASED SELECTION STRATEGY

The selection strategy in adaptive routing algorithms significantly impacts performance [1, 18, 42, 52]. An efficient strategy should ideally satisfy two goals: high adaptivity and dynamic isolation.

5.4.1.1 Congestion information propagation network
NoCs can take advantage of abundant wiring to employ a dedicated network to exchange congestion status. First we explain the design of the low-cost congestion information propagation network, which enables a router to leverage both local

and nonlocal status. Both NoP and RCA utilize such a low-bandwidth monitoring network [1, 22]. We focus on obtaining global information; the propagation network in RCA serves as the best comparison point.

At each hop in RCA's congestion propagation network, the local status is aggregated with information from remote nodes and then propagated to upstream routers [22]. This implementation has two limitations. First, the aggregation logic combines local and distant information during transmission, making it impossible to filter out superfluous information. Second, the aggregation logic adds an extra cycle of latency per hop, leading to stale information at distant routers. On the basis of these two observations, we propose a novel propagation network, which consumes only one cycle per tile, giving DBSS timelier network status. More importantly, the design makes it feasible to filter out information on the basis of the packet destination.

Figure 5.6 shows the congestion propagation network for the third row of an 8×8 mesh; the same structure is present in each row and column. Along a dimension, each router has a register (*congestion_X* or *congestion_Y*) to store the incoming congestion information. The incoming information along with the local status is forwarded to the neighboring nodes in the next cycle via congestion propagation channels.

We use free VC count as the congestion metric. To cover the range of free VCs, the width of each congestion propagation channel along one direction is $\log(numVCs)$. However, a coarser approximation is sufficient. For neighboring routers, making a fine distinction has little impact. On the other hand, since the incoming congestion information is weighted according to the distance, it is also unnecessary to have accurate numbers for distant routers. As we show in Section 5.6.2, one wire per node is sufficient to achieve high performance; the router forwards congestion information in an on/off manner. The threshold for indicating congestion is four VCs (out of eight VCs); when five or more VCs are available, no congestion is signaled. Coarse-grained

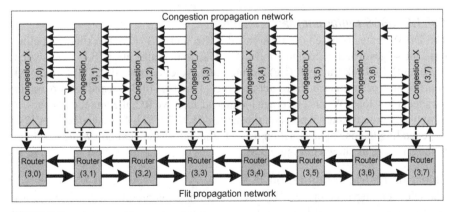

FIGURE 5.6

The congestion information propagation network along one dimension. Bold arrows represent channels of multiple bits, and thin arrows represent channels of 1 bit.

0	1	2	3	4	5	6	7	8
E(3,2)	E(3,1)	E(3,0)	W(3,2)	W(3,3)	W(3,4)	W(3,5)	W(3,6)	W(3,7)

FIGURE 5.7

The format of *congestion_X*. E, east; W, west.

congestion signals will toggle infrequently, resulting in a low activity factor and low power consumption.

With use of this coarse-grained monitoring, the *congestion_X* and *congestion_Y* registers are both $n + 1$ bits wide for an $n \times n$ network (e.g., 9 bits for an 8×8 network). Incoming congestion information from routers in the same dimension is stored in 7 bits and the other 2 bits store the conditions of two ports of the local router. The router weights the incoming congestion information on the basis of the distance from the current router; the weight is halved for each additional hop. This ratio is chosen on the basis of prior work [22] and implementation complexity. Adjacent bit positions of a register inherently maintain a step ratio of 0.5; we implement this weighting by putting the incoming information into the appropriate positions in the register.

Figure 5.7 shows the format of *congestion_X* in router (3,2). Bit 0 stores the east input port status of the current router. Bits 1 and 2 store the incoming congestion information from its nearest and one hop farther west neighbors: (3,1) and (3,0) respectively. These 3 bits are forwarded to the east neighbor: (3,3). Bit 3 stores the west input port status of the current router, and the following 5 bits sequentially store the west input port status of the remaining east side routers on the basis of the distance. These 6 bits are forwarded to the west neighbor: (3,1). The format of *congestion_Y* is similar.

5.4.1.2 DBSS router microarchitecture

The DBSS router is based on a canonical VC router [12, 17]. The router pipeline has four stages: routing computation (RC), VA, switch allocation (SA), and switch traversal (ST). Link traversal (LT) requires one cycle to forward the flit to the next hop. For high performance, the router uses speculative SA [48]; VA and SA proceed in parallel at low load. We also leverage lookahead adaptive RC; the router calculates at most two alternative output ports for the next hop [20, 22, 31]. Advanced bundles [24, 32] encoding the packet destination traverse the link to the next hop while the flit is in the ST stage as shown in Figure 5.8.

Selection metric computation

The selection metric computation (SMC) and dimension preselection (DP) modules are added as illustrated in Figure 5.9. SMC computes the dimension of the optimal output port for every possible destination using the congestion information stored in *congestion_X* and *congestion_Y*. An additional register, *out_dim*, stores the results of SMC. With minimal routing, there are at most two admissible ports (one per

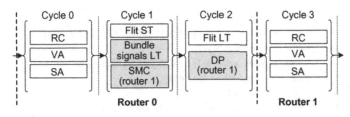

FIGURE 5.8

The DBSS router pipeline. DP, dimension preselection; SMC, selection metric computation.

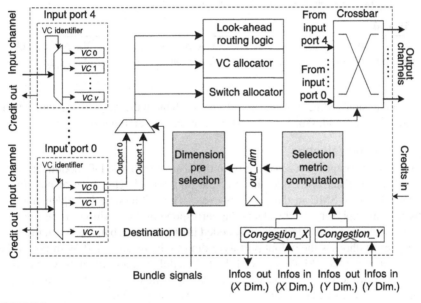

FIGURE 5.9

The DBSS router architecture.

dimension) for each destination. Because of this property, the *out_dim* register uses 1 bit to represent the optimal output port. If the value is "0," the optimal output port is along dimension X; otherwise, it is along dimension Y.

Figure 5.10 shows the pseudocode of SMC to compute the optimal output dimension for a packet whose destination is the *pos*th bit position of the *out_dim* register. Packets forwarded to the local node are excluded from this logic. Along each dimension, only those bit positions in *congestion_X* and *congestion_Y* storing congestion information for nodes inside the quadrant defined by the current and the *pos*th nodes are chosen. The values chosen are the congestion status metric for each dimension. According to the relative magnitude of the congestion status for the X

```
1:    if ( pos_x < cur_x )
2:        tmp_x[0:cur_x-pos_x-1] ← congestion_X[1:cur_x-pos_x];
3:    else if (pos_x > cur_x )
4:        tmp_x[0:pos_x-cur_x-1] ← congestion_X[cur_x+2:pos_x+1];
5:    else { out_dim[pos] ← 1;   return;}
6:    if ( pos_y < cur_y )
7:        tmp_y[0:cur_y-pos_y-1] ← congestion_Y[1:cur_y-pos_y];
8:    else if (pos_y > cur_y )
9:        tmp_y[0:pos_y-cur_y-1] ← congestion_Y[cur_y+2:pos_y+1];
10:   else { out_dim[pos] ← 0;   return;}
11:   if( tmp_x < tmp_y ) out_dim[pos] ←1;
12:   else if( tmp_x > tmp_y ) out_dim[pos] ← 0;
```

FIGURE 5.10

The pseudocode of SMC. Here, (*cur_y, cur_x*) and (*pos_y, pos_x*) are the positions of the current and the *pos*th router respectively. The initial values of *tmp_x* and *tmp_y* are 7-bit 0s.

and Y dimensions, SMC sets the value of the *pos*th bit in the *out_dim* register. If their magnitudes are equal, DBSS randomly chooses an output dimension.

Dimension preselection

To remove the output port selection from the critical path, the DP module accesses the *out_dim* register one cycle ahead of the flit's arrival. The value of *out_dim* is computed out by SMC in the previous cycle. The DP module is implemented as a 64-to-1 multiplexer, which selects the corresponding bit position of the *out_dim* register according to the destination encoded in an advanced bundle. When the head flit arrives, it chooses the output port according to the result of DP. With use of the logical effort model [48], the delay of the DP module is approximately 8.1 fan-outs of 4 (FO4). If the DP were added to the VA stage, the critical path would increase from 20 FO4 to 28.1 FO4. Advanced bundles serve to avoid this increase.

5.4.2 ROUTING FUNCTION DESIGN

The focus of this work is a novel selection strategy; however, the efficacy of the selection strategy is tightly coupled to other aspects of routing algorithm design, including adaptivity and VC reallocation. We compare several adaptive routing functions. Previous analysis [44] ignored the effect of VC reallocation. The majority of NoC packets are short [23, 39], making VC reallocation more important. Thus, it is necessary to re-evaluate routing functions by considering both the offered path diversity and VC reallocation.

5.4.2.1 Offered path diversity

Partially adaptive routing functions avoid deadlock by forbidding certain turns [6, 19, 21]. Owing to the prohibition of east→south and north→west turns, negative-first routing can utilize all minimal paths when the X and Y positions of the destination are both positive or negative with respect to the source. If only one is positive, it provides one minimal path [21]. Similarly, west-first and north-last routing utilize all minimal paths when the destination is east and south of the source respectively. Otherwise, only one path is allowed [21]. The odd-even routing applies different turn restrictions on odd and even columns for even adaptiveness.

In contrast to turn models, fully adaptive routing provides all minimal physical paths but places restrictions on VCs. Most fully adaptive routing functions are designed on the basis of Duato's theory [15], in which the VCs are classified into adaptive and escape VCs. There is no restriction on the routing of adaptive VCs; escape VCs can be utilized only if the output port adheres to a deadlock-free algorithm, usually DOR.

We compare the path diversity offered by a fully adaptive routing function (Duato) with several turn models in Figure 5.11. The path diversity is measured as the ratio of the number of times that the routing function provides two admissible ports versus one. We vary the VC count. For turn models, one VC is enough to avoid deadlock;

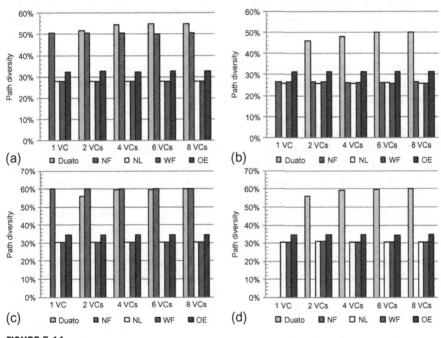

FIGURE 5.11

The path diversity offered by several routing functions. Duato, a fully adaptive function based on Duato's theory; NF, negative-first; NL, north-last; WF, west-first; OE, odd-even. Duato is not shown with one VC as it requires at least two VCs for deadlock freedom. NF shows zero path diversity for transpose-2. (a) Bit reverse, (b) bit complement, (c) transpose-1, and (d) transpose-2.

the Duato model needs at least two VCs. A local selection strategy is utilized. When each port is configured with four or fewer VCs, the selection is based on the buffer availability of neighbors. The VC availability is utilized if the VC count is greater than four. Adjusting the congestion metric according to the VC count can more stringently reflect the network status. The experiments are conducted on a 4×4 mesh; larger networks show similar trends.

As can be seen, at least approximately 25% of RCs produce two admissible ports; a selection strategy is needed to make a choice. For Duato and negative-first, more than 50% of RCs utilize the selection strategy under bit reverse and transpose-1 patterns. This ratio increases with larger network size. These results emphasize the importance of designing an effective selection strategy.

Duato generally shows the highest path diversity. The only exception is for transpose-1. Since all traffic is between the north-east and south-west quadrants, negative-first has the highest path diversity. The limitation on escape VCs yields slightly lower diversity for Duato. Once a packet enters an escape VC, it can only use escape VCs in subsequent hops; this packet loses path diversity. For the same reason, the diversity of Duato decreases with smaller VC counts. There is almost no difference for Duato with six and eight VCs; six VCs are enough to mitigate the escape VC limitation. The path diversity of partially adaptive routing is not sensitive to VC count. Negative-first offers no path diversity for transpose-2 since all traffic is between the north-west and south-east quadrants. Negative-first offers unstable path diversity for two symmetric transpose patterns. The four patterns evaluated are symmetric around the center of the mesh; west-first and north-last perform the same under such patterns.

5.4.2.2 VC reallocation scheme

The VC reallocation is an important yet often overlooked limitation for fully adaptive routing functions. Owing to cyclic channel dependencies, only empty VCs can be reallocated [15, 16]. If nonempty VCs are reallocated, a deadlock configuration is easily formed [15, 16].

Since partially adaptive functions prohibit cyclic channel dependencies, nonempty VCs can be reallocated [10]. They reallocate an output VC when the tail flit of the last packet goes through the ST stage of the *current* router, which is called aggressive VC reallocation. Instead, fully adaptive functions reallocate an output VC only after the tail flit of the last packet has gone through the ST stage of the *next-hop* router, which is called conservative VC reallocation [12, 19]. The difference between these schemes is shown in Figure 5.12.

In cycle 0, packet P_0 resides in VC_0. P_0 waits for VC_2 and VC_3, which are occupied by P_1 and P_2 respectively. Both VC_2 and VC_3 have some free slots. The aggressive VC reallocation scheme forwards the header flit of P_0 to VC_2 in cycle 1, as shown in Figure 5.12b. However, for fully adaptive routing, P_0 must wait for $3 + i$ cycles to be forwarded (Figure 5.12c), where i denotes the delay due to contention. Assuming round-robin arbitration [12] for both VA and SA and no contention from other input ports, it takes six cycles for VC_2 to become empty.

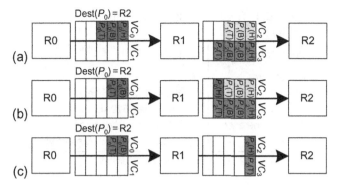

FIGURE 5.12

The difference between two VC reallocation schemes. $P_i(H)$, $P_i(B)$, and $P_i(T)$ are the header, body, and tail flit of P_i respectively. (a) Original state of cycle 0, (b) cycle 1 for aggressive VC reallocation, and (c) cycle $3 + i$ for conservative VC reallocation.

Then, P_0 can be forwarded. More cycles are needed with contention from other input ports. The aggressive VC reallocation has better VC utilization, leading to improved performance. Allowing multiple packets to reside in one VC may result in head-of-line (HoL) blocking [29]. However, as we will show in Section 5.5.1, with limited VCs, making efficient use of VCs strongly outweighs the negative effect of HoL blocking.

5.5 EVALUATION

We modify BookSim [28] to model the microarchitecture discussed in Section 5.4.1.2. The evaluation consists of three parts: we first consider the routing functions. Negative-first, north-last, west-first, and odd-even models and a fully adaptive routing function based on Duato's theory are evaluated with a local selection strategy. Then, the fully adaptive function is chosen for the selection strategy evaluation. A local strategy (LOCAL), NoP, RCA-1D,[3] and DBSS are implemented for meshes of different size. Finally, we extend the evaluation to CMeshes [2, 33].

Several synthetic traffic patterns [6, 12] are used. Each VC is configured with five flit buffers, and the packet length is uniformly distributed between one and six flits. Workload consolidation scenarios are evaluated with multiple regions configured inside a network. The injection procedure for each region is independently controlled as if a region is a whole network. The destinations of all traffic generated from one region stay within the same region. The latency and throughput are measured for each region. The routing algorithm is configured at the network level.

---

[3]RCA-1D is referred to as RCA throughout the evaluation.

Table 5.1 The Full System Simulation Configuration

Parameter	Value
No. of cores	16
L1 cache (D and I)	Private, 4-way, 32 kB each
L2 cache	Private, 8-way, 512 kB each
Cache coherence	MOESI distributed directory
Topology	4 × 4 mesh, 4 VNs, 8 VCs per VN

L1, first level; L2, second level; VN, virtual network.

To measure full-system performance, we leverage FeS2 [47] for x86 simulation and BookSim for NoC simulation. FeS2 is implemented as a module for Virtutech Simics [41]. We run PARSEC benchmarks [4] with 16 threads on a 16-core chip multiprocessor, organized in a 4 × 4 mesh. Workload consolidation scenarios are also evaluated. An 8 × 8 mesh with four 4 × 4 regions is configured in BookSim. Region R0 delivers the traffic generated by FeS2, while the remaining regions run uniform random patterns. A mix of real applications and synthetic traffic was used owing to scalability problems with the simulator and operating system [5]. Prior research showed the frequency of simple cores can be optimized to 5-10 GHz, while the frequency of NoC routers is limited by the allocator speed [14]. We assume cores are clocked four times faster than the network. Cache lines are 64 bytes, and the network flit width is 16 bytes. All benchmarks use the *simsmall* input sets. The total run time is used as the performance metric. Table 5.1 presents the system configuration.

5.5.1 EVALUATION OF ROUTING FUNCTIONS

Figure 5.13 illustrates the results of routing function evaluation; the saturation throughput[4] achieved is the performance metric. From the comparison of partially adaptive functions in Figures 5.11 and 5.13, generally higher path diversity leads to higher saturation throughput. However, owing to the conservative VC reallocation, Duato's performance with two VCs is lower than that of some partially adaptive functions even though it offers higher path diversity. For example, with bit reverse traffic, negative-first and odd-even show 47.1% and 30.6% higher performance than Duato respectively. With two VCs, improving the VC utilization greatly outweighs the negative effect of HoL blocking induced by aggressive VC reallocation.

Duato's performance increases significantly with four VCs. More VCs improve the possibility of finding empty VCs; the difference between aggressive and conservative VC reallocation schemes declines. For example, Duato achieves the highest performance for bit reverse traffic. Most partially adaptive functions have significant improvement in performance when the VC count increases from one to two, but the

---

[4]The saturation point is when the average latency is three times the zero-load latency.

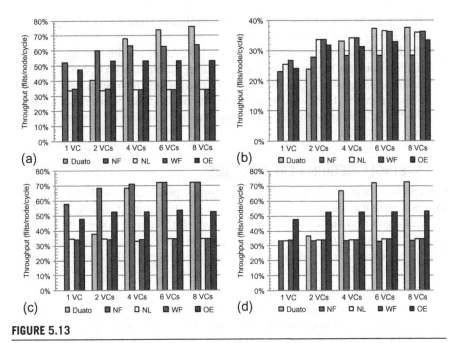

FIGURE 5.13

The saturation throughput of several routing functions in a 4 × 4 mesh. (a) Bit reverse, (b) bit complement, (c) transpose-1, and (d) transpose-2.

performance gain increases only slightly when the VC count increases to four. More VCs mitigate the negative effect of HoL blocking. Two VCs are enough to reduce the HoL effect, thus improving the importance of path diversity. Moreover, with aggressive VC reallocation, configuring two VCs fully utilizes the physical channel. These two factors make path diversity the performance bottleneck with four VCs.

Duato's performance steadily increases with six VCs. Six VCs help to stress the physical channel with conservative VC reallocation. Also, Duato provides enough path diversity, preventing it from rapidly becoming the bottleneck. The performance gain when the VC count increases from six to eight is small; with eight VCs, the physical path congestion becomes the limiting factor. A noteworthy phenomenon is observed for bit complement traffic with eight VCs: although Duato has twice the path diversity of the three turn models, its performance is only slightly better. Most bit-complement traffic traverses the center area of the mesh, and Duato's high path diversity can only slightly mitigate this region's congestion problem.

In summary, our evaluation of routing functions gives the following insights:

(1) With limited VCs, providing efficient VC utilization greatly outweighs the negative effect of HoL blocking. The efficient VC utilization provided by aggressive VC reallocation compensates for the limited path diversity of partially adaptive functions, resulting in higher performance than for fully adaptive ones.

(2) With more VCs, the offered path diversity of routing functions becomes the dominating factor for performance.

(3) Configuring the appropriate VC count for NoCs must consider the applied routing function. For partially adaptive routing functions with aggressive VC reallocation, generally two VCs are enough to provide high performance. More VCs not only add hardware overhead, but do not improve performance much. For the fully adaptive routing functions, at most eight VCs are enough for performance gains.

5.5.2 SINGLE-REGION PERFORMANCE

From previous analysis, Duato achieves the highest performance with more than six VCs; we chose it as the routing function for the selection strategy analysis. Synthetic traffic evaluation is conducted with eight VCs, which are enough to stress the physical channel. We first evaluate the performance of two single-region configurations: 4×4 and 8×8 meshes. There is only one traffic pattern throughout the whole network.

5.5.2.1 Synthetic traffic results

Figures 5.14 and 5.15 give the latency results for transpose-1, bit reverse, shuffle, and bit complement traffic patterns in 4×4 and 8×8 meshes respectively. In the 4×4 mesh, DBSS has the best performance for these four traffic patterns as RCA suffers from intraregion interference. There is one exception: for bit complement traffic, RCA's saturation point is 2.1% higher. Bit complement traffic has the largest AHP, with four hops in a 4×4 network (Figure 5.3); this AHP mitigates the intraregion interference. LOCAL and NoP perform the worst for bit complement traffic owing to their limited knowledge. The small AHP (2.5 hops) of transpose-1 traffic leads to RCA performing the worst among all four adaptive algorithms. DBSS, LOCAL, and NoP offer similar performance for transpose-1 traffic, with approximately 13% improvement in saturation throughput versus RCA. DBSS has a significant improvement of 21.9% relative to RCA for bit reverse traffic.

DBSS improves saturation throughput by 10.2% and 8.5% over LOCAL for shuffle and bit complement traffic. These patterns cause global congestion, and the shortsightedness of the locally adaptive strategy makes it unable to avoid congested areas. The saturation throughput improvements of DBSS over NoP are 17.7% and 11.1% for bit reverse and bit complement traffic respectively. NoP overlooks the status of neighbors. Comparing the performance of LOCAL against NoP further illuminates this limitation. This phenomenon validates our weighting mechanism placing more emphasis on closer nodes.

LOCAL outperforms RCA on a 4×4 mesh; intraregion interference leads RCA to make inferior decisions. However, in the 8×8 mesh, DBSS and RCA offer the best performance, while LOCAL has inferior performance. RCA's improvement comes from the weighting mechanism in the congestion propagation network. The weight of the congestion information halves for each hop; the effect of intraregion interference from distant nodes diminishes. This interference reduction is a result of the high

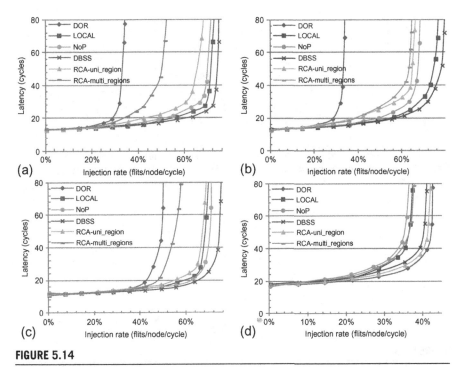

FIGURE 5.14

The routing algorithm performance for a 4 × 4 mesh network (region). *RCA-uni_region* and *RCA-multi_region* give the performance of RCA for a single region and multiple regions respectively. (a) Transpose-1, (b) bit reverse, (c) shuffle, and (d) bit complement.

AHP of 5.58 for these patterns. However, the AHP on the 4 × 4 network is 2.63, which is not large enough to hide the negative effect of interference. Although the weighted aggregation mechanism mitigates some interference in the 8 × 8 mesh, DBSS still outperforms RCA by 11.1% for bit reverse traffic. Compared with the 4 × 4 network, DBSS further improves the saturation throughput for shuffle and bit complement traffic versus LOCAL by 12.4% and 16.5%. The shortsightedness of LOCAL has a stronger impact in a larger network. Similar trends are seen for NoP. For most traffic, DOR's rigidity prevents it from avoiding congestion, resulting in the poorest performance.

5.5.2.2 Application results

Figure 5.16 shows the speedups relative to DOR for eight PARSEC workloads. The workloads can be classified into two groups: network-insensitive applications and network-sensitive applications. Sophisticated routing algorithms can improve the network saturation throughput, but the full-system performance improvements depend on the load and traffic pattern created by each application [51]. Applications

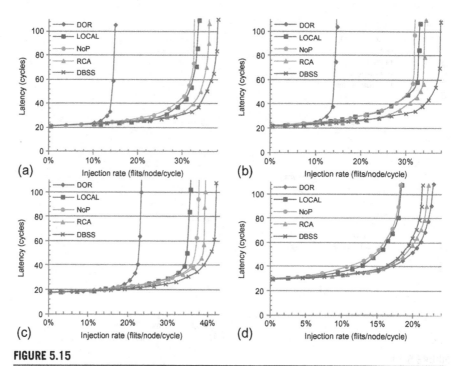

FIGURE 5.15

The routing algorithm performance for an 8 × 8 mesh with a single region. (a) Transpose-1, (b) bit reverse, (c) shuffle, and (d) bit complement.

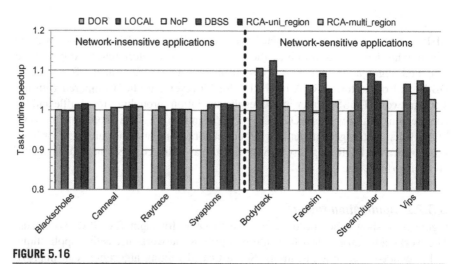

FIGURE 5.16

System speedups for PARSEC workloads.

with a high network load and significant bursty traffic receive benefit from advanced routing algorithms. For `blackscholes`, `canneal`, `raytrace`, and `swaption`, their working sets fit into the private second-level caches, resulting in low network load. Different routing algorithms have similar performance for this group. The other four applications have lots of bursty traffic, emphasizing network-level optimization techniques; their performance benefits from routing algorithms supporting higher saturation throughput. For these four applications, DBSS and LOCAL have the best performance owing to the small network size. DBSS achieves an average speedup of 9.6% and a maximum speedup of 12.5% over DOR. RCA performs better than NoP. NoP is worse than DOR for `facesim`. Its ignorance of neighboring nodes results in suboptimal selections.

5.5.3 MULTIPLE-REGION PERFORMANCE

We evaluate two multiple-region configurations: one regular-region configuration (Figure 5.4) and one irregular-region configuration (Figure 5.17a). In both configurations, we focus on the performance of region R0.

5.5.3.1 Results for a small regular region

In this configuration (Figure 5.4), regions R1, R2, and R3 run uniform random traffic with injection rates of 4%, while we vary the pattern in region R0. LOCAL, NoP, and DBSS do not have inter-region interference, since they consider only the congestion status of nodes belonging to the same region when making selections. Thus, in this configuration, all algorithms except RCA have the same performance as in Figure 5.14. RCA's performance suffers from inter-region interference. The *RCA-multi_regions* curves in Figure 5.14 show RCA's performance for the multiple-region configuration.

Compared with a single region, RCA's performance declines; RCA suffers not only from intraregion interference, but also from inter-region interference. Transpose-1 and shuffle traffic see 22.7% and 16.9% drops in saturation throughput. For bit reverse traffic, the performance degradation is minor; the intraregion interference has already significantly degraded RCA's performance and hides the effect of inter-region interference. DBSS maintains its performance for this configuration, thus revealing a clear advantage. The average saturation throughput improvement is 25.2%, with a maximum improvement of 46.1% for transpose-1 traffic. Figure 5.16 (*RCA-multi_regions*) shows that RCA's performance decreases for these four network-sensitive applications compared with the single-region configuration (R1, R2, and R3 run uniform random traffic with injection rates of 4%). The maximum performance decrease (7.3%) occurs for `bodytrack`.

Routers at the region boundaries strongly affect R0's performance, since some of their input ports are never used. For example, the west input VCs of router (0,4) are always available since no packets arrive from the west. The interference from internal nodes of R1 and R2 is partially masked by RCA's weighting mechanism at these boundary nodes with eight free VCs. This explains why R0's saturation point decreases from 50% to only 47% when the injection rate of R1 increases from 4% to 64% in Figure 5.5.

5.5.3.2 Irregular-region results

Figure 5.17a shows nonrectangular regions. The isolation boundaries of R0 and R1 are the minimal rectangles surrounding them; traffic from both regions shares some network links, so the traffic from R1 affects routing in R0 in some cases. Figure 5.17b shows the performance of R0 while the injection rate of R1 is varied from low load (4%) to high load (55%); the injection rates of R2 and R3 are fixed at 4%. All regions run uniform random traffic.

For both high and low loads in R1, DBSS has the best performance. As the load in R1 increases, the performance of all algorithms declines. For low load in R1, RCA has the second-highest saturation throughput. Two rows of R0 have five routers; LOCAL and NoP are not sufficient to avoid congestion. When R1 has a high injection

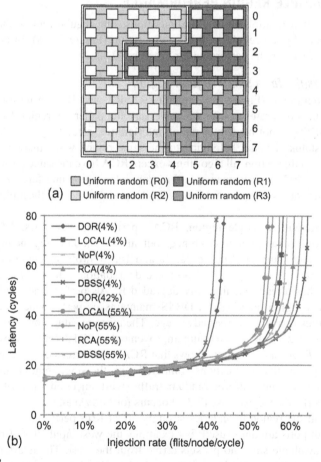

FIGURE 5.17

The irregular-region configuration and its performance. (a) Configuration and (b) performance.

rate, the saturation points decline for DOR, LOCAL, NoP, RCA, and DBSS by 7%, 7%, 6.8%, 6.7%, and 4% respectively; DBSS has the least performance degradation since it offers the best isolation between R0 and R1. As these two regions are not completely isolated, some traffic for R0 and R1 shares common network links; in this case, traffic from R1 should affect routing in R0. Only traffic between the routers in the first two columns and routers in the last two columns is completely isolated; DBSS correctly accounts for this dependency across the regions, which makes it a flexible and powerful technique.

5.5.3.3 Summary
In workload consolidation scenarios, different applications will be mapped to different region sizes according to their intrinsic parallelism. However, with small regions, RCA suffers from interference, while LOCAL and NoP are limited by shortsightedness for large regions. Table 5.2 lists the average saturation throughput improvement of DBSS against other strategies. DBSS provides the best performance for all configurations evaluated and it shows the smallest performance degradation in multiple irregular regions. DBSS can provide more predictable performance when running multiple applications. DBSS is well suited to workload consolidation.

5.5.4 CMesh EVALUATION
5.5.4.1 Configuration
Here, we extend the analysis to the CMesh topology [2, 33]. As a case study, we use radix-4 CMeshes [2, 33]; four cores are concentrated around one router, with two cores in each dimension, as shown in Figure 5.18a. Here, Core0, Core1, Core4, and Core5 are concentrated on router (0,0). Each core has its own injection/ejection channels to the router. Based on a CMesh latency model, the network channel has a two-cycle delay, while the injection/ejection channel has a one-cycle delay [33]. The router pipeline is the same as discussed in Section 5.4.1.2. As shown in Figure 5.18, we evaluate 16-core and 64-core platforms. For the 64-core platform, both single-region and multiple-region experiments are conducted. For the multiple-region configuration (Figure 5.18b), regions R1, R2, and R3 run uniform random traffic with an injection rate of 4% and we vary the pattern in region R0.

5.5.4.2 Performance
Figure 5.19 shows the results. In the 2×2 CMesh, LOCAL, DBSS, and RCA have the same performance as there are only two routers along each dimension;

Table 5.2 The Average Throughput Improvement of DBSS

Network	LOCAL	NoP	RCA	RCA_multi
4×4	7.2%	8.8%	10.4%	25.2%
8×8	12.6%	14.9%	4.7%	–
Irregular	16.5%	14.3%	–	6.8%

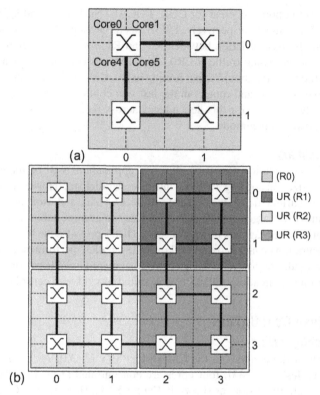

FIGURE 5.18

The CMesh configurations. (a) 16-Core platform and (b) 64-core platform.

these selection strategies utilize the same congestion information. For example, in Figure 5.18a, when a packet is routed from router (0,1) to (1,0), LOCAL, DBSS, and RCA all make selections by comparing the congestion status of the east input port of (0,0) and the north input port of (1,1). NoP performs differently; it compares the status of the north and south input ports of (1,0). This difference results in poorer performance for NoP (Figure 5.19a and b), since it ignores the status of the nearest nodes. DOR has the lowest performance for transpose-1 traffic (Figure 5.19a). With bit reverse traffic in a 4 × 4 mesh, DOR's performance is about one third of the performance of the best adaptive algorithm (Figure 5.14a). But in a 2 × 2 CMesh with bit reverse traffic, the performance difference between DOR and adaptive routing declines (Figure 5.19b). Concentration reduces the AHP; adaptive routing has less opportunity to balance network load.

In the multiple-region scenario of a 4 × 4 CMesh, RCA's performance drops owing to the inter-region interference. There is a 26.2% performance decline with transpose-1 traffic (*RCA-uni_region* vs. *RCA-multi_region* in Figure 5.19a). In a 4 ×

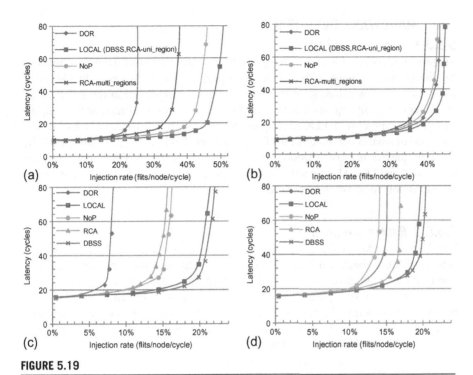

FIGURE 5.19

The routing algorithm performance for CMeshes. *RCA-uni_region* and *RCA-multi_region* give the performance of RCA for a single region and multiple regions respectively. (a) Transpose-1 (2 × 2 CMesh), (b) bit reverse (2 × 2 CMesh), (c) transpose-1 (4 × 4 CMesh), and (d) bit reverse (4 × 4 CMesh).

4 mesh region with bit reverse traffic, the inter-region interference only slightly affects RCA since the intraregion interference already strongly deteriorates its performance (Figure 5.14b). However, here with bit reverse traffic, the inter-region interference brings a 9.8% performance drop as there is no intraregion interference in a 2 × 2 CMesh (Figure 5.19b).

With one region configured in a 4 × 4 CMesh, the performance trend (Figure 5.19c and d) is generally similar to that of a 4 × 4 mesh (Figure 5.14a and b); DBSS and LOCAL show the highest performance, while RCA suffers from intraregion interference. RCA's performance is 29.4% lower than that of DBSS in a 4 × 4 CMesh (Figure 5.19c), while the performance gap is only 7.4% in a 4 × 4 mesh (Figure 5.14a). The effect of intraregion interference increases owing to the change of the latency ratio between the network and injection/ejection channel, resulting in judicious selection becoming more important in a 4 × 4 CMesh. These results indicate that considering the appropriate congestion information is also important in CMeshes; DBSS still offers high performance.

5.5.5 HARDWARE OVERHEAD

5.5.5.1 Wiring overhead

Adaptive routers require some wiring overhead to transmit congestion information. Assuming an 8×8 mesh with eight VCs per port, DBSS introduces eight additional wires for each dimension. RCA uses eight wires in each direction for a total of 16 per dimension. NoP requires $4 \times \log(numVCs) = 12$ wires per direction; there are 24 wires for one dimension. LOCAL requires $\log(numVCs) = 3$ wires in each direction for a total of 6 wires per dimension. Given a state-of-the-art NoC [23] with 128-bit flit channels, the overhead of DBSS is just 3.125% versus 6.25%, 9.375%, and 2.34% for RCA, NoP, and LOCAL respectively. DBSS has a modest overhead; abundant wiring on chip is able to accommodate these wires.

5.5.5.2 Router overhead

DBSS adds the SMC, DP, and three registers to the canonical router. In an 8×8 mesh network, SMC utilizes two 7-bit shifters to select and align the congestion information of the two dimensions. The 64-to-1 multiplexer of DP can be implemented in tree format. *Congestion_X* and *congestion_Y* are 9 bits wide, while *out_dim* is 64 bits wide. Considering the wide (128-bit) datapath of our mesh network (five ports, eight VCs per ports, and five slots per VC), the overhead of these registers is only 0.3% compared with the existing buffers.

5.5.5.3 Power consumption

We leverage an existing NoC power model [43], which divides the total power consumption into three main components: channels, input buffers, and router control logics. We also model the power consumption of the congestion propagation network. The activity of these components is obtained from BookSim. We use a 32 nm technology process with a clock frequency of 1 GHz.

Figure 5.20a shows the average power for transpose-1 traffic in an 8×8 mesh. Since DOR cannot support injection rates higher than 20%, there are no results for injection rates of 30% and 35%. The increased hardware complexity, especially the congestion propagation network of adaptive routing, results in a higher average power than the simple DOR. Among these adaptive routers, LOCAL and DBSS have the lowest power consumption owing to their wiring overhead being the lowest. NoP has the highest power consumption. LOCAL needs fewer additional wires than DBSS, but these wires have a higher activity factor. For an injection rate of 20%, the activity of DBSS's congestion propagation network is 15.8% versus 17.5% for LOCAL. This smaller activity factor mitigates the increased power consumption of DBSS's congestion propagation network. For an injection rate of 35%, LOCAL consumes more power than DBSS. The adaptive routing accelerates packet transmission, showing a significant energy-delay product (EDP) advantage. As shown in Figure 5.20b, DBSS provides the smallest EDP for medium (20%) and high (30% and 35%) injection rates.

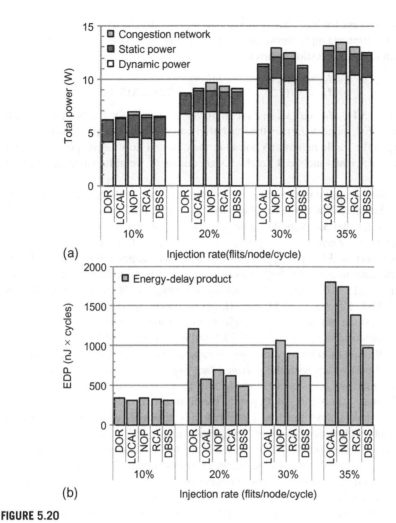

FIGURE 5.20

Power consumption results for transpose-1 traffic. (a) Power and (b) energy-delay product.

5.6 ANALYSIS AND DISCUSSION

5.6.1 IN-DEPTH ANALYSIS OF INTERFERENCE

Here, we delve into the cause of interference. We find that it is strongly related to the detailed implementation of the congestion metric. There are many metrics, such as free VCs, free buffers, and crossbar demands [22]. We use free buffers as an example.

There are two implementation choices. One is to choose the output port with the *maximal free buffers*, and the other is to use *minimal occupied buffers*. These two choices appear equivalent at first glance. Indeed, for LOCAL, they are the same.

This is not the case for RCA owing to the interference. To clarify this point, consider the packet routing example shown in Figure 5.2c. The following equation shows the calculation of the congestion metrics:

$$\begin{cases} M_{\text{w}} & = E(0,1) + 0.5 \times E(0,0), \\ M_{\text{s}} & = N(1,2) + 0.5 \times N(2,2) + 0.25 \times N(3,2). \end{cases} \tag{5.1}$$

In Equation (5.1), M_{w} and M_{s} are the congestion metrics for the west and south output ports. $E(i,j)$ and $N(i,j)$ are the free buffer counts (or the occupied buffer counts according to the implementation choice) in the east and north input ports of router (i,j). The coefficients, such as 0.5 and 0.25, are due to the weighting mechanism used by RCA.

Since fully adaptive routing uses the conservative VC reallocation scheme, and the packet length in NoCs is generally short, most of the buffer slots are unoccupied. Let us consider that an example range of occupied buffers $occupied_{range}$ for each input port is $0 \leqslant occupied_{range} \leqslant 6$. Then the range of free buffers $free_{range}$ for each input port is $34 \leqslant free_{range} \leqslant 40$ in our experimental configuration. If we implement *minimal occupied buffers* as the congestion metric, then the range of M_{w} in Equation (5.1) is $0 \leqslant M_{\text{w}} \leqslant 9$. Similarly, the range of M_{s} in Equation (5.1) is $0 \leqslant M_{\text{s}} \leqslant 10.5$. The ranges of M_{w} and M_{s} are nearly the same.

However, if we use *maximal free buffers* as the congestion metric, then the range of M_{w} in Equation (5.1) is $51 \leqslant M_{\text{w}} \leqslant 60$. Similarly, the range of M_{s} is $59.5 \leqslant M_{\text{s}} \leqslant 70$. Unlike the situation with the *minimal occupied buffers*, the ranges of M_{w} and M_{s} have almost no overlap with the *maximal free buffers* implementation choice. For almost all situations, M_{s} is larger than M_{w}; this packet loses significant adaptivity as RCA will always choose the south port as the optimal one.

This example shows that the interference is strongly related to the detailed implementation choices of the same congestion metric. We find that if the output port with the *maximal free buffers* is chosen as the optimal one in RCA, its saturation throughput is about half that of LOCAL for most synthetic traffic patterns. But if the output port with the *minimal occupied buffers* is chosen as the optimal one, RCA's performance is almost the same as LOCAL's performance.

Even with aggressive VC reallocation in partially adaptive routing, *maximal free buffers* and *minimal occupied buffers* show similar properties as well. The reason is that the network saturates when some resources are saturated [12]. In particular, these central buffers, VCs, and links may easily be saturated, while other resources may have low utilization [45]. Kim *et al.* [30] observed that even with one VC per physical channel, the average buffer utilization is below 40% at saturation; more resources are in a free status than in an occupied status. For this reason, other metrics, such as "free VCs," have similar properties; the "maximum free VCs" implementation used in previous evaluations has inferior performance compared with the "minimal occupied VCs" metric. Although RCA achieves good performance by carefully implementing the congestion metric, its instability with different implementation choices easily leads to an inferior design. However, DBSS shows stable performance with either *maximal free buffers* (VCs) or *minimal occupied buffers* (VCs) since it eliminates the interference.

5.6.2 DESIGN SPACE EXPLORATION

5.6.2.1 Number of propagation wires

We evaluate the saturation throughput of DBSS with 1-, 2-, and 3-bit-wide propagation networks. The detailed results are given in Ma *et al.* [37]. The increase in wiring yields only minor performance improvements, and these performance gains decrease as the network scales. Making a fine distinction about the available VCs has little practical impact as the selection strategy is interested only in the relative difference between two routing candidates. To demonstrate this point, we simultaneously keep the 1-, 2-, and 3-bit status information and record the fraction of times the selection strategy using 1 bit or 2 bits makes a different selection than the 3-bit status in a 4×4 mesh. At very low network load (2% injection rate) with uniform random traffic, the 1-bit status and the 2-bit status make a different selection 8.8% and 7.6% of the time. This difference mainly comes from the randomization when facing two equal output port statuses. These fractions decrease with higher network loads. The fractions are 1.8% and 1.1% respectively at saturation. This minor difference verifies that making a fine distinction is not necessary.

5.6.2.2 DBSS scalability

The cost of scaling DBSS to a larger network increases linearly as N 1-bit congestion propagation wires are needed for each row (column) in an $N \times N$ network. For a 16×16 mesh, this represents a 6.25% overhead with 128-bit flit channels. The size of the added registers in Figure 8.10 increases linearly. The latency of the DP module increases logarithmically with network radix; however, this delay is not on the critical path so it will not increase the cycle time. DBSS is a cost-effective solution for many-core networks.

5.6.2.3 Congestion propagation delay

In addition to eliminating interference, our novel congestion network operates with only a one cycle per hop delay compared with two cycles per hop in RCA. To isolate this effect from interference effects, we compare DBSS with a one cycle per hop and a two cycles per hop congestion propagation network. The timeliness of the one cycle per hop network improves saturation throughput by up to 5% over the two-cycle design (for shuffle traffic).

5.7 CHAPTER SUMMARY

The shortsightedness of locally adaptive routing limits the performance for large networks, while globally adaptive routing suffers from interference for multiple regions. Interference (or false dependency) comes from utilizing network status across region boundaries. By leveraging a novel congestion propagation network, DBSS provides both high adaptivity for congestion avoidance and dynamic isolation to eliminate interference. Although DBSS does not provide strict isolation, it is a powerful technique that can dynamically adapt to changing region configurations as threads are migrated or rescheduled. Experimental results show that DBSS offers high performance in all network configurations evaluated. The wiring overhead of DBSS is small. We have provided additional insights into the design of routing functions.

REFERENCES

[1] G. Ascia, V. Catania, M. Palesi, D. Patti, Implementation and analysis of a new selection strategy for adaptive routing in networks-on-chip, IEEE Trans. Comput. 57(6) (2008) 809–820. 144, 145, 148, 149

[2] J. Balfour, W. Dally, Design tradeoffs for tiled CMP on-chip networks, in: Proc. of the International Conference on Supercomputing (ICS), 2006, pp. 187–198. 155, 163

[3] S. Bell, B. Edwards, J. Amann, R. Conlin, K. Joyce, V. Leung, J. MacKay, M. Reif, L. Bao, J. Brown, et al., TILE64 processor: a 64-core SoC with mesh interconnect, in: Proc. of the International Solid-State Circuits Conference Digest of Technical Papers (ISSCC), 2008, pp. 88–598. 142

[4] C. Bienia, S. Kumar, J.P. Singh, K. Li, The PARSEC benchmark suite: characterization and architectural implications, in: Proc. of the International Conference on Parallel Architectures and Compilation Techniques (PACT), 2008, pp. 72–81. 156

[5] S. Boyd-Wickizer, A.T. Clements, Y. Mao, A. Pesterev, M.F. Kaashoek, R. Morris, N. Zeldovich, An analysis of Linux scalability to many cores, in: Proc. of the USENIX Conference on Operating Systems Design and Implementation (OSDI), 2010, pp. 1–8. 156

[6] G.-M. Chiu, The odd-even turn model for adaptive routing, IEEE Trans. Parallel Distrib. Syst. 11(7) (2000) 729–738. 144, 147, 152, 155

[7] C.-L. Chou, R. Marculescu, Run-time task allocation considering user behavior in embedded multiprocessor networks-on-chip, IEEE Trans. Comput. Aided Des. Integr. Circuits Syst. 29(1) (2010) 78–91. 143, 148

[8] C.-L. Chou, U. Ogras, R. Marculescu, Energy- and performance-aware incremental mapping for networks on chip with multiple voltage levels, IEEE Trans. Comput. Aided Des. Integr. Circuits Syst. 27(10) (2008) 1866–1879. 143, 148

[9] W. Dally, Virtual-channel flow control, IEEE Trans. Parallel Distrib. Syst. 3(2) (1992) 194–205. 142

[10] W. Dally, C. Seitz, Deadlock-free message routing in multiprocessor interconnection networks, IEEE Trans. Comput. C-36(5) (1987) 547–553. 144, 154

[11] W. Dally, B. Towles, Route packets, not wires: on-chip interconnection networks, in: Proc. of the Design Automation Conference (DAC), 2001, pp. 684–689. 142

[12] W. Dally, B. Towles, Principles and Practices of Interconnection Networks, first ed., Morgan Kaufmann Publishers Inc., San Francisco, CA, 2003. 143, 145, 147, 150, 154, 155, 168

[13] W.J. Dally, H. Aoki, Deadlock-free adaptive routing in multicomputer networks using virtual channels, IEEE Trans. Parallel Distrib. Syst. 4 (1993) 466–475. 144, 145

[14] R. Das, A.K. Mishra, C. Nicopoulos, D. Park, V. Narayanan, R. Iyer, M.S. Yousif, C.R. Das, Performance and power optimization through data compression in network-on-chip architectures, in: Proc. of the International Symposium on High-Performance Computer Architecture (HPCA), 2008, pp. 215–225. 156

[15] J. Duato, A new theory of deadlock-free adaptive routing in wormhole networks, IEEE Trans. Parallel Distrib. Syst. 4(12) (1993) 1320–1331. 144, 153, 154

[16] J. Duato, A necessary and sufficient condition for deadlock-free adaptive routing in wormhole networks, IEEE Trans. Parallel Distrib. Syst. 6(10) (1995) 1055–1067. 144, 154

[17] N. Enright Jerger, L. Peh, On-Chip Networks, first ed., Morgan & Claypool, San Rafael, CA, 2009. 150

[18] W.-C. Feng, K.G. Shin, Impact of selection functions on routing algorithm performance in multicomputer networks, in: Proc. of the International Conference on Supercomputing (ICS), 1997, pp. 132–139. 144, 148

[19] B. Fu, Y. Han, J. Ma, H. Li, X. Li, An abacus turn model for time/space-efficient reconfigurable routing, in: Proc. of the International Symposium on Computer Architecture (ISCA), 2011, pp. 259–270. 144, 152, 154

[20] M. Galles, Spider: a high-speed network interconnect, IEEE Micro 17(1) (1997) 34–39. 150

[21] C. Glass, L. Ni, The turn model for adaptive routing, in: Proc. of the International Symposium on Computer Architecture (ISCA), 1992, pp. 278–287. 144, 152, 153

[22] P. Gratz, B. Grot, S. Keckler, Regional congestion awareness for load balance in networks-on-chip, in: Proc. of the International Symposium on High-Performance Computer Architecture (HPCA), 2008, pp. 203–214. 142, 144, 145, 149, 150, 167

[23] P. Gratz, C. Kim, K. Sankaralingam, H. Hanson, P. Shivakumar, S.W. Keckler, D. Burger, On-chip interconnection networks of the TRIPS chip, IEEE Micro 27(5) (2007) 41–50. 152, 166

[24] P. Gratz, K. Sankaralingam, H. Hanson, P. Shivakumar, R. McDonald, S. Keckler, D. Burger, Implementation and evaluation of a dynamically routed processor operand network, in: Proc. of the International Symposium on Networks-on-Chip (NOCS), 2007, pp. 7–17. 150

[25] Y. Hoskote, S. Vangal, A. Singh, N. Borkar, S. Borkar, A 5-GHz mesh interconnect for a teraflops processor, IEEE Micro 27(5) (2007) 51–61. 142

[26] J. Hu, R. Marculescu, DyAD—smart routing for networks-on-chip, in: Proc. of the Design Automation Conference (DAC), 2004, pp. 260–263. 144, 145

[27] J. Hu, R. Marculescu, Energy- and performance-aware mapping for regular NoC architectures, IEEE Trans. Comput. Aided Des. Integr. Circuits Syst. 24(4) (2005) 551–562. 143

[28] N. Jiang, D.U. Becker, G. Michelogiannakis, J. Balfour, B. Towles, D. Shaw, J. Kim, W. Dally, A detailed and flexible cycle-accurate network-on-chip simulator, in: Proc. of the International Symposium on Performance Analysis of Systems and Software (ISPASS), 2013, pp. 86–96. 155

[29] M. Karol, M. Hluchyj, S. Morgan, Input versus output queueing on a space-division packet switch, IEEE Trans. Commun. 35(12) (1987) 1347–1356. 155

[30] G. Kim, J. Kim, S. Yoo, FlexiBuffer: reducing leakage power in on-chip network routers, in: Proc. of the Design Automation Conference (DAC), 2011, pp. 936–941. 168

[31] J. Kim, D. Park, T. Theocharides, N. Vijaykrishnan, C.R. Das, A low latency router supporting adaptivity for on-chip interconnects, in: Proc. of the Design Automation Conference (DAC), 2005, pp. 559–564. 144, 145, 150

[32] A. Kumar, P. Kundu, A. Singh, L.-S. Peh, N. Jha, A 4.6 Tbits/s 3.6 GHz single-cycle NoC router with a novel switch allocator in 65 nm CMOS, in: Proc. of the International Conference on Computer Design (ICCD), 2007, pp. 63–70. 150

[33] P. Kumar, Y. Pan, J. Kim, G. Memik, A. Choudhary, Exploring concentration and channel slicing in on-chip network router, in: Proc. of the International Symposium on Networks-on-Chip (NOCS), 2009, pp. 276–285. 155, 163

[34] T. Lei, S. Kumar, A two-step genetic algorithm for mapping task graphs to a network on chip architecture, in: Proc. of the Euromicro Symposium on Digital System Design (DSD), 2003, pp. 180–187. 143

[35] M. Li, Q.-A. Zeng, W.-B. Jone, DyXY—a proximity congestion-aware deadlock-free dynamic routing method for network on chip, in: Proc. of the Design Automation Conference (DAC), 2006, pp. 849–852. 144, 145

[36] D. llitzky, J. Hoffman, A. Chun, B. Esparza, Architecture of the scalable communications core's network on chip, IEEE Micro 27(5) (2007) 62–74. 142

[37] S. Ma, N. Enright Jerger, Z. Wang, DBAR: an efficient routing algorithm to support multiple concurrent applications in networks-on-chip, in: Proc. of the International Symposium on Computer Architecture (ISCA), 2011, pp. 413–424. 141, 169

[38] S. Ma, N. Enright Jerger, Z. Wang, DBAR: an efficient routing algorithm to support multiple concurrent applications in networks-on-chip, ACM SIGARCH Comput. Archit. News 39(3) (2011) 413–424. 141

[39] S. Ma, N. Enright Jerger, Z. Wang, Whole packet forwarding: efficient design of fully adaptive routing algorithms for networks-on-chip, in: Proc. of the International Symposium on High-Performance Computer Architecture (HPCA), 2012, pp. 467–478. 144, 152

[40] S. Ma, N. Enright Jerger, Z. Wang, L. Huang, M. Lai, Holistic routing algorithm design to support workload consolidation in NoCs, IEEE Trans. Comput. 63 (3) (2012) 529–542. 141

[41] P.S. Magnusson, M. Christensson, J. Eskilson, D. Forsgren, G. Hallberg, J. Hogberg, F. Larsson, A. Moestedt, B. Werner, Simics: a full system simulation platform, Computer 35 (2002) 50–58. 156

[42] J.C. Martínez et al., On the influence of the selection function on the performance of networks of workstations, in: Proc. of the International Symposium High Performance Computing (ISHPC), 2000, pp. 292–299. 144, 148

[43] G. Michelogiannakis, D. Sanchez, W.J. Dally, C. Kozyrakis, Evaluating bufferless flow control for on-chip networks, in: Proc. of the International Symposium on Networks-on-Chip (NOCS), 2010, pp. 9–16. 166

[44] M. Mirza-Aghatabar, S. Koohi, S. Hessabi, M. Pedram, An empirical investigation of mesh and torus NoC topologies under different routing algorithms and traffic models, in: Proc. of the Euromicro Symposium on Digital System Design (DSD), 2007, pp. 19–26. 152

[45] A.K. Mishra, N. Vijaykrishnan, C.R. Das, A case for heterogeneous on-chip interconnects for CMPs, in: Proc. of the International Symposium on Computer Architecture (ISCA), 2011, pp. 389–400. 168

[46] O. Mutlu, T. Moscibroda, Parallelism-aware batch scheduling: enhancing both performance and fairness of shared DRAM systems, in: Proc. of the International Symposium on Computer Architecture (ISCA), 2008, pp. 63–74. 142

[47] N. Neelakantam, C. Blundell, J. Devietti, M.M. Martin, C. Zilles, FeS2: a full-system execution-driven simulator for x86, in: Poster Presented at ASPLOS 2008, 2008. 156

[48] L.-S. Peh, W. Dally, A delay model and speculative architecture for pipelined routers, in: Proc. of the International Symposium on High-Performance Computer Architecture (HPCA), 2001, pp. 255–266. 150, 152

[49] R.S. Ramanujam, B. Lin, Destination-based adaptive routing on 2D mesh networks, in: Proc of the ACM/IEEE Symposium on Architectures for Networking and Communications Systems (ANCS), 2010, pp. 1–12. 144

[50] S. Rodrigo, J. Flich, J. Duato, M. Hummel, Efficient unicast and multicast support for CMPs, in: Proc. of the International Symposium on Microarchitecture (MICRO), 2008, pp. 364–375. 144

[51] D. Sanchez, G. Michelogiannakis, C. Kozyrakis, An analysis of on-chip interconnection networks for large-scale chip multiprocessors, ACM Trans. Archit. Code Optim. 7(1) (2010) 4:1–4:28. 159

[52] L. Schwiebert, R. Bell, Performance tuning of adaptive wormhole routing through selection function choice, J. Parallel Distrib. Comput. 62 (2002) 1121–1141. 144, 148

[53] S. Zhuravlev, S. Blagodurov, A. Fedorova, Addressing shared resource contention in multicore processors via scheduling, in: Proc. of the International Conference on Architectural Support for Programming Languages and Operating Systems (ASPLOS), 2010, pp. 129–142. 142

Flow control for fully adaptive routing[†]

[†]Part of this research was first presented at the 18th IEEE International Symposium on High Performance Computer Architecture (HPCA-2012) [32]. The journal extension edition of the HPCA-2012 paper was published in *IEEE Transactions on Parallel and Distributed Systems* [33].

Networks-on-Chip. http://dx.doi.org/10.1016/B978-0-12-800979-6.00006-8

6.1 INTRODUCTION

The performance of networks-on-chip (NoCs) is sensitive to the routing algorithm, as it defines the network latency and saturation throughput [7, 35]. Many novel NoC routing algorithms have been proposed to deliver high performance [18, 20, 23, 28, 31, 46, 55]. In addition to performance considerations, the routing algorithm has correctness implications; any routing algorithm must be deadlock-free at both the network level and the protocol level. The guarantee of network-level deadlock freedom is generally based on deadlock avoidance theories. There are many theories for fully adaptive routing design [12, 13, 17, 29, 44, 48, 52, 53] and partially adaptive routing design [4, 6, 18, 19]. Although most theories were originally proposed for off-chip networks, they are widely used in today's NoCs [18, 20, 23, 28, 31, 46, 55].

However, NoC packets are quite different from off-chip network packets. Abundant wires lead to wider flits, which decreases the flit count per packet; short packets dominate the NoC traffic. In contrast, the wires in off-chip networks are limited by pin counts; the flit width of a typical off-chip router is of the order of 32 bits (e.g., Alpha 21364 router [37]), while the typical NoC flit width is between 128 bits [21] and 256 bits [10]. With such wide flits, coherence messages with an address and control information but no data are encoded as single-flit packets (SFPs). Figure 6.1 shows that, on average, 78.7% of packets are single-flit control ones for PARSEC workloads [3]; other data packets are five flits long with a 64-bit full cache line.

Another noteworthy difference is that the NoC buffers are more expensive than the off-chip network ones owing to the tight area and power budgets [16, 22]; thus, NoCs generally use flit-based wormhole flow control [9]. Although buffers are limited, several physical networks or virtual networks (VNs) are needed for delivering different messages to avoid protocol-level deadlock. Table 6.1 lists the number of physical networks and VNs in some off-chip and on-chip networks. We also show the VN counts for some coherence protocols in the GEMS simulator [36]. Typically, four or five VNs are needed to avoid protocol-level deadlock. Considering the expense of NoC buffers and large allocators, each VN will have only a small number of virtual channels (VCs) [5]. For example, the TILE64 [54] and TRIPS [21] processors have one VC per VN. Thus, an NoC routing algorithm generally runs with limited VCs.

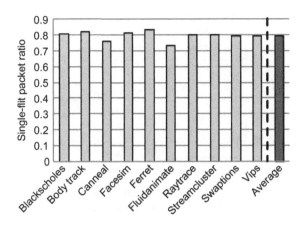

FIGURE 6.1

The SFP ratio for PARSEC benchmarks. The flit width is 16 bytes.

Table 6.1 Number of Physical Networks (PNs)/Virtual Networks (VNs)

Industrial products	
Alpha 21364 [37]	1 PN (7 VNs)
TILE64 [54]	5 PNs (1 VN/PN)
TRIPS [21]	2 PNs (4 VNs for on-chip network, 1 VN for operand network)
Cache coherence protocols in the GEMS simulator [36]	
MESI directory	5 VNs
MOESI directory	4 VNs
MOESI token	4 VNs

In VC-limited NoCs with short packets dominating traffic, the design of fully adaptive routing faces new challenges. In a wormhole network, fully adaptive routing algorithms based on existing theories require conservative VC reallocation: only empty VCs can be reallocated [12, 13, 17, 29, 44, 48, 52, 53]. This scheme prevents network-level deadlock, but it is very restrictive with many short packets [18]. Figure 6.2 shows the performance of three algorithms.[1] Despite its flexibility and load-balancing capability, the fully adaptive algorithm is inferior to the deterministic and partially adaptive ones, since the latter algorithms apply aggressive VC reallocation. It is imperative to enhance the design of fully adaptive routing.

This chapter focuses on improving fully adaptive routing algorithms by designing novel flow control. We propose two mechanisms. First, whole packet forwarding

---

[1] See Section 6.5 for the experimental configuration and description.

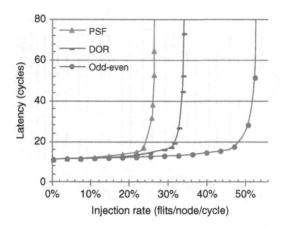

FIGURE 6.2

The routing algorithm performance with bit reverse traffic. PSF, fully adaptive routing; DOR, deterministic routing; odd-even, partially adaptive routing.

(WPF) allows a nonempty VC which has enough buffers for an entire packet to be reallocated. WPF can be viewed as applying packet-based flow control in a wormhole network. We prove that a fully adaptive routing algorithm using WPF is deadlock-free if this algorithm is deadlock-free with conservative VC reallocation. The proposed WPF is an important extension to previous deadlock avoidance theories. Second, we extend Duato's theory [12] to apply aggressive VC reallocation on escape VCs (EVCs) without deadlock. This extension further improves the VC utilization and the network performance.

The novel flow control enables the design of a fully adaptive routing algorithm with high VC utilization and maximal routing flexibility. Compared with existing fully adaptive routing algorithms, our design provides an average gain of 88.9% for synthetic traffic, and an average speedup of 21.3% and a maximal speedup of 37.8% for PARSEC workloads that heavily load the network. It is also superior to partially adaptive and deterministic algorithms.

In this chapter we make the following main contributions:

- We propose WPF to improve the performance of fully adaptive routing algorithms.
- We demonstrate WPF is an important extension to existing deadlock avoidance theories.
- We extend Duato's theory to apply aggressive VC reallocation on EVCs without deadlock.
- We propose an efficient design which maintains high routing flexibility with low overhead.

6.2 BACKGROUND

Here, we discuss related work in deadlock avoidance theories and designs for fully adaptive routing.

6.2.1 DEADLOCK AVOIDANCE THEORIES

Since NoCs typically use wormhole flow control [7, 9, 35, 41], we focus on theories for wormhole networks. Dally and Seitz [6] proposed a seminal deadlock avoidance theory to design deterministic and partially adaptive routing algorithms. Duato [12, 13] introduced the routing subfunction, and gave an efficient design method. The message flow model [29] and channel waiting graph [44] were used to analyze deadlock. Taktak *et al.* [48] and Verbeek and Schmaltz [52] leveraged the decision procedure to establish deadlock freedom. This method can also be applied to our designs. Verbeek and Schmaltz [51, 53] gave the first static necessary and sufficient condition for deadlock-free routing. These theories [12, 13, 17, 29, 44, 48, 52, 53] can be used to design fully adaptive routing algorithms.

The theories of fully adaptive routing algorithms require only empty VCs to be reallocated. This requirement guarantees that all blocked packets can reach VC heads to access "deadlock-free" paths. Yet, it limits performance with many short packets. Some deadlock-recovery designs [1] or theories [14] remove this limitation. They allow the formation of deadlock, and then invoke some recovery mechanisms. In contrast, we extend deadlock avoidance theories to prohibit the formation of deadlock.

There are several partially adaptive routing algorithms based on turn models [4, 18, 19]. They allow aggressive VC reallocation: a VC can be reallocated once the tail flit of the last packet has arrived [6, 8]. This property can be deduced from Dally and Seitz's theory [6] since the channel dependency graphs of these algorithms are acyclic. However, partially adaptive routing algorithms suffer from limited adaptivity: packets cannot use all minimal paths between the source and the destination.

6.2.2 FULLY ADAPTIVE ROUTING ALGORITHMS

Duato's theory [12] is widely used to design fully adaptive routing algorithms. This theory classifies the network VCs into EVCs and adaptive VCs (AVCs). EVCs apply more restrictive algorithms, typically dimensional order routing (DOR), to form deadlock-free paths. An EVC can only be used when the port adheres to DOR.

Many fully adaptive routing algorithms based on Duato's theory [20, 31, 37, 55] select the physical port first. Once a port has been selected, packets can request only VCs of this chosen port. This requirement imposes a limitation: if a packet enters an EVC, it can only use EVCs until it is delivered. Otherwise, EVCs may be involved in deadlock. In VC-limited NoCs, this limitation easily induces adaptivity loss. However, Duato's theory supports the design of algorithms which can use AVCs after using an EVC if packets can always request EVCs [12]. On the basis of these observations, we propose a design which maintains high routing flexibility with low hardware cost.

6.3 MOTIVATION

In this section, we analyze the requirements and limitations of fully adaptive routing algorithms. We also illustrate how these requirements negatively affect performance.

6.3.1 VC REALLOCATION

Fully adaptive routing has the limitation that only empty VCs can be reallocated. This requirement is reasonable since VCs may form cycles in fully adaptive routing. For example, Duato's theory arbitrarily uses AVCs [12]; if multiple packets reside in one VC, a deadlock configuration appears, as shown in Figure 6.3. Here each VN has two VCs: an AVC and an EVC. Configuring more VCs cannot eliminate this deadlock scenario since cycles exist among AVCs.

Figure 6.3 involves eight packets: P_0-P_7. P_0's head flit is behind P_1's tail flit in AVC_1. The same is true for P_1, P_2, P_4, P_5, and P_6. Although the head flits of P_3 and P_7 are at VC heads, they cannot move as both AVC_0 and EVC_0 are occupied. This deadlock scenario occurs because some head flits are not at VC heads, resulting in some packets being unable to access the "deadlock-free" path. Also, the tail flits of these packets reside in other VCs, prohibiting other packets from reaching VC heads. These following packets then cyclically block the aforementioned packets. For example, P_0's tail flit resides in AVC_0, blocking P_3 from using this VC, which cyclically blocks P_0. If the packet length is greater than the VC depth, putting multiple packets in one VC may induce deadlock. Yet, we notice that moving entire short packets into nonempty VCs will not prevent following packets from reaching VC heads. This is an opportunity for optimization, especially with many short NoC packets.

6.3.2 ROUTING FLEXIBILITY

This section discusses routing flexibility in VC-limited NoCs. Many fully adaptive algorithms based on Duato's theory consist of the routing function and selection strategy [20, 31, 37, 55]; they are called *port-selection-first* (PSF) algorithms since once the selection strategy has chosen a port, the packet requests VCs of only this particular port. If we have a separable VC allocator with two stages of arbiters [2, 38, 42], a PSF algorithm only needs V:1 arbiters in the first stage as shown in Figure 6.4a.

A limitation of these algorithms is that once a packet enters an EVC, it must continue to use EVCs; the packet loses adaptivity in subsequent hops. Violating this limitation induces deadlock as shown in Figure 6.5. If P_1 and P_2 select the south port, they can request only AVC_2 since EVC_2 can be requested only when the port adheres to DOR. Similarly, P_4 and P_5 select the north port; they request only AVC_0. No packet can move. Thus, requiring that if a packet enters an EVC it can use only EVCs

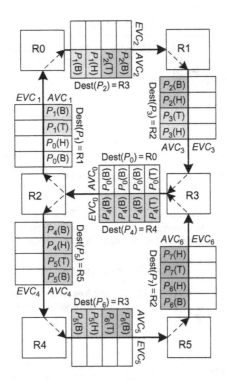

FIGURE 6.3

Deadlock in a fully adaptive algorithm violating conservative VC reallocation. $P_i(H)$, $P_i(B)$, and $P_i(T)$ are the head, body, and tail flits of packet P_i, and Dest(P_i) is the destination of packet P_i.

subsequently if necessary for PSF algorithms. However, this requirement results in significant adaptivity loss with limited VCs.

Duato's theory supports algorithms using AVCs after using EVCs, if they guarantee packets are always able to request EVCs. To achieve this target, a packet could request all permissible output VCs, since at least one port must adhere to DOR and the packet can use the EVC of this port [12]. However, this naive design has additional overhead. As shown in Figure 6.4b, the VC allocator uses $2V{:}1$ arbiters to cover the at most two permissible ports for minimal routing algorithms in the first stage. We propose a low-cost design to maintain high routing flexibility.

6.4 FLOW CONTROL AND ROUTING DESIGNS

First, we present WPF and prove that it is deadlock-free. Then, we extend Duato's deadlock avoidance theory to apply aggressive VC reallocation on EVCs. Next, we provide a low-cost design to maintain routing flexibility. Finally, we describe the implementation and overhead.

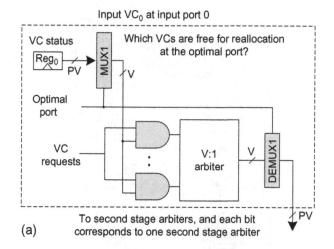

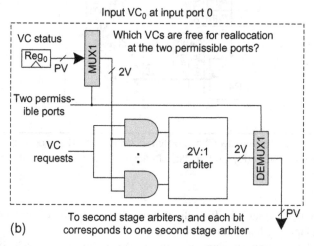

FIGURE 6.4

The first stage arbiter. Each VN has V VCs and P ports. (a) PSF design and (b) naive design.

6.4.1 WHOLE PACKET FORWARDING

We propose WPF. Suppose a packet P_k with $length(P_k)$ resides in VC_i, and VC_j is downstream of VC_i. Assume the routing algorithm allows P_k to use VC_j. With conservative VC reallocation, VC_j can be reallocated to P_k only if the tail flit of its last allocated packet is sent out, i.e., it is empty [8]. For WPF, VC_j can be reallocated if it has received the tail flit of the last allocated packet, and its free buffer count is no less than $length(P_k)$.

Figure 6.6 shows a WPF example. The algorithm allows P_1 to use VC_2. VC_2 received P_2's tail flit and has two free slots; this space is enough for P_1's two flits.

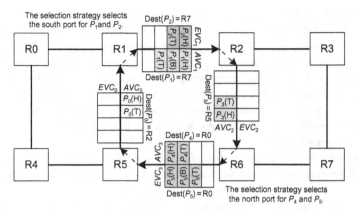

FIGURE 6.5

A deadlock configuration if packets in EVCs can request AVCs in PSF algorithms.

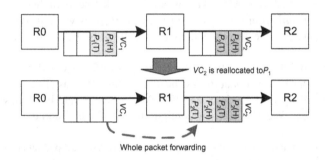

FIGURE 6.6

An example of WPF. P_i(H) and P_i(T) are the head and tail flits of packet P_i.

Thus, WPF reallocates VC_2 to P_1. WPF works similarly to packet-based flow control such as store-and-forward flow control [15] and virtual cut-through (VCT) flow control [26]. VCT enables the design of high-frequency routers [43]. Yet, our design uses wormhole flow control with empty VCs, which does not require the empty VC to be large enough for entire packets; this reduces the buffering requirement compared with VCT flow control, and provides more flexibility.

Our contention is that if the routing algorithm with conservative VC reallocation is deadlock-free, then applying WPF will not induce deadlock. Intuitively, this contention is true, since short packets will be either at VC heads or behind some packets whose head flits are at VC heads. Thus, some packets can access the "deadlock-free" path. Yet, the "deadlock-free" path is coupled to special theories. We provide a general proof. We label the algorithm with conservative VC reallocation as Alg; $Alg + WPF$ is Alg with WPF to forward entire packets into nonempty VCs. The proof needs a simple assumption about the router.

Assumption 6.1. *Assume a packet can use multiple VCs. If any permissible VC is eventually available, the packet must have some possibility to request this VC.*

This assumption provides *weak fairness* among permissible VCs. It can be fulfilled in the following manner. The router continually monitors the status of all downstream VCs. If one permissible VC is available, the packet requests it immediately. If the algorithm first selects a port and limits the packet to request VCs of the selected port, the selection strategy should guarantee that any permissible port has some possibility to be selected. As shown in Section 6.4.4, common NoC routers adhere to Assumption 6.1.

Theorem 6.1. *If Alg is deadlock-free, then Alg + WPF is also deadlock-free.*

Informal description: We prove this theorem by contradiction. We prove that if *Alg + WPF* has a deadlock configuration, then *Alg* has a deadlock configuration as well. Using *Config₀* in Figure 6.7 as an example, we remove packets whose head flits are not at VC heads, and get a new configuration *Config₁*. We prove that *Alg* can achieve *Config₁*, and *Config₁* is a deadlock configuration. However, *Alg* is deadlock-free; thus, there is no such configuration.

Proof: By contradiction. If $Alg + WPF$ is not deadlock-free, then there is a deadlock configuration ($Config_0$) in which a set of packets, P_{set_0}, is waiting on VCs held by other packets in P_{set_0}. We prove that there is a deadlock configuration for Alg in three steps.

Step 1: We build a new configuration based on $Config_0$. Consider each packet P_i in P_{set_0}. If P_i's head flit is not at the VC head, then this VC was allocated to P_i using WPF; all flits of P_i must reside in one VC. We remove P_i and label these removed packets as P_{subset_0}. We label the new configuration as $Config_1$, and the set of packets in $Config_1$ as P_{subset_1}.

Step 2: We prove that when the network is routed by Alg, all packets in P_{subset_1} can move into their current VCs in $Config_1$. For each packet P_j in P_{subset_1}, we consider each hop hop_k of P_j when it is routed by $Alg + WPF$. We assume that P_j's head flit moves into VC_k during hop_k. There are two situations:

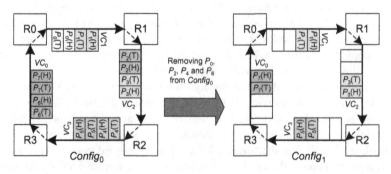

FIGURE 6.7

The construction of a new configuration.

(1) VC_k is empty when P_j reaches it; VC_k is allocated to P_j with conservative VC reallocation. Thus, when it is routed by Alg, P_j can use VC_k.

(2) VC_k is not empty when P_j reaches it; VC_k is allocated to P_j with WPF. The algorithm allows P_j to use VC_k. Yet, when it is routed by Alg, P_j cannot move into VC_k until it is empty. Since Alg is deadlock-free, the packet currently in VC_k must move out in a limited time. Then VC_k is available for conservative VC reallocation. On the basis of Assumption 6.1, P_j has some possibility to request VC_k. Then, P_j can move into VC_k. Thus, when it is routed by Alg, P_j can use VC_k.

If we consider situations 1 and 2 together, and take into account that the head flits of all packets in $Config_1$ are at VC heads, if a VC is used by P_j during any hop when it is routed by $Alg + WPF$, this VC can also be used by P_j when it is routed by Alg. Thus, P_j can be routed to its current VC(s) in $Config_1$ by Alg.

Step 3: We prove that $Config_1$ is a deadlock configuration for Alg. For each P_i in the removed set P_{subset_0}, all flits of P_i reside in one VC but P_i's head is not at the VC head. Thus, removing P_i does not create empty VCs. Alg uses conservative VC reallocation, which reallocates only empty VCs; all packets in P_{subset_1} still wait for VCs held by other packets in P_{subset_1}. $Config_1$ is a deadlock configuration for Alg. But Alg is deadlock-free, so there is no deadlock configuration. Thus, $Alg + WPF$ is deadlock-free as well. ∎

This proof needs only a simple assumption. WPF can be used with many fully adaptive routing algorithms if they are deadlock-free with conservative VC reallocation. It is an important extension to existing theories [12, 13, 17, 29, 44, 48, 52, 53]. In the Appendix, we further prove that if $Alg + WPF$ is deadlock-free, Alg is also deadlock-free.

6.4.2 AGGRESSIVE VC REALLOCATION FOR EVCs

We apply WPF to fully adaptive algorithms designed on the basis of Duato's theory [12]. Yet, using WPF directly may bring about fairness issues for long packets, as shown in Figure 6.8. Both three-flit packet P_3 and SFP P_0 are waiting to move forward. The free buffers in EVC_2 and AVC_2 allow P_0 to move. P_3 must wait for EVC_2 or AVC_2 to have three free slots, or for them to be empty. If P_4 moves forward in the next cycle, a similar situation happens again: the free buffers in EVC_2 or AVC_2 allow only P_1 to move. Under the extreme case, P_3 will be permanently waiting if SFPs are continuously injected into AVC_1.[2]

A design to address this fairness issue is to deploy a time counter and a priority allocator. Once the counter crosses the threshold value for the blocked packet P_3, a high priority is assigned to P_3's VC allocation so as to prevent SFPs from occupying the buffers in EVC_2 and AVC_2. This design involves additional costs, including those for the time counter and the priority allocator. Also, an appropriate threshold value needs to be configured to provide the desired fairness.

---

[2]This extreme case never appears in our simulation as short and long packets are randomly injected.

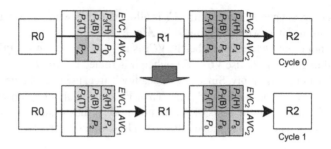

FIGURE 6.8

A fairness issue example.

Instead of using this complex design, we leverage a more elegant observation. We notice the EVCs in Duato's theory can apply aggressive VC reallocation without deadlock. With aggressive reallocation, P_3 and P_0 can equally use EVC_2; P_3 cannot be blocked forever. Under the worst case with a stream of continuous short packets using AVC_2, P_3 cannot acquire AVC_2; this may cause long packets to have less adaptivity than short ones. Yet, two factors mitigate the possible negative effect. First, network streams generally consist of both short and long packets. Second, our routing design in Section 6.4.3 can request AVCs after using EVCs; thus, this limitation causes only minor adaptivity loss. Also, our design has less overhead than the complex design, and aggressive VC reallocation further improves the VC utilization and performance. This design is coupled to Duato's theory; we reiterate some definitions from Duato's paper [12, 13] to make this section self-contained.

Definition 6.1 (network path). *A network path consists of a set of VCs, VC_0, VC_1, ..., VC_i. Along this path, packets can be sent from VC_0 to VC_i.*

Definition 6.2 (connected routing function). *A routing function is connected if it can always establish a path for every packet from its current VC to its destination.*

Definition 6.3 (routing subfunction). *A routing subfunction is based on a routing function. The input VCs of the routing subfunction are the same as those of the routing function. The output VCs supplied by the routing subfunction are a subset of those of the routing function.*

Definition 6.4 (direct dependency). *If a packet can use VC_j immediately after using VC_i, there is a direct dependency from VC_i to VC_j.*

Definition 6.5 (indirect dependency). *There is an indirect dependency between two EVCs, EVC_i and EVC_j, if there is a network path consisting of EVC_i, AVC_0, ..., $AVC_{k(k \geqslant 0)}$, EVC_j. In other words, EVC_i and EVC_j are the first and last VCs in this path, and all intermediate VCs are AVCs.*

Definition 6.6 (extended VC dependency graph). *The extended VC dependency graph D_E is defined for EVCs. The vertices of D_E are EVCs. The arcs of D_E are the pairs of EVCs (EVC_i, EVC_j) such that there is either a direct dependency or an indirect dependency from EVC_i to EVC_j.*

Duato's necessary and sufficient theory [13] defines the direct cross and indirect cross dependency. Our research is mainly based on Duato's necessary theory [12]; thus, we omit these two types of dependency.

Theorem 6.2 (Duato's necessary theory). *An adaptive routing function is deadlock-free if there exists a routing subfunction that is connected and has no cycles in its extended VC dependency graph [12].*

The routing subfunction is defined on EVCs. As the extended VC dependency graph is acyclic, we can assign an order among EVCs so that if there is a direct or an indirect dependency from EVC_i to EVC_j, then $EVC_i > EVC_j$ [12]. Moreover, since the routing subfunction is connected, any packet can move to its destination by using EVCs [12].

Theorem 6.1 applies WPF to both AVCs and EVCs. In this section we further apply aggressive VC reallocation to EVCs. We label the *deadlock-free* algorithm based on Duato's theory with conservative VC reallocation as Alg; $Alg + WPF + Agg$ is Alg with WPF for AVCs and aggressive VC reallocation for EVCs. We prove Theorem 6.3.

Theorem 6.3. *If Alg is a deadlock-free routing algorithm designed on the basis of Duato's theory [12], then Alg + WPF + Agg is also deadlock-free.*

Informal description: The structure of our proof is similar to the proof of Theorem 2 in Duato's paper [12]. The difference is that we prove that even with WPF on AVCs and aggressive VC reallocation on EVCs, there is still a movable packet in a hypothetical deadlock configuration. Two key points are the routing subfunction on EVCs is connected, and there is an order among EVCs.

Proof: Suppose $Alg + WPF + Agg$ has a deadlock configuration ($Config_0$), in which no packet head has already reached the destination. There are two cases:

(1) All EVCs are empty. $Config_0$ consists only of AVCs. Let AVC_i be an AVC in $Config_0$, and P_i be the packet whose flit is at AVC_i's head. There are two situations:

    (a) The flit at AVC_i's head is P_i's head flit. Since the routing subfunction on EVCs is connected, there is an EVC (EVC_i) that P_i can use. As all EVCs are empty, P_i can move into EVC_i.

    (b) The flit at AVC_i's head is not P_i's head flit; P_i spans multiple VCs. Since WPF moves entire packets, P_i is not forwarded by WPF. P_i's head flit must reside at another AVC's head. Similarly to the first situation, P_i can move into an empty EVC as well.

(2) $Config_0$ involves EVCs. There is an order among EVCs; let EVC_i be the nonempty EVC in $Config_0$ such that all EVCs with an order less than that of EVC_i are empty. Let P_i be the packet whose flit is at EVC_i's head. There are two situations:

    (a) The flit at EVC_i's head is P_i's head flit. Since the routing subfunction is connected, there is an EVC EVC_j that P_i can use. This implies there is a direct dependency from EVC_i to EVC_j; thus, the order of EVC_j is less than that of EVC_i. EVC_j is empty; P_i can move into EVC_j.

(b) The flit at EVC_i's head is not P_i's head flit. Aggressive VC reallocation may make P_i's head flit reside in another EVC (EVC_j). In such a case, there is a direct or an indirect dependency from EVC_i to EVC_j. The order of EVC_j is less than that of EVC_i. EVC_j should be empty; P_i's head flit cannot reside in EVC_j.

As P_i spans multiple VCs, it was not forwarded by WPF. P_i's head flit must be at the head of an AVC. Since the routing subfunction is connected, there is an EVC (EVC_k) that P_i can use, which implies there is an indirect dependency from EVC_i to EVC_k. Thus, EVC_k is empty. P_i can move into EVC_k.

In all cases, a packet can move. There is no deadlock configuration for $Alg + WPF + Agg$. ∎

Theorem 6.3 extends Dauto's theory to apply aggressive VC reallocation on EVCs. It not only addresses the fairness issue of WPF on long packets, but also yields additional VC utilization and performance gains.

6.4.3 MAINTAIN ROUTING FLEXIBILITY

In this section we propose a low-cost design to maintain high routing flexibility. This design is based on Duato's theory [12]. The design should allow the use of AVCs after the use of EVCs. Otherwise, once a packet enters EVCs, it will lose adaptivity in subsequent routing. The design must guarantee that a packet can request an EVC at any time [12]. Once this condition is satisfied, packets can always find a path which is not involved in cyclic dependencies, since the extended dependency graph of EVCs is acyclic [12].

In PSF algorithms, the only time a packet cannot use EVCs is when the selected port violates DOR. We make a simple modification by violating the selection in this case; the packet requests the EVC of the nonselected permissible port in addition to AVCs of the selected one. We use P_1 in Figure 6.5 as an example. If the south port is selected, our design allows P_1 to request the EVC of the east port. If there is only one permissible port, this port must adhere to DOR, and the packet can request its EVC. This design guarantees that a packet can always request an EVC. It allows a packet to move back into AVCs after using an EVC. Also, it needs only $V{:}1$ arbiters in the first stage of the VC allocator. Large arbiters result in more hardware overhead and introduce additional delay in the critical path.

6.4.4 ROUTER MICROARCHITECTURE

The pipeline of a canonical NoC router consists of routing computation, VC allocation, switch allocation, and switch traversal [8, 15, 38, 42]. We apply several optimizations. The speculative allocation parallelizes switch allocation with VC allocation at low loads [42]. Lookahead routing calculates at most two permissible ports one hop ahead and the selection strategy chooses the optimal one [20, 27, 31]. The router delay is two cycles plus one cycle for link traversal.

This baseline router adheres to Assumption 6.1 in Section 6.4.1. First, a common selection strategy guarantees that any permissible port can be selected. For example, in Section 6.5 we use a strategy which prefers the port with more free buffers. This strategy will not prohibit any permissible port from being selected since once the permissible ports have equal free buffers, it randomly selects one from them. Second, the router continually monitors the status of all downstream VCs using credits [8]. If one permissible VC of the selected port is eventually available, the packet can find out this situation, and requests this VC immediately. Therefore, an NoC router generally provides *weak fairness* among permissible VCs, even when selecting the port first.

Our design requires only simple modifications to the VC allocator. Any type of allocator can be used; we assume a low-cost separable allocator which consists of two stages of arbiters [2, 38, 42]. We modify the first-stage arbiters to apply WPF on AVCs and aggressive VC reallocation on EVCs. WPF requires calculating whether the free buffers are enough for an entire packet, which has some overhead. Yet, considering cache coherence packets exhibit a bimodal distribution and the majority are SFPs, we apply WPF only on SFPs.

Figure 6.9 shows the proposed VC allocator. Reg_0 and Reg_1 record whether a downstream VC can be reallocated. If a downstream VC is an AVC, the corresponding bit in Reg_0 records whether it can be reallocated with conservative reallocation. If a downstream VC is an EVC, the corresponding bit in Reg_0 records whether it can be reallocated with aggressive reallocation. The criterion is that the VC holds the tail flit of its last allocated packet and still has free buffers. Reg_1 records whether a VC can be reallocated with WPF for SFPs. Its criterion is the same as applying aggressive VC reallocation on SFPs. Thus, we use Reg_0 or Reg_1 on the basis of the incoming packet type. If it is an SFP, we apply WPF, sending Reg_1 to MUX1. Otherwise, Reg_0 is sent to MUX1. Updates to Reg_0 and Reg_1 are off the critical path as the router monitors the status of downstream VCs. The increased delay is an additional two-input multiplexer: MUX0.

To maintain routing flexibility, we modify MUX1 and DEMUX1, as shown in Figure 6.9. MUX1 needs two additional signals: *DOR* and *the other output port*. The *DOR* signal indicates if the selected port obeys DOR. *The other output port* signal records the nonselected permissible port. The routing logic produces these signals. If *DOR* is "0," the selected port violates DOR. Then, the status of the EVC for *the other output port* rather than the selected one is sent to the $V{:}1$ arbiter. This is accomplished with a two-input multiplexer whose control signal is *DOR*. DEMUX1 also needs these signals. If *DOR* is "0," the result from the $V{:}1$ arbiter is demultiplexed to the second-stage arbiter for the EVC of *the other output port*. This is accomplished with a two-input demultiplexer. The increased delay is an additional two-input multiplexer and demultiplexer.

We implement the three VC allocators (Figures 6.4 and 6.9) in RTL Verilog for an open-source NoC router [2] and synthesize in Design Compiler with a Taiwan Semiconductor Manufacturing Company 65 nm standard cell library. The designs operate at 500 MHz under normal conditions (1.2 V, 25 °C). We use round-robin

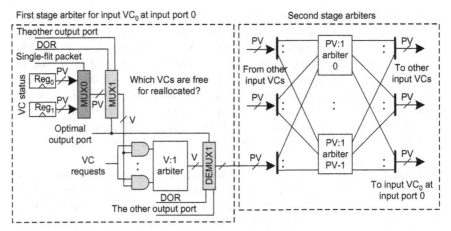

FIGURE 6.9

The proposed VC allocator for one VN.

Table 6.2 The Critical Path Delay and Area Results

Design	Delay (ns)	Area (μm^2)
PSF design (Figure 6.4a)	1.78	49,437.4
Naive design (Figure 6.4b)	1.92	56,045.4
Proposed design (Figure 6.9)	1.79	49,512.6

arbiters [8]. This router has five ports ($P = 5$) and four VNs; each VN has two VCs ($V = 2$). Table 6.2 gives the area and critical path delay estimates. The naive design uses *4:1* arbiters in the first stage, resulting in a 7.9% longer critical path and 13.4% more area than the PSF design. Our design uses *2:1* arbiters in the first stage and increases the critical path by only 0.5% and the area by only 0.2%. An allocator's power consumption is largely decided by the arbiter size [2, 55]; given the small arbiters in our design, there should be negligible power overhead compared with the PSF design.

6.5 EVALUATION ON SYNTHETIC TRAFFIC

We modify BookSim [24] for evaluation. Fully adaptive routing includes the high flexibility design (FULLY) in Section 6.4.3 and the PSF design. FULLY and PSF are evaluated with conservative reallocation (named FULLY and PSF), and with WPF on AVCs and aggressive reallocation on EVCs (named FULLY + WA and PSF + WA). The deterministic routing is DOR. West-first, negative-first, and odd-even represent partially adaptive routing. We use a local selection strategy for all adaptive routing;

when there are two permissible ports, it chooses the one with more free buffers. If the ports have equal free buffers, it randomly chooses one of them.

Our evaluation for synthetic traffic uses one VN since each VN is independent. We use a 4 × 4 mesh with two VCs that are each four flits deep. The packet lengths exhibit a bimodal distribution; there are SFPs and five-flit packets. The SFP ratio is 80%. The simulator is warmed up for 10,000 cycles and then the performance is measured over another 100,000 cycles. We also perform sensitivity studies on several configuration parameters.

6.5.1 PERFORMANCE OF SYNTHETIC WORKLOADS

Figure 6.10 gives the performance with four synthetic traffic patterns [8]. Across the four patterns, PSF and FULLY perform worst. Although PSF and FULLY offer adaptiveness for all traffic, conservative VC reallocation significantly limits their performance. In contrast, DOR and partially adaptive routing use aggressive VC reallocation. PSF is inferior to FULLY. PSF is further limited by its poor flexibility: once a packet enters an EVC, the packet can be routed only by DOR using EVCs until it is delivered.

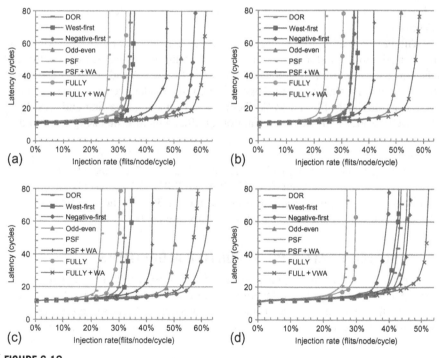

FIGURE 6.10

The routing algorithm performance for the baseline configuration. (a) Bit reverse, (b) transpose-1, (c) transpose-2, and (d) hotspot.

For bit reverse, a node with bit address $\{s_3, s_2, s_1, s_0\}$ sends traffic to node $\{s_0, s_1, s_2, s_3\}$. Of this traffic, 62.5% is between the north-east and south-west quadrants; negative-first offers adaptiveness for this traffic. Further, 37.5% of the traffic is eastbound; west-first offers adaptiveness for this traffic, and performs worse than negative-first. Although WPF and aggressive VC reallocation improve the VC utilization for PSF + WA, PSF + WA is inferior to odd-even and negative-first. PSF + WA is limited by poor flexibility. FULLY + WA provides high VC utilization and routing flexibility, leading to the highest saturation throughput.[3]

Transpose-1 sends a message from node (i, j) to node $(3 - j, 3 - i)$. Negative-first deteriorates to DOR for this pattern. West-first still offers adaptiveness for 37.5% of the traffic; thus, it is superior to negative-first. Odd-even offers greater adaptiveness than the other two partially adaptive algorithms and achieves higher performance. FULLY + WA offers adaptiveness for all traffic, performing 15.7% better than odd-even.

Transpose-2 is favorable to negative-first; node (i, j) communicates with node (j, i). Negative-first offers adaptiveness for all traffic and performs best. Although FULLY + WA offers adaptiveness for all traffic, it is limited by the restriction on EVCs: only if the port adheres to DOR can the EVC be used. The performance of FULLY + WA and odd-even with transpose-2 is similar to their performance with transpose-1 since the two patterns are symmetric and these designs offer the same adaptiveness for them.

With hotspot traffic, four nodes are chosen as hotspots and receive an extra 20% traffic in addition to uniform random traffic. This pattern mimics memory controllers receiving a disproportionate amount of traffic. FULLY + WA and odd-even are worse than negative-first and west-first. Owing to the limited adaptiveness offered by odd-even, it is inferior to FULLY+WA. DOR outperforms negative-first and west-first, since DOR more evenly distributes uniform traffic, which is the background in this pattern.

Table 6.3 gives average throughput gains of FULLY + WA over other designs. The 88.9% gap between FULLY + WA and FULLY reflects the effect of novel flow control. The gap between FULLY + WA and PSF + WA represents the effect of routing flexibility; high flexibility brings a gain of 31.3%. The next section provides further insights into performance trends by analyzing the buffer utilization.

6.5.2 BUFFER UTILIZATION OF ROUTING ALGORITHMS

This section provides further insights into the performance trends shown in Section 6.5.1 by analyzing the buffer utilization of all evaluated designs. Figure 6.11 illustrates the average buffer utilization of all network VCs at saturation for eight synthetic traffic patterns. The maximum and minimum rates are given by the error

---

[3]The saturation point is measured as the injection rate at which the average latency is three times the zero-load latency.

Table 6.3 The Average Saturation
Throughput Improvement

Algorithm	Improvement (%)
FULLY	88.9
DOR	64.5
West-first	58.6
Negative-first	26.6
Odd-even	16.3
PSF	130.9
PSF + WA	31.3

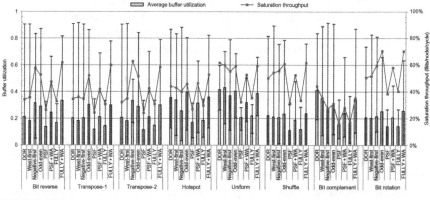

FIGURE 6.11

The buffer utilization and throughput at saturation for several routing algorithms.

bars. Figure 6.11 also shows the saturation throughput supported by routing algorithms. There are several important insights.

First, buffers are used to support high throughput. The saturation throughput of routing algorithms is related to the buffer utilization. Higher average buffer utilization generally leads to higher saturation throughput. This trend is obvious for bit complement. Bit complement sends traffic from node $\{s_3,s_2,s_1,s_0\}$ to node $\{\neg s_3,\neg s_2,\neg s_1,\neg s_0\}$, which has the largest average hop count [31], and puts the highest pressure on the buffers among all evaluated traffic patterns.

Second, for the same routing algorithm in different traffic patterns, there may be fluctuations between the buffer utilization and the saturation throughput. For example, even though DOR shows much higher buffer utilization for bit complement than for shuffle, its saturation throughput for bit complement is lower than for shuffle. These fluctuations are because the same routing algorithm uses different VCs for different patterns; the network becomes saturated in different statuses. Similarly, since some algorithms use different VCs in the same traffic pattern, there may be slight fluctuations as well. For example, even though the average buffer utilization of west-first is slightly lower than that of DOR for bit reverse, its saturation throughput is higher than that of DOR.

Third, since all fully adaptive routing algorithms, including PSF, PSF + WA, FULLY, and FULLY + WA, use the same set of VCs for the same traffic pattern, their saturation throughput is roughly proportional to the buffer utilization. For instance, for bit reverse, PSF + WA's average buffer utilization is 78.3% higher than that of PSF, making it perform 79.5% better than PSF. Similarly, PSF + WA has 67.4% higher average buffer utilization than PSF in shuffle, and its saturation throughput is 67.2% higher than that of PSF.

Fourth, the network becomes saturated when some resources become saturated [8]. For most patterns and most algorithms, the bottleneck is VCs; networks become saturated when some VCs have more than 80% utilization, including DOR, west-first, negative-first, and odd-even in most patterns. When the algorithm provides abundant adaptiveness, it distributes traffic more uniformly among VCs, preventing them from becoming the bottleneck. Instead, the bottleneck may be the crossbar or network interface [50]. For example, since negative-first offers adaptiveness for all traffic in transpose-2, the network becomes saturated even when the maximum buffer utilization is only 49.2%. Similarly, at saturation, the maximum buffer utilization of PSF+WA and FULLY + WA is lower than that of partially adaptive routing and DOR.

Fifth, PSF and FULLY are limited by conservative VC reallocation; their maximum buffer utilization for the evaluated patterns is less than 40%. Applying WPF on AVCs and aggressive VC reallocation on EVCs greatly increases buffer utilization, which proportionally improves performance. FULLY supports higher routing flexibility than PSF, which brings higher buffer utilization and performance.

6.5.3 SENSITIVITY TO NETWORK DESIGN

This section includes a sensitivity study for network configuration parameters. Except for the parameter analyzed, the other parameters are the same as for the baseline configuration. Table 6.4 lists the baseline parameter values and their variations. In this section we make a comparison with the performance of the baseline configuration shown Section 6.5.1.

6.5.3.1 SFP ratio

SFP ratios depend on the cache hierarchy, the coherence protocol, and the application. To test the robustness of our design, we evaluate SFP ratios of 60% and 40% for transpose-1. As shown in Figure 6.12, DOR, west-first, negative-first, and odd-even exhibit nearly identical performance for different SFP ratios. The aggressive VC reallocation makes their performance insensitive to the packet length distribution. However, the performance of PSF and FULLY improves as the SFP ratio shrinks. Their conservative VC reallocation favors long packets,

Table 6.4 The Baseline Configuration and Variations

Characteristic	Baseline	Variations
Topology (mesh)	4 × 4	8 × 8
VCs per VN	2	4
Flit buffers per VC	4	3, 2
SFP ratio	80%	60%, 40%

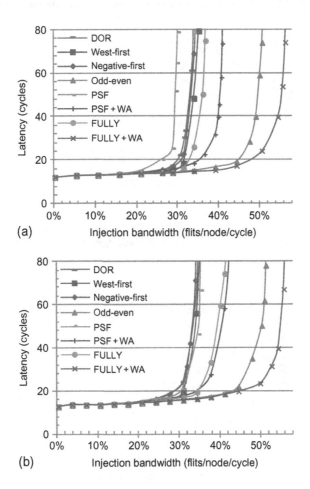

FIGURE 6.12

The performance with different SFP ratios for transpose-1 traffic. (a) 60% SFP ratio and (b) 40% SFP ratio.

which utilize buffers more efficiently than short ones. As the SFP ratio decreases, so does the possibility of applying WPF. Thus, the performance gap between FULLY + WA and FULLY (or PSF + WA and PSF) decreases. However, even with an SFP ratio of 40%, FULLY + WA achieves a 53.1% saturation throughput gain over FULLY.

6.5.3.2 VC depth

Different NoCs may use different VC depths. To test the flexibility, we evaluate three-flit-deep and two-flit-deep VCs with bit reverse traffic. In a comparison of four flits per VC (Figure 6.10a) and three flits per VC (Figure 6.13a), DOR and west-first perform similarly, while FULLY and PSF show minor performance degradation. DOR and

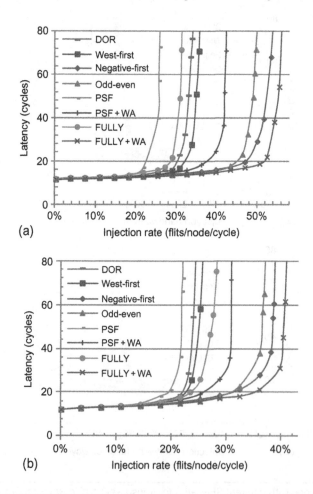

(a)

(b)

FIGURE 6.13

The performance with different VC depths for bit reverse traffic. (a) Three-flit-deep VCs and (b) two-flit-deep VCs.

west-first offer no or very limited adaptiveness, which is a major factor in their performance. Thus, reducing the VC depth from four to three has little effect. The bottleneck of FULLY and PSF is conservative VC reallocation. When we consider the majority of short packets, reducing the VC depth from four to three affects performance only slightly. However, the performance of FULLY + WA, PSF + WA, odd-even, and negative-first declines with shallower VCs since the VC depth is their bottleneck. Shallow VCs increase the number of hops that a blocked packet spans, which increases the effect of chained blocking [50].

When we compare three fits per VC and two flits per VC, the performance drops for all algorithms. FULLY outperforms DOR and west-first with two flits per VC. As the VC depth decreases, the difference between aggressive and conservative VC reallocation declines; FULLY exhibits a relative gain. Even with two flits per VC, WPF still optimizes the performance since short packets dominate the traffic. In Figure 6.13b, FULLY + WA performs 46.2% better than FULLY. Novel flow control leads to superior performance even with half of the buffers, which enables the design of a low-cost NoC. With two flits per VC (Figure 6.13b), FULLY + WA's saturation throughput is 40.3%, while FULLY's saturation throughput is 32.3% with four flits per VC (Figure 6.10a). The same is true for PSF + WA.

6.5.3.3 VC count

Semiconductor scaling and coherence protocol optimization may allow a VN to be configured with more VCs. When we compare four VCs per VN (Figure 6.14a) and two VCs per VN (Figure 6.10a), the performance of DOR, west-first, and odd-even is almost the same. These algorithms offer limited adaptiveness; although additional results show increasing the VC count from one to two improves performance, increasing the VC count from two to four cannot reduce the physical path congestion and does not further improve performance. Negative-first has a modest gain. In contrast, PSF, FULLY, PSF + WA, and FULLY + WA all have significant gains; more VCs mitigate the negative effects of conservative VC reallocation. The gap between PSF and FULLY (or PSF + WA and FULLY + WA) decreases with more VCs; more VCs reduce the possibility of using EVCs in PSF, which forces packets to lose adaptivity.

Figure 6.14b shows the performance of transpose-2, which is a favorable pattern for negative-first. FULLY + WA performs similarly to negative-first; with more VCs, the effect of restricting the use of EVCs in FULLY + WA declines. More VCs reduce the gap between FULLY and FULLY + WA (or PSF and PSF + WA), since the possibility of facing empty VCs increases. Furthermore, using WPF to forward entire packets into nonempty VCs may result in head-of-line blocking [8] and limit FULLY + WA (or PSF + WA). Nevertheless, FULLY + WA still shows an average 19.8% gain over FULLY for these two patterns with four VCs; providing high VC utilization outweighs the negative effect of head-of-line blocking in a VC-limited NoC. Similarly to VC depth, with only two VCs, FULLY + WA performs similarly or even better (Figure 6.10a and b) than FULLY with four VCs (Figure 6.14). WPF provides similar or higher performance with half as many VCs.

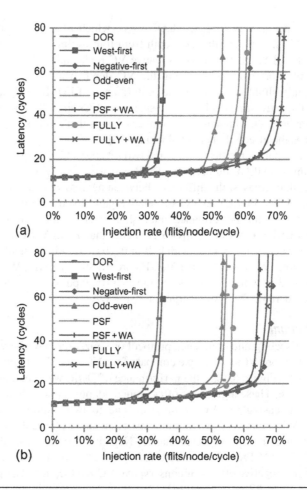

FIGURE 6.14

The performance with four VCs per VN. (a) Bit reverse and (b) transpose-2.

6.5.3.4 Network size

Figure 6.15 explores the scalability for an 8×8 mesh. The trends across different algorithms are similar to those for the 4×4 mesh (Figure 6.10). Communication is mostly determined by the traffic pattern. Since a larger network leads to higher average hop counts [31], it puts more pressure on VCs than a smaller one. Novel flow control achieves more gains in a larger network. The average improvement for these two patterns of FULLY + WA over FULLY is 108.2%, while it is 93.1% in a 4×4 mesh. As packets undergo more hops in a larger network, the possibility of entering an EVC increases. For PSF and PSF + WA, once the packet enters an EVC, it loses adaptivity in subsequent hops. Therefore, the gap between FULLY + WA and PSF + WA (or FULLY and PSF) increases with a larger network; providing routing flexibility becomes more important with a larger network.

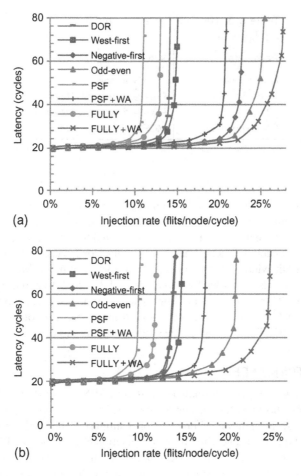

FIGURE 6.15

The performance for an 8 × 8 mesh. (a) Bit reverse and (b) transpose-1.

6.6 EVALUATION OF PARSEC WORKLOADS

In this section we evaluate routing algorithms using PARSEC workloads.

6.6.1 METHODOLOGY AND CONFIGURATION

To measure full-system performance, we leverage FeS2 [39] for x86 simulation and
BookSim for NoC simulation. FeS2 is implemented as a module for Simics [34].
We run PARSEC benchmarks [3] with 16 threads on a 16-core chip multiprocessor,
which is organized as a 4 × 4 mesh. Prior research shows the frequency of simple

Table 6.5 The Full System Simulation Configuration

Parameter	Value
No. of cores	16, 4 × 4 mesh
L1 cache (D and I)	Private, 4-way, 32 kB each
L2 cache	Private, 8-way, 512 kB each
Cache coherence	MOESI distributed directory

L1, first level; L2, second level.

cores in a many-core platform can be optimized to 5-10 GHz, while the frequency of an NoC router is limited by the allocator speed [11]. We assume cores are clocked five times faster than the network. Each core is connected to private, inclusive first-level and second-level caches. Cache lines are 64 bytes; long packets are five flits long with a 16-byte flit width. We use a distributed, directory-based MOESI coherence protocol which needs four VNs for protocol-level deadlock freedom. Each VN has two VCs; each VC is four flits deep. The router pipeline is the same as described in Section 6.4.4. All benchmarks use the *simsmall* input sets. The total run time is the performance metric. Table 6.5 gives the system configuration.

6.6.2 PERFORMANCE

Figure 6.16 shows the speedups relative to PSF for PARSEC workloads. We divide the 10 applications into two classes. For blackscholes, fluidanimate, raytrace, and swaptions, different algorithms perform similarly. The working sets of these applications fit into the caches, and their computation phases consist of few synchronization points, leading to a lightly loaded network. The system performance of these applications is unaffected by techniques that improve the network throughput, such as sophisticated routing algorithms.

However, the routing algorithms affect the other six applications. These applications have heavier loads than the previous four applications. Yet, their average aggregate loads during the running periods are lower than the network saturation points. The highest average aggregate injection rate for canneal is 12.3% of flits per node per cycle, which is below the saturation points for all designs. Two factors bring performance improvements for these six applications. First, these applications exhibit period bursty communication and the synchronization primitives create one hotspot inside the network. The bursty communication and the hotspot make the network operate past saturation at times during the running period. Thus, these applications benefit from routing algorithms with higher throughputs. Second, short packets are critical in cache-coherent many-core platforms; they carry time-critical control messages that are often on the application's critical path. These short packets can affect the execution time significantly. The novel flow control reduces the

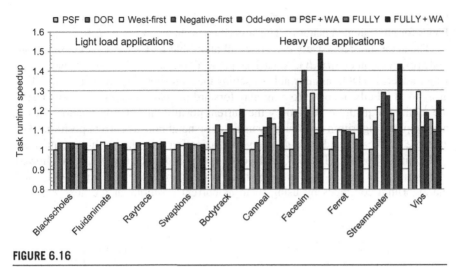

FIGURE 6.16

System speedups for PARSEC benchmarks.

packet latency, especially under heavy loads, which brings performance gains. For example, FULLY + WA has speedups of 48.5% and 43.0% over PSF for `facesim` and `streamcluster`.

Since most of the `vips` application's bursty communication is eastbound, west-first performs best for `vips`. For `facesim` and `streamcluster`, negative-first offers higher adaptiveness than odd-even, thus achieving better performance. For all heavy-load applications except `vips`, FULLY + WA performs best. Across these applications, FULLY + WA achieves an average speedup of 21.3% and a maximum speedup of 37.8% over FULLY. With sufficient flexibility, FULLY + WA has an average speedup of 12.1% over PSF + WA. The average speedups of FULLY + WA are 29.3%, 15.0%, 10.1%, 9.9%, and 10.4% over PSF, DOR, west-first, negative-first, and odd-even respectively.

6.7 DETAILED ANALYSIS OF FLOW CONTROL

In this section we perform a detailed analysis for the flow control mechanisms utilized in fully adaptive routing algorithms.

6.7.1 THE DETAILED BUFFER UTILIZATION

6.7.1.1 Allowable EVCs

Here, we analyze aggressive VC reallocation and WPF by measuring the buffer utilization for network VCs. Since EVCs can be used only when the port adheres to DOR, not all EVCs are allowable for a particular traffic pattern. To clearly understand the system bottleneck, we leverage a recursive algorithm to calculate the allowable EVCs for all evaluated patterns. This algorithm is shown in Figure 6.17.

The algorithm recursively calculates all allowable EVCs for a packet sent from node (cx, cy) to node (dx, dy). At each step, the algorithm marks the EVC at the port which adheres to DOR as allowable by setting the corresponding element of the *EVCs* array to "1." Fully adaptive routing may forward the packet to all neighbors inside the minimum quadrant defined by the current position and the destination. At the next step, the algorithm continues by using these neighbors as current positions. For example, if $cx > dx$ and $cy > dy$, then the destination is in the north-west quadrant of the current position. The EVC at the east input port of node $(cx - 1, cy)$ is allowed to be used. The packet may move to node $(cx - 1, cy)$ and node $(cx, cy - 1)$, and the algorithm continues by using them as current positions at the next step. By

```
void AllowableEVCs(int cx,int cy,int dx,int dy,
                   int*** EVCs){
    if( cx == dx && cy == dy ){  // destination
    return;
    } else if( cx < dx && cy == dy ){    // east
    EVCs[cx+1][cy][1] = 1;
    AllowableEVCs( cx+1, cy, dx, dy, EVCs );
    } else if( cx > dx && cy == dy ){    // west
    EVCs[cx-1][cy][0] = 1;
    AllowableEVCs( cx-1, cy, dx, dy, EVCs );
    } else if( cx == dx && cy < dy ){    // north
    EVCs[cx][cy+1][3] = 1;
    AllowableEVCs( cx, cy+1, dx, dy, EVCs );
    } else if( cx == dx && cy > dy ){    // south
    EVCs[cx][cy-1][2] = 1;
    AllowableEVCs( cx, cy-1, dx, dy, EVCs );
    } else if( cx>dx && cy>dy ){    // north-west
    EVCs[cx-1][cy][0] = 1;
    AllowableEVCs( cx-1, cy, dx, dy, EVCs );
    AllowableEVCs( cx, cy-1, dx, dy, EVCs );
    } else if( cx>dx && cy<dy ){    // south-west
    EVCs[cx-1][cy][0] = 1;
    AllowableEVCs( cx-1, cy, dx, dy, EVCs );
    AllowableEVCs( cx, cy+1, dx, dy, EVCs );
    } else if( cx<dx && cy>dy ){    // north-east
    EVCs[cx+1][cy][1] = 1;
    AllowableEVCs( cx+1, cy, dx, dy, EVCs );
    AllowableEVCs( cx, cy-1, dx, dy, EVCs );
    } else if( cx<dx && cy<dy ){    // south-east
    EVCs[cx+1][cy][1] = 1;
    AllowableEVCs( cx+1, cy, dx, dy, EVCs );
    AllowableEVCs( cx, cy+1, dx, dy, EVCs );
    }
}
```

FIGURE 6.17

The AllowableEVCs algorithm. The initial value of all elements of EVCs is 0. cx and cy are the X and Y positions of the current node. dx and dy are the X and Y positions of the destination. Port encoding: east, 0; west, 1; south, 2; north, 3.

Table 6.6 Allowable EVC Counts for Synthetic Traffic (Total EVC Count of 48)

Traffic Pattern	Allowable EVCs
Bit reverse	36
Transpose-1	36
Transpose-2	36
Hotspot	48
Uniform random	48
Shuffle	36
Bit complement	48
Bit rotation	40

considering all source and destination pairs defined in the traffic pattern, we get the allowable EVCs in a 4 × 4 mesh network, As shown in Table 6.6, some traffic patterns, such as bit reverse, cannot use all EVCs, while other patterns, including bit complement and uniform random, can use all EVCs.

6.7.1.2 Performance analysis

Figure 6.18 gives the saturation throughput and buffer utilization for fully adaptive designs. PSF + WPF and FULLY + WPF apply WPF on both AVCs and EVCs. The difference between PSF + WA and PSF + WPF (or FULLY + WA and FULLY + WPF) is that PSF + WA (or FULLY + WA) applies aggressive VC reallocation on EVCs. The AVC utilization and EVC utilization are shown by the "average AVC utilization" and "average EVC utilization" bars. The "average allowable EVC utilization" bar gives the average utilization of allowable EVCs.

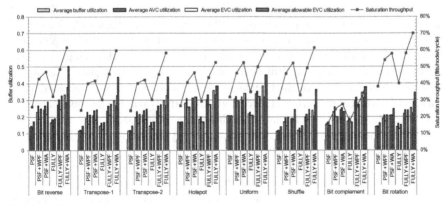

FIGURE 6.18

The buffer utilization and saturation throughput of fully adaptive routing algorithms.

There are several insights from Figure 6.18. First, the performance gain of FULLY + WPF over FULLY (or PSF + WPF over PSF) is due to the improvement of buffer utilization for both AVCs and EVCs. The gain of FULLY + WA over FULLY + WPF (or PSF + WA over PSF + WPF) is due to the improvement of buffer utilization for EVCs; aggressive VC reallocation more efficiently utilizes EVCs than does WPF. For example, for bit reverse, the EVC utilization of FULLY + WA and FULLY + WPF is 0.377 and 0.247.

Second, the efficiency of aggressive VC reallocation depends on routing flexibility. FULLY can use AVCs after using EVCs; aggressive VC reallocation offers more gains for FULLY + WA than for PSF + WA. For bit reverse, FULLY + WA performs 26.6% better than FULLY + WPF; the 52.9% EVC utilization improvement of FULLY + WA over FULLY + WPF brings this gain. In contrast, PSF + WA achieves only 16.3% higher EVC utilization than PSF + WPF, which brings a 9.7% performance gain for PSF + WA.

Third, PSF + WPF requests an EVC only when the selected port adheres to DOR. Meanwhile, it always applies WPF on AVCs. These two factors induce higher pressure on AVCs than EVCs for PSF + WPF. For example, for transpose-1, the AVC utilization of PSF + WPF is 0.228, while the EVC utilization is 0.158 and the allowable EVC utilization is 0.210. FULLY can always request an EVC. Its allowable EVC utilization is higher than the AVC utilization for most patterns; FULLY puts more pressure on EVCs than on AVCs.

Fourth, the aggressive VC reallocation in FULLY + WA further increases the pressure on EVCs; the allowable EVC utilization of most patterns is more than 60% higher than the AVC utilization. FULLY + WA is limited by allowable EVCs, and mechanisms to improve the capacity of EVCs, such as dynamically sharing buffers between EVCs and AVCs, will be helpful to FULLY + WA.

6.7.2 THE EFFECT OF FLOW CONTROL ON FAIRNESS

As shown in Section 6.4.2, aggressive VC reallocation on EVCs can address the fairness issue of WPF on long packets. To illustrate this point, Figures 6.19 and 6.20 show the latency distribution of short and long packets for FULLY + WPF and FULLY + WA with bit complement traffic; the overall network average latencies are both 52 cycles. Other patterns have similar trends. The short packet latency of FULLY + WPF has three distinct peaks at 11, 17, and 23 cycles, corresponding to packets which take three, five, and seven hops respectively without any queuing delay. Since each flit of a long packet may face different switch allocation congestion, the long packet latency only exhibits one peak at 27 cycles, corresponding to packets with seven hops.

The short and long packet latencies of FULLY + WA both have one peak, at 23 and 27 cycles respectively. To illustrate the difference between FULLY + WA and FULLY + WPF, we subtract the data in Figure 6.19 from the data in Figure 6.20, as shown in Figure 6.21. The ratio of short packets with a latency between nine and 23 cycles for FULLY + WA is 13.2% less than that for FULLY + WPF. On the other hand,

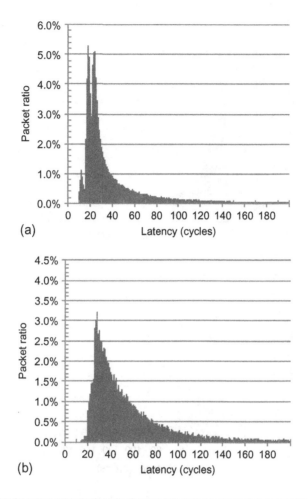

FIGURE 6.19

The latency distribution of FULLY + WPF when the average latency is 52 cycles for bit complement traffic. (a) Short packets and (b) long packets.

FULLY + WA has 8.8% more long packets with a latency between 19 and 42 cycles than FULLY + WPF. Compared with FULLY + WPF, FULLY + WA accelerates long packets, while sacrificing short packets.

The effect on fairness can also be observed from the average latency of packets injected from different sources, as shown in Figure 6.22. In FULLY + WPF, middle nodes have very high latency as their injected long packets are always inferior to short packets from edge nodes. FULLY + WA reduces the peak latency for these middle nodes from 134 cycles to 104 cycles, and increases the latency for the edge nodes, achieving more even latency distribution throughout the network.

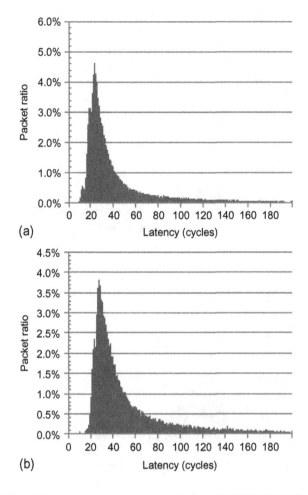

FIGURE 6.20

The latency distribution of FULLY + WA when the average latency is 52 cycles for bit complement traffic. (a) Short packets and (b) long packets.

The latencies shown in Figures 6.19–6.22 are snapshots when the network average latencies are 52 cycles. With decreasing injection rate, the latency differences between FULLY + WA and FULLY + WPF become insignificant as the network contention becomes less likely. Conversely, the differences become more pronounced as the network approaches saturation. For example, when the average network latencies of FULLY + WPF and FULLY + WA are both 75 cycles, FULLY + WA has 10.3% more long packets with a latency between 21 and 48 cycles than FULLY + WPF, while FULLY + WPF has 15.7% more short packets with a latency between 11 and 25 cycles than FULLY + WA.

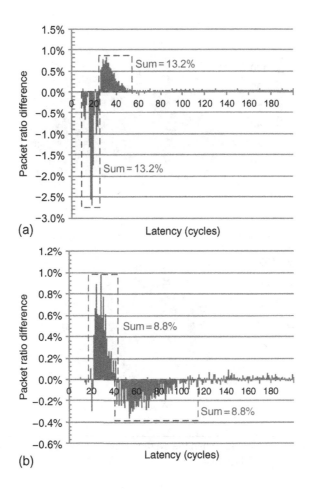

FIGURE 6.21

The latency distribution difference between FULLY + WPF and FULLY + WA for bit comple-ment traffic. (a) Short packets and (b) long packets.

6.8 FURTHER DISCUSSION

6.8.1 PACKET LENGTH

Packet lengths for cache coherence traffic typically have a bimodal distribution. However, optimizations such as cache line compression [11, 25] create packet distributions that are not bimodal; the packet length may be distributed between a single flit and the maximum number of flits per packet supported by the architecture. To apply WPF on such NoCs, more downstream VC status registers are needed for the first-stage arbiters shown in Figure 6.9. An important consideration is how many different packet lengths to apply WPF to. The longest packet length that can use WPF

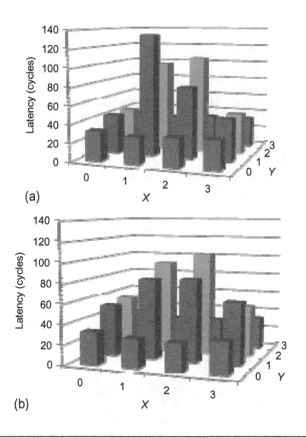

FIGURE 6.22

The latency of each source node when the average latency is 52 cycles for bit complement traffic. (a) FULLY + WPF and (b) FULLY + WA.

is one flit shorter than the VC depth. Designers can ignore long packets, since there are few opportunities to apply WPF on long packets. This tradeoff depends on the packet length distribution, VC depth, hardware overhead, and expected performance gain.

6.8.2 DYNAMICALLY ALLOCATED MULTIQUEUE AND HYBRID FLOW CONTROLS

Previous research proposed dynamically allocated multiqueue (DAMQ) designs for both off-chip networks [49] and on-chip networks [40, 55] to improve VC utilization. Even with DAMQ, allowing multiple packets to reside in one VC may lead to deadlock similar to that in Figure 6.3 for fully adaptive routing in wormhole networks. WPF is complementary to DAMQ as it ensures deadlock freedom. WPF can be viewed as a hybrid mechanism combining wormhole flow control and VCT flow control. There are some hybrid flow controls [30, 45, 47]. Hybrid switching [45] and buffered wormhole [47] remove a blocked packet to release held physical channels by using either the processing node memory [45] or a central buffer [47]. Layered switching divides a long packet into several groups and tries to keep switch allocation grants for a whole group [30]. Our research is different; we focus on improving the performance while avoiding deadlock.

6.9 CHAPTER SUMMARY

An abundance of short packets and a limited VC budget increase the importance of flow control in the performance of cache-coherent NoCs. We proposed two novel flow control designs for fully adaptive routing in wormhole networks. WPF is an important extension to several deadlock avoidance theories. It reallocates a nonempty VC if the VC has enough buffers for an entire packet. Then, we extended Duato's theory to apply aggressive VC reallocation on EVCs. On the basis of the proposed flow control, we designed a low-cost routing algorithm to maintain maximal routing flexibility. Compared with existing fully adaptive routing, our design performs similarly or even better with half of the buffers or VCs, enabling the design of a low-cost NoC. Our design also outperforms several partially adaptive and deterministic routing algorithms.

APPENDIX: LOGICAL EQUIVALENCE OF *ALG* AND *ALG* + *WPF*

In Section 6.4.1 we proposed Theorem 6.1, which declares that if a fully adaptive routing algorithm with conservative VC reallocation (named *Alg*) is deadlock-free, then applying WPF on this routing algorithm (named *Alg* + *WPF*) is also deadlock-free. In this section we prove that if *Alg* + *WPF* is deadlock-free, then *Alg* is deadlock-free as well. In addition to Assumption 6.1 in Section 6.4.1, this proof needs another simple assumption about the routing algorithm.

Assumption 6.2. *The definition of a routing algorithm is independent of the packet length.*

Assumption 6.2 implies that if two packets are injected from the same source and destined for the same node, they can use the same set of VCs even though these packets have different lengths. This assumption was generally used in many previous theories [4, 6, 12, 13, 17–19, 29, 44, 48, 52, 53]. We prove the following theorem.

Theorem 6.4. *If Alg + WPF is deadlock-free, then Alg is also deadlock-free.*

Informal description: This proof is similar to the proof of Theorem 6.1. We prove that if *Alg* has a deadlock configuration, then *Alg* + *WPF* has a deadlock configuration as well. We use *Config*$_0$ in Figure 6-A.1 as an example. We replace packet P_i with packet P_i' to completely fill all VCs. The source and destination of P_i' are the same as those of P_i, while P_i' may have more flits than P_i to completely fill the VCs. We prove that *Alg* + *WPF* can achieve *Config*$_0$, and that *Config*$_0$ is a deadlock configuration. However, *Alg* + *WPF* is deadlock-free; thus, there is no such configuration.

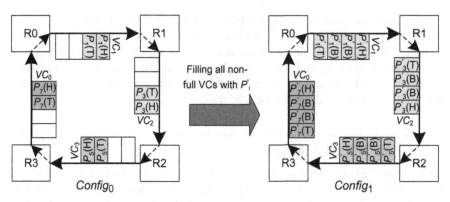

FIGURE 6-A.1

The construction of a new configuration.

Proof: By contradiction. If *Alg* is not deadlock-free, then there is a deadlock configuration (*Config₀*) in which a set of packets, P_{set_0}, is waiting for VCs held by other packets in P_{set_0}. We prove that there is a deadlock configuration for *Alg + WPF* in three steps.

Step 1: We build a new configuration based on *Config₀*. Consider each packet P_i in P_{set_0}. If P_i does not completely fill all VCs where its flits reside, we replace P_i with packet P_i' to completely fill all VCs occupied by P_i. The source and destination of P_i' are the same as those of P_i, and the length of P_i' is larger than that of P_i. If P_i completely fills all VCs where its flits reside, P_i' is the same as P_i. We label the new configuration as *Config₁*, and the set of packets in *Config₁* as P_{set_1}.

Step 2: We prove that when the network is routed by *Alg + WPF*, all packets in P_{set_1} can move to their current VCs in *Config₁*, and the head flits of these packets are at VC heads. For each packet P_i' in P_{set_1}, we consider its corresponding packet P_i in *Config₀*. We further consider each hop hop_k of P_i when it is routed by *Alg*. Assume the head flit of P_i moves to VC_k's head during hop_k. The routing algorithm allows P_i to use VC_k. Although P_i' and P_i may have different packet lengths, they have the same source and destination information. On the basis of Assumption 6.2, the routing algorithm allows P_i' to use VC_k as well.

Since *Alg + WPF* is deadlock-free, VC_k must become available at some time. On the basis of Assumption 6.1, P_i' has some possibility to request VC_k. There are two cases when P_i' requests VC_k:

(1) VC_k is empty. VC_k can be reallocated to P_i' with conservative VC reallocation. Thus, the head flit of P_i' will move to VC_k's head.

(2) VC_k is not empty. But VC_k has already received the tail flit of the last allocated packet and still has enough buffers for P_i'. WPF can reallocate VC_k to P_i'. Yet, when P_i' reaches VC_k, its head flit is not at VC_k's head. Considering *Alg + WPF* is deadlock-free, the packets in VC_k before P_i' must be sent out in limited time. Then, the head flit of P_i' is at VC_k's head.

When we consider cases 1 and 2 together, if the head flit of P_i moves to the head of a VC during any hop when it is routed by Alg, the head flit of P'_i can also move to the same VC head when it is routed by $Alg + WPF$. Thus, when it is routed by $Alg + WPF$, P'_i can be routed to its current VC(s) in $Config_1$, and the head flit of P'_i is at the VC head.

Step 3: We prove that $Config_1$ is a deadlock configuration for $Alg + WPF$. Since all VCs in $Config_1$ are completely filled, all packets in P_{set_1} are still waiting for VCs held by other packets in P_{set_1}, even with WPF applied. $Config_1$ is a deadlock configuration for $Alg + WPF$. But $Alg + WPF$ is deadlock-free, so there is no deadlock configuration. Thus, Alg is deadlock-free as well. ∎

Theorem 6.1 declares that if Alg is deadlock-free, then $Alg + WPF$ is deadlock-free as well. Theorem 6.4 declares that if $Alg + WPF$ is deadlock-free, then Alg is deadlock-free as well. If we consider these two theorems together, the deadlock freedom of $Alg + WPF$ is logically equivalent to the deadlock freedom of Alg. Therefore, the WPF optimization of traditional wormhole flow control does not have any influence on the deadlock freedom.

REFERENCES

[1] K. Anjan, T. Pinkston, An efficient, fully adaptive deadlock recovery scheme: DISHA, in: Proc. of the International Symposium on Computer Architecture (ISCA), 1995, pp. 201–210. 179

[2] D.U. Becker, W.J. Dally, Allocator implementations for network-on-chip routers, in: Proc. of the Conference on High Performance Computing Networking, Storage and Analysis (SC), 2009, pp. 1–12. 180, 189, 190

[3] C. Bienia, S. Kumar, J.P. Singh, K. Li, The PARSEC benchmark suite: characterization and architectural implications, in: Proc. of the International Conference on Parallel Architectures and Compilation Techniques (PACT), 2008, pp. 72–81. 176, 199

[4] G.-M. Chiu, The odd-even turn model for adaptive routing, IEEE Trans. Parallel Distrib. Syst. 11(7) (2000) 729–738. 176, 179, 209

[5] W. Dally, Virtual-channel flow control, IEEE Trans. Parallel Distrib. Syst. 3(2) (1992) 194–205. 176

[6] W. Dally, C. Seitz, Deadlock-free message routing in multiprocessor interconnection networks, IEEE Trans. Comput. C-36(5) (1987) 547–553. 176, 179, 209

[7] W. Dally, B. Towles, Route packets, not wires: on-chip interconnection networks, in: Proc. of the Design Automation Conference (DAC), 2001, pp. 684–689. 176, 179

[8] W. Dally, B. Towles, Principles and Practices of Interconnection Networks, first ed., Morgan Kaufmann Publishers Inc., San Francisco, CA, 2003. 179, 182, 188, 189, 190, 191, 194, 197

[9] W.J. Dally, C.L. Seitz, The torus routing chip, Distrib. Comput. 1(4) (1986) 187–196. 176, 179

[10] R. Das, S. Eachempati, A.K. Mishra, V. Narayanan, C.R. Das, Design and evaluation of a hierarchical on-chip interconnect for next-generation CMPs, in: Proc. of the International Symposium on High-Performance Computer Architecture (HPCA), 2009, pp. 175–186. 176

[11] R. Das, A.K. Mishra, C. Nicopoulos, D. Park, V. Narayanan, R. Iyer, M.S. Yousif, C.R. Das, Performance and power optimization through data compression in network-on-chip architectures, in: Proc. of the International Symposium on High-Performance Computer Architecture (HPCA), 2008, pp. 215–225. 200, 207

[12] J. Duato, A new theory of deadlock-free adaptive routing in wormhole networks, IEEE Trans. Parallel Distrib. Syst. 4(12) (1993) 1320–1331. 176, 177, 178, 179, 180, 181, 185, 186, 187, 188, 209

[13] J. Duato, A necessary and sufficient condition for deadlock-free adaptive routing in wormhole networks, IEEE Trans. Parallel Distrib. Syst. 6(10) (1995) 1055–1067. 176, 177, 179, 185, 186, 187, 209

[14] J. Duato, T. Pinkston, A general theory for deadlock-free adaptive routing using a mixed set of resources, IEEE Trans. Parallel Distrib. Syst. 12(12) (2001) 1219–1235. 179

[15] N. Enright Jerger, L. Peh, On-Chip Networks, first ed., Morgan & Claypool, San Rafael, CA, 2009. 183, 188

[16] C. Fallin, C. Craik, O. Mutlu, CHIPPER: a low-complexity bufferless deflection router, in: Proc. of the International Symposium on High-Performance Computer Architecture (HPCA), 2011, pp. 144–155. 176

[17] E. Fleury, P. Fraigniaud, A general theory for deadlock avoidance in wormhole-routed networks, IEEE Trans. Parallel Distrib. Syst. 9 (1998) 626–638. 176, 177, 179, 185, 209

[18] B. Fu, Y. Han, J. Ma, H. Li, X. Li, An abacus turn model for time/space-efficient reconfigurable routing, in: Proc. of the International Symposium on Computer Architecture (ISCA), 2011, pp. 259–270. 176, 177, 179

[19] C. Glass, L. Ni, The turn model for adaptive routing, in: Proc. of the International Symposium on Computer Architecture (ISCA), 1992, pp. 278–287. 176, 179, 209

[20] P. Gratz, B. Grot, S. Keckler, Regional congestion awareness for load balance in networks-on-chip, in: Proc. of the International Symposium on High-Performance Computer Architecture (HPCA), 2008, pp. 203–214. 176, 179, 180, 188

[21] P. Gratz, C. Kim, K. Sankaralingam, H. Hanson, P. Shivakumar, S.W. Keckler, D. Burger, On-chip interconnection networks of the TRIPS chip, IEEE Micro 27(5) (2007) 41–50. 176, 177

[22] M. Hayenga, N. Enright Jerger, M. Lipasti, SCARAB: a single cycle adaptive routing and bufferless network, in: Proc. of the International Symposium on Microarchitecture (MICRO), 2009, pp. 244–254. 176

[23] J. Hu, R. Marculescu, DyAD—smart routing for networks-on-chip, in: Proc. of the Design Automation Conference (DAC), 2004, pp. 260–263. 176

[24] N. Jiang, D.U. Becker, G. Michelogiannakis, J. Balfour, B. Towles, D. Shaw, J. Kim, W. Dally, A detailed and flexible cycle-accurate network-on-chip simulator, in: Proc. of the International Symposium on Performance Analysis of Systems and Software (ISPASS), 2013, pp. 86–96. 190

[25] Y. Jin, K.H. Yum, E.J. Kim, Adaptive data compression for high-performance low-power on-chip networks, in: Proc. of the International Symposium on Microarchitecture (MICRO), 2008, pp. 354–363. 207

[26] P. Kermani, L. Kleinrock, Virtual cut-through: a new computer communication switching technique, Comput. Netw. 3(4) (1979) 267–286. 183

[27] J. Kim, D. Park, T. Theocharides, N. Vijaykrishnan, C.R. Das, A low latency router supporting adaptivity for on-chip interconnects, in: Proc. of the Design Automation Conference (DAC), 2005, pp. 559–564. 188

[28] M. Li, Q.-A. Zeng, W.-B. Jone, DyXY—a proximity congestion-aware deadlock-free dynamic routing method for network on chip, in: Proc. of the Design Automation Conference (DAC), 2006, pp. 849–852. 176

[29] X. Lin, P. McKinley, L. Ni, The message flow model for routing in wormhole-routed networks, IEEE Trans. Parallel Distrib. Syst. 6(7) (1995) 755–760. 176, 177, 179, 185, 209

[30] Z. Lu, M. Liu, A. Jantsch, Layered switching for networks on chip, in: Proc. of the Design Automation Conference (DAC), 2007, pp. 122–127. 209

[31] S. Ma, N. Enright Jerger, Z. Wang, DBAR: an efficient routing algorithm to support multiple concurrent applications in networks-on-chip, in: Proc. of the International Symposium on Computer Architecture (ISCA), 2011, pp. 413–424. 176, 179, 180, 188, 193, 198

[32] S. Ma, N. Enright Jerger, Z. Wang, Whole packet forwarding: Efficient design of fully adaptive routing algorithms for networks-on-chip, in: Proc. of the International Symposium on High-Performance Computer Architecture (HPCA), 2012, pp. 467–478. 175

[33] S. Ma, Z. Wang, N. Enright Jerger, L. Shen, N. Xiao, Novel flow control for fully adaptive routing in cache-coherent NoCs, IEEE Trans. Parallel Distrib. Syst. 25 (9) (2014) 2397–2407. 175

[34] P.S. Magnusson, M. Christensson, J. Eskilson, D. Forsgren, G. Hallberg, J. Hogberg, F. Larsson, A. Moestedt, B. Werner, Simics: a full system simulation platform, Computer 35 (2002) 50–58. 199

[35] R. Marculescu, U.Y. Ogras, L.-S. Peh, N. Enright Jerger, Y. Hoskote, Outstanding research problems in NoC design: system, microarchitecture, and circuit perspectives, IEEE Trans. Comput. Aided Des. Integr. Circuits Syst. 28 (2009) 3–21. 176, 179

[36] M.M.K. Martin, D.J. Sorin, B.M. Beckmann, M.R. Marty, M. Xu, A.R. Alameldeen, K.E. Moore, M.D. Hill, D.A. Wood, Multifacet's general execution-driven multiprocessor simulator (GEMS) toolset, ACM SIGARCH Comput. Archit. News 33 (2005) 92–99. 176, 177

[37] S.S. Mukherjee, P. Bannon, S. Lang, A. Spink, D. Webb, The alpha 21364 network architecture, in: Proc. of the the Symposium on High Performance Interconnects (HOTI), 2001, pp. 113–118. 176, 177, 179, 180

[38] R. Mullins, A. West, S. Moore, Low-latency virtual-channel routers for on-chip networks, in: Proc. of the International Symposium on Computer Architecture (ISCA), 2004, pp. 188–197. 180, 188, 189

[39] N. Neelakantam, C. Blundell, J. Devietti, M.M. Martin, C. Zilles, FeS2: a full-system execution-driven simulator for x86, in: Poster Presented at ASPLOS 2008, 2008. 199

[40] C.A. Nicopoulos, D. Park, J. Kim, N. Vijaykrishnan, M.S. Yousif, C.R. Das, ViChaR: a dynamic virtual channel regulator for network-on-chip routers, in: Proc. of the International Symposium on Microarchitecture (MICRO), 2006, pp. 333–346. 208

[41] U.Y. Ogras, J. Hu, R. Marculescu, Key research problems in NoC design: a holistic perspective, in: Proc. of the International Conference on Hardware/Software Codesign and System Synthesis (CODES+ISSS), 2005, pp. 69–74. 179

[42] L.-S. Peh, W. Dally, A delay model and speculative architecture for pipelined routers, in: Proc. of the International Symposium on High-Performance Computer Architecture (HPCA), 2001, pp. 255–266. 180, 188, 189

[43] A. Roca, J. Flieh, F. Silla, J. Duato, VCTlite: towards an efficient implementation of virtual cut-through switching in on-chip networks, in: Proc. of the International Conference on High-Performance Computing (HiPC), 2010, pp. 1–12. 183

[44] L. Schwiebert, D.N. Jayasimha, A necessary and sufficient condition for deadlock-free wormhole routing, J. Parallel Distrib. Comput. 32 (1996) 103–117. 176, 177, 179, 185, 209

[45] K. Shin, S. Daniel, Analysis and implementation of hybrid switching, in: Proc. of the International Symposium on Computer Architecture (ISCA), 1995, pp. 211–219. 209

[46] A. Singh, W.J. Dally, A.K. Gupta, B. Towles, GOAL: a load-balanced adaptive routing algorithm for torus networks, in: Proc. of the International Symposium on Computer Architecture (ISCA), 2003, pp. 194–205. 176

[47] C.B. Stunkel, D.G. Shea, B. Abali, M.G. Atkins, C.A. Bender, D.G. Grice, P. Hochschild, D. Joseph, B.J. Nathanson, R.A. Swetz, R.F. Stucke, M. Tsao, P.R. Varker, The SP2 high-performance switch, IBM Syst. J. 34 (1995) 185–204. 209

[48] S. Taktak, E. Encrenaz, J. Desbarbieux, A polynomial algorithm to prove deadlock-freeness of wormhole networks, in: Proc. of the Euromicro International Conference on Parallel, Distributed and Network-Based Processing (PDP), 2010, pp. 121–128. 176, 177, 179, 185, 209

[49] Y. Tamir, G. Frazier, High-performance multiqueue buffers for VLSI communication switches, in: Proc. of the International Symposium on Computer Architecture (ISCA), 1988, pp. 343–354. 208

[50] A. Vaidya, A. Sivasubramaniam, C. Das, Impact of virtual channels and adaptive routing on application performance, IEEE Trans. Parallel Distrib. Syst. 12(2) (2001) 223–237. 194, 197

[51] F. Verbeek, J. Schmaltz, A comment on "A necessary and sufficient condition for deadlock-free adaptive routing in wormhole networks", IEEE Trans. Parallel Distrib. Syst. 22(10) (2011) 1775–1776. 179

[52] F. Verbeek, J. Schmaltz, Automatic verification for deadlock in networks-on-chips with adaptive routing and wormhole switching, in: Proc. of the International Symposium on Networks-on-Chip (NOCS), 2011, pp. 25–32. 176, 177, 179, 185, 209

[53] F. Verbeek, J. Schmaltz, On necessary and sufficient conditions for deadlock-free routing in wormhole networks, IEEE Trans. Parallel Distrib. Syst. 22(12) (2011) 2022–2032. 176, 177, 179, 185, 209

[54] D. Wentzlaff, P. Griffin, H. Hoffmann, L. Bao, B. Edwards, C. Ramey, M. Mattina, C.-C. Miao, J.F. Brown III, A. Agarwa, On-chip interconnection architecture of the TILE processor, IEEE Micro 27(5) (2007) 15–31. 176, 177

[55] Y. Xu, B. Zhao, Y. Zhang, J. Yang, Simple virtual channel allocation for high throughput and high frequency on-chip routers, in: Proc. of the International Symposium on High-Performance Computer Architecture (HPCA), 2010, pp. 1–11. 176, 179, 180, 190, 208

Deadlock-free flow control for torus networks-on-chip†

7

CHAPTER OUTLINE

---

†Part of this research was first published in *IEEE Transactions on Computers* [39].

Networks-on-Chip. http://dx.doi.org/10.1016/B978-0-12-800979-6.00007-X

7.1 INTRODUCTION

Optimizing networks-on-chip (NoCs) [18] on the basis of coherence traffic is necessary to improve the efficiency of many-core coherence protocols [41]. The torus is a good NoC topology candidate [53, 54]. The wraparound links convert plentiful on-chip wires into bandwidth [18], and reduce hop counts and latencies [53]. Its node symmetry helps to balance network utilization [53, 54]. Several products [21, 28, 31] use a ring or 1D torus NoC. Also, the 2D or higher-dimensional torus is widely used in off-chip networks [4, 20, 44, 51].

Despite the many desirable properties of a torus, additional effort is needed to handle deadlock owing to cyclic dependencies introduced by wraparound links. A deadlock avoidance scheme should support high performance with low overhead. Requiring minimum virtual channels (VCs) [16] is preferable, because more VCs not only increase the router complexity, but also limit the frequency. Buffers are a precious resource [24, 46]; an efficient design should maximize performance with limited buffers. There is a gap between existing proposals and these requirements.

A conventional design [19] uses two VCs to remove cyclic dependencies; this introduces large allocators and negatively affects the router frequency. Optimizations [10, 13] for virtual cut-through (VCT) networks [32] avoid deadlock by preventing the use of the last free packet-size buffer inside rings; only one VC is needed. However, with variable-size packets, each packet must be regarded as a maximum-length packet [4]. This restriction prevents deadlock, but results in poor buffer utilization and performance, especially for short-packet-dominating coherence traffic.

In addition to the majority short packets, cache-coherent NoCs also deliver long packets. Even though multiple virtual networks (VNs) [19] may be configured to avoid protocol-level deadlock, these two types of packets still coexist in a single VN. For example, short read requests and long write-back requests are sent in VN0 of AlphaServer GS320, while long read responses and short write-back acknowledgments (ACKs) are sent in VN1; both VNs carry variable-size packets [25]. Similarly, all VNs in DASH [36], Origin 2000 [35], and Piranha [6] deliver variable-size packets.

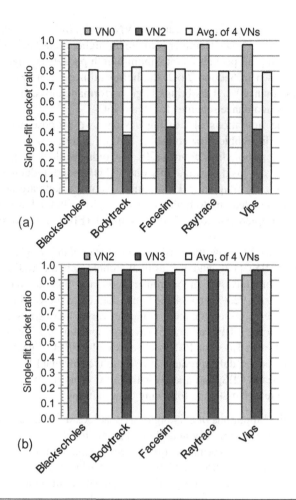

FIGURE 7.1

Single-flit packet ratios. In the MOESI directory, VN0 has a read request (one flit), clean write-back (one flit), and dirty write-back (five flits), and VN2 has an ACK (one flit) and a read response (five flits). In AMD's Hammer, VN2 has an ACK (one flit) and a read response (five flits), and VN3 has unblock (one flit), clean write-back (one flit), and dirty write-back (five flits). (a) MOESI directory. (b) AMD's Hammer.

With a typical 128-bit NoC flit width [24, 38, 46], the majority control packets have one flit; the remaining data packets contain a 64-byte cache line and have five flits. Figure 7.1 shows the packet length distribution of some representative PARSEC workloads [8] with two coherence protocols.[1] Both protocols use four VNs. For each protocol, two VNs carry variable-size packets, and the other two have only short

---

[1] See Section 7.5 for the experimental configuration and description.

packets. The single-flit packet (SFP) ratios of VN0 in the MOESI directory [45], and VN2 and VN3 in the AMD's Hammer [15] are all higher than 90%. With such high SFP ratios, regarding all packets as maximum-length packets strongly limits buffer utilization. As shown in Section 7.6.2, the buffer utilization of existing designs at saturation is less than 40%. This brings large performance loss. It is imperative to improve deadlock avoidance schemes for torus NoCs.

To address the limitations of existing designs, we propose a novel deadlock avoidance theory, flit bubble flow control (FBFC), for torus cache-coherent NoCs. FBFC leverages wormhole flow control [20]. It avoids deadlock by maintaining one free *flit-size* buffer slot inside a ring. Only one VC is needed, reducing the allocator size and improving the frequency. Furthermore, short packets are not regarded as long packets in FBFC, leading to high buffer utilization. On the basis of this theory, we provide two implementations: FBFC-localized (FBFC-L) and FBFC-critical (FBFC-C).

Experimental results show that FBFC outperforms dateline [19], the localized bubble scheme (LBS) [10], and the critical bubble scheme (CBS) [13]. FBFC achieves an approximately 30% higher router frequency than dateline. For synthetic traffic, FBFC performs 92.8% and 34.2% better than LBS and CBS in a 4×4 torus. FBFC's advantage is more significant in larger networks; these gains are 107.2% and 40.1% in an 8×8 torus. FBFC achieves an average speedup of 13.0% and a maximal speedup of 22.7% over LBS for PARSEC workloads. FBFC's gains increase with fewer buffers. The power-delay product (PDP) results show that FBFC is more power efficient than LBS and CBS, and a torus with FBFC is more power efficient than a mesh. In this chapter we make the following contributions:

- We analyze the limitations of existing torus deadlock avoidance schemes, and show that they perform poorly in cache-coherent NoCs.
- We demonstrate that in wormhole torus networks, maintaining one free *flit-size* buffer slot can avoid deadlock, and propose the FBFC theory.
- We present two implementations of FBFC; both show substantial performance and power efficiency gains over previous proposals.

7.2 LIMITATIONS OF EXISTING DESIGNS

Here, we analyze existing designs. Avoiding deadlock inside a ring combined with dimensional order routing (DOR) is the general way to avoid deadlock in tori. We use the ring for discussion.

7.2.1 DATELINE

As shown in Figure 7.2, dateline [19] avoids deadlock by leveraging two VCs: VC_{0i} and VC_{1i}. It forces packets to use VC_{1i} after crossing the dateline to form acyclic channel dependency graphs [17, 19]. Dateline can be used in both packet-based VCT

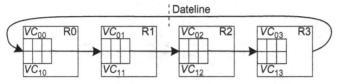

FIGURE 7.2

Dateline uses two VCs.

and flit-based wormhole networks. It uses two VCs, which results in larger allocators and lower router frequency.

7.2.2 LOCALIZED BUBBLE SCHEME

Bubble flow control [10, 50] is a deadlock avoidance theory for VCT torus networks. It forbids the use of the last free packet-size amount of buffers (a packet-size bubble); only one VC is needed. Theoretically, any free packet-size bubble in a ring can avoid deadlock [10, 50]. However, owing to difficulties in gathering global information and coordinating resource allocation for all nodes, previous designs applied a localized scheme; a packet may be injected only when the receiving VC has two free packet-size bubbles [10, 50]. Figure 7.3 gives an example. Here, three packets, P_0, P_1, and P_2, are waiting. Theoretically, they can all be injected. Yet, with a localized scheme, only P_0 can be injected since only VC_1 has two free packet-size bubbles. LBS requires each VC to be deep enough for two maximum-length packets.

7.2.3 CRITICAL BUBBLE SCHEME

CBS [13] marks at least one packet-size bubble in a ring as critical. A packet can be injected only if its injection will not occupy a critical bubble. Control signals between routers track the movement of critical bubbles. CBS reduces the minimum buffer requirement to one packet-size bubble. In the example in Figure 7.4, the bubble at VC_2 is marked as critical; P_2 can be injected. P_1 cannot be injected since its injection would result in it occupying the critical bubble. Requiring that critical bubbles can be occupied only by packets in a ring guarantees that there is at least one free bubble

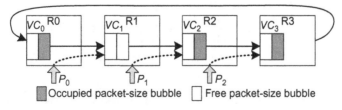

FIGURE 7.3

LBS uses one VC with two packet-size bubbles.

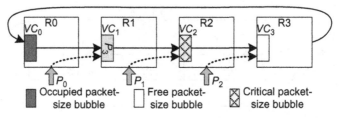

FIGURE 7.4

CBS uses one VC with one packet-size bubble.

to avoid deadlock. When P_3 advances into VC_2, the critical bubble moves into VC_1. Now, VC_1 maintains one free bubble.

7.2.4 INEFFICIENCY WITH VARIABLE-SIZE PACKETS

LBS and CBS are proposed for VCT networks; they are efficient for constant-size packets. Yet, as observed by the Blue Gene/L team, LBS deadlocks with variable-size packets owing to bubble fragmentation [4]. Figure 7.5 shows an example with SFPs and two-flit packets. A free full-size (two-slot) bubble exists in VC_2 at cycle 0. When P_0 moves into VC_2, the bubble is fragmented across VC_1 and VC_2. VCT reallocates a VC only if it has enough space for an entire packet. Since VC_2's free buffer size is less than P_1's length, P_1 cannot advance, and deadlock results. CBS has a similar problem. To handle this issue, Blue Gene/L regards each packet as a maximum-length packet [4]. Now, there is no bubble fragmentation. However, this reduces buffer utilization, especially in coherence traffic, for which short packets are in the majority.

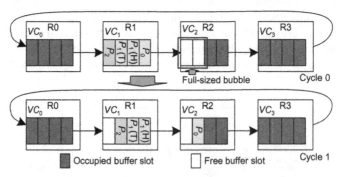

FIGURE 7.5

Deadlock with variable-size packets. $P_i(H)$ and $P_i(T)$ are head and tail flits of P_i. Starting with this figure, each box represents one buffer slot, while it represents a packet-size amount of buffers in Figures 7.3 and 7.4.

7.3 FLIT BUBBLE FLOW CONTROL

We first propose the FBFC theory. Then, we give two implementations. Finally, we discuss starvation.

7.3.1 THEORETICAL DESCRIPTION

We notice that maintaining one free *flit-size* buffer slot can avoid deadlock in wormhole networks. This insight leverages a property of wormhole flow control: advancing a packet with wormhole flow control does not require the downstream VC to have enough space for the entire packet [20]. To show this, in Figure 7.6, a free buffer slot exists in VC_2 at cycle 0; P_0 advances at cycle 1. A free slot is created in VC_1 owing to P_0's movement. Similarly, P_3's head flit moves to VC_1 at cycle 2, creating a free slot in VC_0. This free buffer slot cycles inside the ring, allowing all flits to move.

The packet movement in a ring does not reduce free buffer amounts since forwarding one flit leaves its previously occupied slot free; only injection reduces free buffer amounts. The theory is declared as follows.

Theorem 7.1. *If packet injection maintains one free buffer slot inside a ring, there is no deadlock with wormhole flow control.*

Informal description: A deadlock configuration in wormhole networks involves a set of cyclically dependent flits where no flit can move [17]. In a ring, a cyclic dependency needs the participation of all VCs. Thus, we need only prove that a flit in any VC can advance.

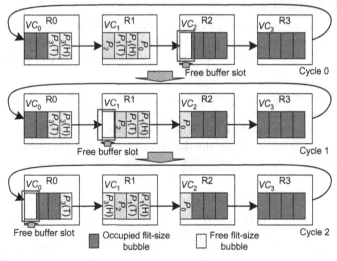

FIGURE 7.6

A wormhole routing example.

Proof: Assume there is only one free buffer slot at VC_{i+1} and all other VCs are full. We label VC_{i+1}'s upstream VC in the ring as VC_i. There are two possible situations for the flit f at VC_i's head:

(1) f is a head flit. If f arrives at the destination, it can be ejected. If f needs to advance into VC_{i+1}, we consider the packet P_k which most recently utilized VC_{i+1}. Again, there are two possible situations:

    (a) P_k was forwarded from VC_i into VC_{i+1}. Since now the head flit f of another packet is at the head of VC_i, P_k's tail flit has already advanced into VC_{i+1}. f can advance with wormhole flow control.

    (b) P_k was injected into VC_{i+1}. Its tail flit must have already advanced into VC_{i+1}. Otherwise, the tail flit will occupy the free buffer slot, which violates the premise that the injection procedure maintains one free buffer slot. f can advance.

(2) f is a body or tail flit. It can be ejected or forwarded. In all cases, a flit can move. There is no deadlock. ∎

Since one free buffer slot (flit bubble)[2] avoids deadlock, we call this theory flit bubble flow control (FBFC). DOR removes the cyclic dependency across dimensions; combining DOR with FBFC avoids deadlock in tori. FBFC has no bubble fragmentation; its bubble is flit-size. Thus, FBFC does not regard each packet as a maximum-length packet. Only one VC is needed; this improves the frequency. FBFC uses wormhole flow control to move packets inside a ring. It requires the injection procedure to leave one slot empty. Later, we show two schemes to satisfy this requirement.

7.3.2 FBFC-LOCALIZED

The key point in implementing FBFC is to maintain a free buffer slot inside each ring. We first give a localized scheme: FBFC-L. When combined with DOR, a dimension-changing packet is treated the same as an injecting packet. The rules of FBFC-L are as follows:

(1) Forwarding of a packet within a dimension is allowed if the receiving VC has one free buffer slot. This is the same as wormhole flow control.

(2) Injecting a packet (or changing its dimension) is allowed only if the receiving VC has one more free buffer slot than the packet length. This requirement ensures that after injection, one free buffer slot is left in the receiving VC to avoid deadlock.

Figure 7.7 shows an example. Three packets are waiting. There are two and four free slots in VC_2 and VC_3 respectively; these are one more than the lengths of P_3 and P_4 respectively. P_3 and P_4 can be injected. After injection, at least two free slots are left in the ring. P_2 cannot be injected since VC_1 has only one free slot, which is equal

---

[2]We use "flit bubble" and "buffer slot" interchangeably.

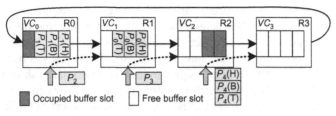

FIGURE 7.7

A FBFC-L example. $P_i(H)$, $P_i(B)$, and $P_i(T)$, the head, body, and tail flits of P_i. P_2 and P_3 have one flit.

to P_2's length. However, according to wormhole flow control, the free slot in VC_1 allows P_1's head flit to advance. In FBFC-L, each VC must have one more buffer slot than the size of the longest packet.

7.3.3 FBFC-CRITICAL

To reduce the minimum buffer requirement, we propose a critical design: FBFC-C. FBFC-C marks at least one free buffer slot as a critical slot, and restricts this slot to be occupied only by packets traveling inside the ring. The rules of FBFC-C are as follows:

(1) Forwarding of a packet within a dimension is allowed if the receiving VC has one free buffer slot, no matter if it is a *normal* or a *critical* slot.
(2) Injecting a packet is allowed only if the receiving VC has enough free *normal* buffer slots for the entire packet. After injection, the critical slot must not be occupied. This requirement maintains one free buffer slot.

Figure 7.8 shows an example. At cycle 0, one critical buffer slot is in VC_1. VC_2 and VC_3 have enough free normal slots to hold P_3 and P_4 respectively; P_3 and P_4

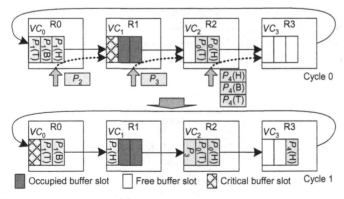

FIGURE 7.8

An FBFC-C example. P_2 and P_3 have one flit.

can be injected. They do not occupy the critical slot, indicating the existence of a free slot (the critical slot) elsewhere in the ring. Since the only free slot in VC_1 is a critical one, P_2 cannot be injected. Yet, this critical slot allows P_1's head flit to move. At cycle 1, P_1's head flit advances into VC_1, moving the critical slot backward into VC_0. This is done by R0 asserting a signal to indicate to R3 that it should mark the newly freed slot in VC_0 as a critical one. More details are provided in Section 7.4. The minimum buffer requirement of FBFC-C is the same as that of CBS; a VC can hold the largest packet. This is one slot less than FBFC-L.

The injection in FBFC-L and FBFC-C is similar to that in VCT; they require enough buffers for packets before injection. After injection, a minimum of one slot is left free for wormhole flow control. They can be regarded as applying VCT for injection (or dimension-changing) in wormhole networks. These hybrid schemes are straightforward ways to address the limitations of existing designs.

7.3.4 STARVATION

FBFC-L and FBFC-C must deal with starvation. The starvation in FBFC-L is intrinsically the same as in LBS [10]: Injected packets need more buffers than inside-ring traveling packets. Figure 7.9 shows a starvation example for FBFC-L. Here, if node R0 continually injects packets, such as P_0, destined for R3, then P_1 cannot be injected. We design a starvation prevention mechanism; if a node detects starvation, it will notify all other nodes in a ring to stop injecting packets. A sideband network conveys the control signal (*starve*). Once blocked cycles of P_1 exceed a threshold value, R1 asserts the "*starve*" signal. R0 stops injecting packets after receiving "*starve*" and forwards it to R3. All nodes except R1 stop injecting packets. Finally, P_1 can be injected. Then, R1 deasserts "*starve*" to resume injection by other nodes. To handle the corner case of multiple nodes simultaneously detecting starvation, "*starve*" carries an "*ID*" field to differentiate the nodes of a ring. Since the sideband network is unblocking, a router can identify the sending time slot of "*starve*" on the basis of the "*ID*" field. The "*ID*" field and the sending time slot order "*starve*" signals. If the incoming "*starve*" has a higher order than the currently serving "*starve*," the router forwards the incoming signal to its neighbor.

FBFC-C has another starvation scenario in addition to the previous one; it is due to the critical bubble stall. CBS has a similar issue (L. Chen and T. Pinkston, personal

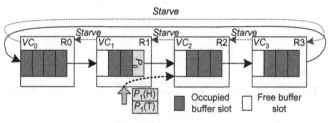

FIGURE 7.9

A starvation example for FBFC-L.

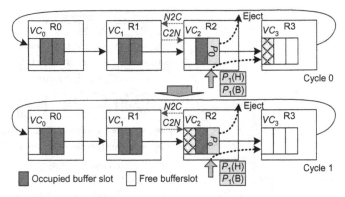

FIGURE 7.10

A starvation example for FBFC-C.

communication, 2012 [11]). Figure 7.10 shows that the critical bubble is in VC_3 at cycle 0. The bubble movement depends on the packet advancement. If all packets in VC_2, such as P_0, are destined for R2, they will be ejected. Since no packet moves to VC_3, the critical bubble stalls at VC_3. P_1 cannot be injected. This can be prevented by proactively transferring the critical bubble backward if the upstream VC has a free normal bubble. As shown in Figure 7.10, a pair of "*N2C*" (NormalToCritical) and "*C2N*" (CriticalToNormal) signals are used. If R2 detects that the critical bubble stall prohibits P_1's injection, it asserts "*N2C*" to R1. If VC_2 has a free normal bubble, R1 will change it into a critical one in cycle 1. "*C2N*" notifies R2 that the critical bubble in VC_3 can now be changed into a normal one. P_1 can be injected. Note that the bubble status is maintained at upstream routers.

7.4 ROUTER MICROARCHITECTURE

In this section, we discuss wormhole routers for FBFC. We also discuss VCT routers for LBS and CBS.

7.4.1 FBFC ROUTERS

The left side of Figure 7.11 shows a canonical wormhole router, which is composed of the input units, routing computation (RC) logic, VC allocator (VA), switch allocator (SA), crossbar, and output units [19, 23]. Its pipeline includes RC, VA, SA, and switch traversal [19, 23]. The output unit tracks downstream VC status. The "*input_vc*" register records the allocated input VC of a downstream VC. The 1-bit "*idle*" register indicates whether the downstream VC receives the tail flit of the last packet. "*Credits*" records credit amounts. Lookahead routing [19] performs RC in parallel with VC allocation. To be fair with VCT routers, wormhole routers try to hold SA grants for entire packets; they prioritize VCs that received switch access previously [34].

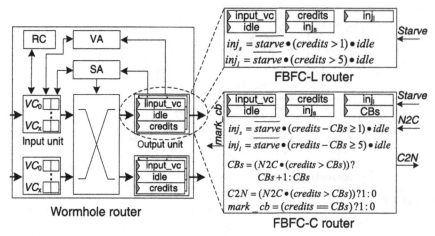

FIGURE 7.11

FBFC routers, assuming short packets have one flit, while long packets have five flits.

FBFC mainly modifies output units. As shown at the top right in Figure 7.11, two 1-bit registers, "inj_s" and "inj_l", are needed for the bimodal length coherence traffic. They record whether a downstream VC is available for injecting (or dimension-changing) short and long packets. When packets are to be injected (or change dimensions), VA checks the appropriate register according to packet lengths. SFPs require at least two credits in a downstream VC. A five-flit packet needs at least six credits. If the incoming "*starve*" signal is asserted to prevent starvation for some other router, both registers are reset to forbid injection. Figure 7.11 shows the logic. This logic can be precalculated and is off the critical path.

The output unit of an FBFC-C router is shown at the bottom right in Figure 7.11. Another register, "*CBs*", records critical flit bubble counts. The logic of "inj_s" and "inj_l" is modified; packet injection is allowed only if the downstream VC has enough free normal slots. Specifically, "$credits - CBs$" is not less than the packet length. FBFC-C routers proactively transfer critical slots to prevent starvation. When there is an incoming "*N2C*" signal, the output unit checks whether there are free normal slots. If there are, "*CBs*" is increased by 1, and "*C2N*" is asserted to inform the neighboring router to change its critical slot into a normal one. The "*mark_cb*" signal is asserted when a flit will occupy a downstream critical slot; it informs the upstream router to mark the newly freed slot as critical. Similarly, this logic is off the critical path.

7.4.2 VCT ROUTERS

We discuss VCT routers for LBS and CBS. A typical VCT router [19, 22] is similar to the wormhole one shown in Figure 7.11. The main difference is VC allocation: VCT reallocates a VC only if it guarantees enough space for an entire packet. The advance of a packet returns one credit, which represents the release of a packet-size amount of buffers. We apply some optimizations to favor LBS and CBS. The SA grant holds for an entire packet. Since VA guarantees enough space for an entire packet, once a head flit moves out, that packet's remaining flits can advance without interruption; all of the packet's occupied buffers will be freed in limited time. Thus, a credit is returned once a head flit moves out. The lookahead credit return allows the next packet to use this VC even if there is only one free slot, overlapping the transmission of an incoming packet and an outgoing packet. This optimization brings an injection benefit for CBS, which we discuss in Section 7.7.2. The LBS router's output unit is similar to that of the FBFC-L router. The difference is that the LBS router needs only one "*inj*" register since all packets are regarded as long packets. The "*credits*" register records free buffer slots in the unit of packets instead of flits. The CBS router's output unit also has these differences. Since CBS starves only because of the critical bubble stall, there is no incoming "*starve*" signal.

7.5 METHODOLOGY

We modify BookSim [29] to implement FBFC-L and FBFC-C and to compare them with dateline, LBS, and CBS. We use both synthetic traffic and real applications. Synthetic traffic uses one VN since each VN is independent. The traffic has randomly injected SFPs and five-flit packets. The baseline SFP ratio is 80%, which is similar to the overall SFP ratio of a MOSEI directory protocol. The warm-up and measurement periods are 10,000 and 100,000 cycles.

Although FBFC works for high-dimensional tori, we focus on 1D and 2D tori as they best match the physical layouts. The routing is DOR; with equal positive and negative distances, the positive direction is used. Buffers are precious in NoCs; most evaluation uses 10 slots at each port per VN. Bubble designs have one VC per VN. Dateline divides 10 slots into 2 VCs; 5 slots per VC cover credit round-trip delays [19]. Instead of injecting packets into VC_{0i} first (Figure 7.2), then switching to VC_{1i} after the dateline [19], we apply a balancing optimization to favor dateline; injected packets choose VCs according to whether they will cross dateline [51]. Packets use VC_{1i}s if they will cross dateline. Otherwise, they use VC_{0i}s. CBS and FBFC-C set one critical bubble for each ring; CBS marks five slots as a packet-size critical bubble, and FBFC-C marks one slot as a flit-size critical bubble. The starvation threshold values (STVs) in FBFC-L and LBS are 30 cycles. The STVs due to critical bubble stall in CBS and FBFC-C are three cycles. We explore this issue in Section 7.7.5.

VA and SA delays determine router frequencies [7, 48]. Dateline uses two VCs per VN, resulting in large allocators and long critical paths. A technology-independent model [48] is used to calculate the delays, as shown in Table 7.1. Separable input-first allocators [19] with matrix arbiters [19] are used. VC allocation is independent for each VN [7], making switch allocation the critical path with multiple VNs. Dateline's SA delay with four VNs is approximately 30% higher than that of bubble

Table 7.1 The Delay Results, Expressed as Fan-out of 4

No. of VN	Ring (3 ports)			Torus (5 ports)		
	Bubble	Dateline	Increase (%)	Bubble	Dateline	Increase (%)
1 VA	8.4	12.2	45	10.0	13.8	38
1 SA	6.9	11.7	69	8.5	13.3	57
2 SA	11.7	16.5	41	13.3	18.1	36
3 SA	14.5	19.3	33	16.1	20.9	30
4 SA	16.5	21.3	29	18.1	22.9	27

One VC per VN for bubble, and two VCs per VN for dateline.

designs. Similarly, with a 45-nm low-power technology, the SA delay with eight VCs is about 15-26% higher than the delay with four VCs for several allocator types [7].

To measure full-system performance, we leverage FeS2 [45] for x86 simulation and BookSim for NoC simulation. FeS2 is a module for Simics [40]. We run PARSEC workloads [8] with 16 threads on a 16-core chip multiprocessor (CMP). Since dateline's frequency can be different with bubble designs, we do not evaluate dateline for real workloads. The frequency of a simple CMP core can be 2-4 GHz, while the frequency of NoC is limited by allocator speeds [27, 42]. We assume cores run two times faster than the NoC. Each core is connected to private, inclusive first-level (L1) and second-level (L2) caches. Cache lines are 64 bytes; long packets are five flits with a 16-byte flit width. We use a MOESI directory protocol to maintain the coherence among L2 caches; it uses four VNs to avoid protocol-level deadlock. Each VN has 10 slots. Workloads use *simsmall* input sets. The task run time is the performance metric. Table 7.2 gives the configuration.

7.6 EVALUATION ON 1D TORI (RINGS)

7.6.1 PERFORMANCE

Our evaluation for synthetic patterns [19] starts with an eight-node ring. As shown in Figure 7.12, FBFC-L is similar to FBFC-C; although FBFC-C needs one less slot for injection, this benefit is minor since 10 slots are used. FBFC obviously outperforms

Table 7.2 The Full System Simulation Configuration

Parameter	Value
Topology	16 cores, 4 × 4 torus
L1 cache (D and I)	Private, 4-way, 32 kB each
L2 cache	Private, 8-way, 512 kB each

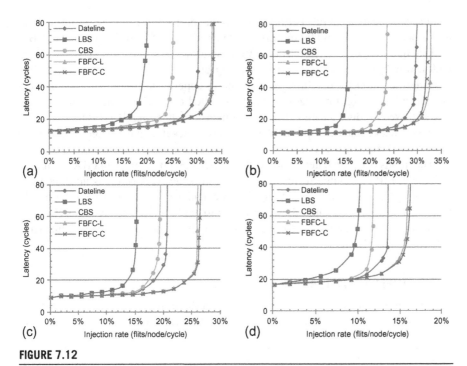

FIGURE 7.12

The performance for an eight-node ring. (a) Uniform random. (b) Bit rotation. (c) Transpose. (d) Tornado.

LBS and CBS. Across all patterns, the average saturation throughput[3] gains of FBFC-C over LBS and CBS are 73.5% and 33.9%. LBS and CBS are limited by regarding short packets as long ones. The advance of a short packet in FBFC-C uses one slot, while it uses five slots in LBS and CBS. LBS is also limited by its high injection buffer requirement; CBS shows an average gain of 29.6% over LBS.

In Figure 7.12, dateline is superior to LBS and CBS. The results are reported in cycles and flits per node per cycle. These metrics do not consider router frequencies. According to Table 7.1, if all routers are optimized to maximum frequencies, dateline is approximately 30% slower than bubble designs. To make a fair comparison, we leverage frequency-independent metrics. The unit seconds is used for latency comparison. Owing to its lower frequency, dateline's cycle in seconds is 30% longer than the bubble design's cycle in seconds. Thus, dateline's zero-load latency in seconds is 30% higher. Flits per node per second is used for throughput comparison. Since dateline's cycle in seconds is longer than the bubble design's cycle in seconds,

---

[3]The saturation point is measured as the injection rate at which the average latency is three times the zero-load latency.

dateline's throughput in flits per node per second drops. For example, its throughput in flits per node per second for uniform random traffic is 7.5% lower than that of CBS.

Dateline divides buffers into two VCs. Shallow VCs make packets span more nodes, which increases the chained blocking effect [56]; this brings a packet forwarding limitation for dateline. Yet, dateline is superior to FBFC for injection and dimension-changing: long packets can be injected or change dimensions with one free slot. We introduce a metric: injection/dimension-changing (IDC) count. The IDC count of a packet includes the number of times a packet is injected plus the number of times it changes dimensions.

The trends between dateline and FBFC depend on hop counts and IDC counts of the traffic. FBFC-C outperforms dateline for all patterns. A ring has no dimension changing; all patterns' IDC counts are no more than one, hiding dateline's merit. The largest gains are 29.2% and 18.8% for transpose and tornado. Transpose's IDC and hop counts are 0.75 and 2.25. Tornado's IDC and hop counts are one and four. They reveal dateline's limitation on packet forwarding.

7.6.2 BUFFER UTILIZATION

To delve into performance trends, the average buffer utilization of all VCs is shown in Figure 7.13. The maximum and minimum rates are given by error bars. Buffers

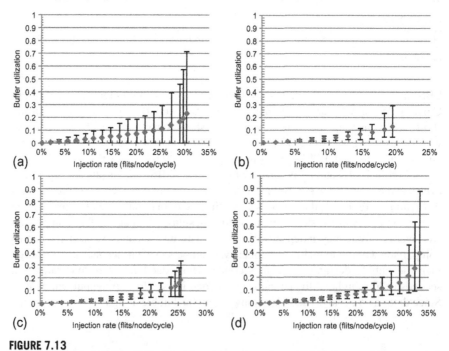

FIGURE 7.13

The buffer utilization with uniform random traffic until network saturation. (a) Dateline. (b) LBS. (c) CBS. (d) FBFC-C.

can support high throughput; higher utilization generally means better performance. For LBS and CBS, the average rates are 13.0% and 19.2% at saturation. They inefficiently use buffers. LBS requires more free buffers for injection; its utilization is lower than that of CBS. Dateline's minimum rate is always zero; one VC is never used. For example, VC_{00} in Figure 7.2 is not used. This scenario combined with chained blocking limits its buffer utilization. Dateline's average and maximum rates at saturation are 23.2% and 71.8%, while these rates are 39.5% and 89.8% for FBFC-C.

7.6.3 LATENCY OF SHORT AND LONG PACKETS

FBFC's injection of long packets requires more buffers than short ones. Figure 7.14 shows latency compositions. "InjVC" and "NI" are delays in injection VCs and network interfaces. "Network" is all other delays. Long and short packets are treated the same in LBS and CBS; they show similar delays at injection VCs and inside the ring. Long packets have four more flits; they spend approximately four more cycles in network interfaces. With low-to-medium injection rates (20% or lower) in FBFC-C, the delays at injection VCs for long and short packets are similar. With higher loads, the difference increases. Even when saturated, they are only 3.4 and 3.2 cycles for the two patterns. FBFC-C does not sacrifice long packets. FBFC-C's acceleration of short packets helps long packets since short and long packets are randomly injected. Indeed, compared with FBFC-C, LBS and CBS sacrifice short packets. For example, with an injection rate of 20% for uniform random traffic, short packets spend 11.7 and 5.4 cycles in injection VCs for LBS and CBS, and 4.3 cycles for FBFC-C.

7.7 EVALUATION ON 2D TORI

Section 7.6 analyzed the performance for 1D tori with buffer utilization and latency composition. This section thoroughly analyzes the performance for 2D tori with several configurations for further insights.

7.7.1 PERFORMANCE FOR A 4 × 4 TORUS

Figure 7.15 shows the performance for a 4 × 4 torus with the baseline configuration. The error bars in Figure 7.15a are average latencies of long and short packets. The average gains of FBFC-C over LBS and CBS are 92.8% and 34.2%. FBFC's high buffer utilization brings these gains. CBS shows an average gain of 45.7% over LBS, and the highest gain is 100% for transpose traffic. Most transpose traffic is between the same row and column; many packets change dimensions at the same router requiring the same port. CBS's low dimension-changing buffer requirement yields high performance.

Compared with the ring (Figure 7.12), the trends between FBFC-C and dateline for a 2D torus change. Dateline performs similarly to FBFC-C for uniform random and transpose traffic. A 2D torus has a dimension-changing step. The IDC counts of

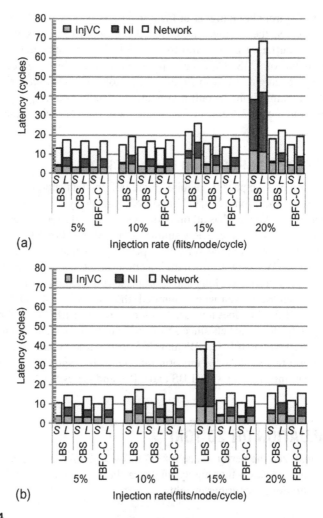

FIGURE 7.14

Latencies of short (*S*) and long (*L*) packets. (a) Uniform random. (b) Bit rotation.

these patterns are both 1.5, and their hop counts are three. As a result, injection and dimension-changing factor significantly into performance. Dateline's low injection and dimension-changing buffer requirement brings gains. Dateline outperforms FBFC-C by 5.7% for hotspot traffic. This pattern sends packets from different rows to the same column of four hotspot nodes, exacerbating FBFC-C's injection limitation. FBFC-C outperforms dateline by 6.4% for bit rotation. Packets of bit rotation change dimensions by requiring different ports; the light congestion mitigates FBFC-C's limitation. These results do not consider frequencies. If routers are optimized to maximum frequencies, dateline's performance will drop.

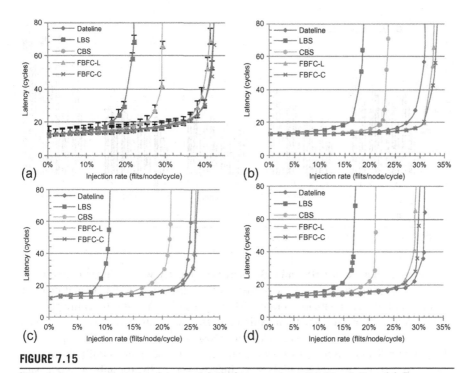

FIGURE 7.15

The performance for a 4 × 4 torus. (a) Uniform random. (b) Bit rotation. (c) Transpose. (d) Hotspot.

As shown in Figure 7.15a, the delay difference between long and short packets is almost constant for LBS and CBS; long packets have approximately four cycles more delay. Dateline's difference is 6.2 cycles at saturation. FBFC-L's difference is 9.8 cycles. This is approximately three cycles more than for the ring; a 2D torus has one dimension-changing step. We measure the behavior after saturation by increasing the load to 1.0 flits per node per cycle. All designs maintain performance after saturation. DOR smoothly delivers injected packets. Adaptive routing may decrease performance after saturation because escape paths drain packets at lower rates than injection rates [13].

7.7.2 SENSITIVITY TO SFP RATIOS

As discussed in Section 7.4.2, VCT routers' lookahead credit return overlaps incoming and outgoing packets; a packet can move to a VC with one free slot. This brings an injection or dimension-changing benefit for CBS over FBFC. CBS's packet injection begins with one free slot. FBFC-C's long packet injection needs five normal slots. Thus, SFP ratios affect trends between CBS and FBFC. Figure 7.16 shows the results of sensitivity studies on SFP ratios.

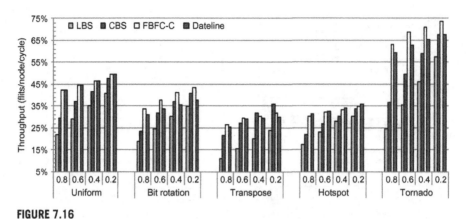

FIGURE 7.16

The performance of several SFP ratios.

The performance of LBS and CBS increases linearly with reduced ratios; more long packets proportionally improve their buffer utilization. FBFC-C and dateline perform slightly better with lower ratios. They try to hold SA grants for entire packets; long packets reduce SA contention. Although FBFC-C's gains over CBS reduce with lower ratios, FBFC-C is always superior for most traffic patterns. FBFC-C performs 10.1% better than CBS for tornado with a 0.2 ratio. Tornado sends traffic from node (i,j) to node $((i + 1)\%4, (j + 1)\%4)$. Each link delivers traffic for only one node pair; there is no congestion. Transpose is different; CBS outperforms FBFC-C with ratios of 0.4 and 0.2. This pattern congests turn ports, emphasizing CBS's injection benefit.

We also experiment with a 64-bit flit width; the short and long packets have one and nine flits. With longer packets, the negative effect of regarding short packets as long ones in LBS and CBS becomes more significant. Meanwhile, FBFC-C needs more slots for long-packet injection. These factors result in trends for a 64-bit configuration similar to those in Figure 7.16.

7.7.3 SENSITIVITY TO BUFFER SIZE

We perform sensitivity studies on buffer sizes. In Figure 7.17a, CBS, FBFC-C, and FBFC-L use minimum buffers that ensure correctness. They use five, five, and six slots. Dateline uses two three-slot VCs. In Figure 7.17b, bubble designs use a 15-slot VC, and dateline uses two eight-slot VCs. Although FBFC-C's injection limitation has higher impact with fewer buffers, dateline performs only 7.6% better than FBFC-C with five slots per VC. Dateline's shallow VCs (three slots) cannot cover the credit round-trip delay (five cycles), making link-level flow control the bottleneck [19]. FBFC-C's gain over CBS increases with fewer buffers. The gains are 26.6% with 15 slots per VC, 41.4% with 10 slots per VC, and 121.8% with five slots per VC. From a comparison of Figures 7.15a and 7.17a, CBS with 10 slots per VC is similar to FBFC-C with 5 slots per VC. With half as many buffers, FBFC-C

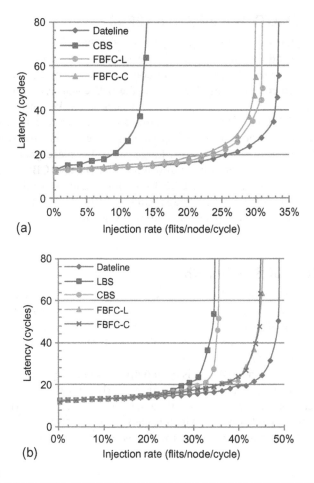

FIGURE 7.17

The performance of uniform random traffic with other buffer sizes. FBFC-L and dateline use six slots per VN in (a) and dateline uses 16 slots per VN in (b).

is comparable to CBS. LBS with 15 slots per VC (Figure 7.17b) has a gain of only 5.2% over FBFC-C with five slots per VC.

LBS almost matches CBS with 15 slots per VC. More buffers mitigate LBS's high injection buffer limitation. Additional results show that with abundant buffers, bubble designs perform similarly; there is little difference among them as many free buffers are available anyway. The convergence points depend on the traffic. For example, owing to congested ports in transpose, at least 30 slots per VC are needed for LBS to match CBS. For uniform random, 50 slots per VC are needed for CBS to match FBFC. Many buffers are required for convergence, which makes high buffer utilization designs, such as FBFC, winners in reasonable configurations.

7.7.4 SCALABILITY FOR AN 8 × 8 TORUS

Figure 7.18 explores scalability to an 8 × 8 torus. Two buffer sizes are evaluated: one assigns 10 slots and the other assigns five or six slots (the same as Figure 7.17a). As the network scales, the traffic's hop count increases, exacerbating dateline's limitation for packet forwarding. Meanwhile, the IDC counts are similar to those of a 4 × 4 torus. Thus, FBFC-C's injection bottleneck is masked in larger networks. With 10 slots, FBFC-C outperforms dateline for all patterns. The largest gain is 26.5%, for tornado, whose average hop count is seven. With five slots, FBFC-C outperforms dateline for four patterns. Larger networks place greater pressure on buffers, worsening inefficient buffer utilization of LBS and CBS. With 10 slots, FBFC-C has a gain of 82.5% over CBS for uniform random, while it is 41.4% in a 4 × 4 torus. With 10 slots, the average gains of FBFC-C over LBS and CBS are 107.2% and 40.1% for the 8 patterns. With five slots, the average gain of FBFC-C over CBS is 78.7%.

7.7.5 EFFECT OF STARVATION

For LBS [10] there is limited discussion of starvation; CBS [13] relies on adaptive routing and therefore does not address starvation. We analyze starvation in a 4 × 4 torus. Reducing buffers makes starvation more likely. We use the same buffer size as in Figure 7.17a (LBS uses 10 slots) with uniform random. The results for larger networks or other patterns are similar.

Starvation in LBS and FBFC-L is essentially the same. Figure 7.19a shows their performance with three STVs. They perform poorly with the three-cycle STV. A small STV causes a router to frequently assert the "*starve*" signal (Figure 7.9) to prohibit injection by other nodes, which negatively affects overall performance. We also evaluate the saturation throughput with several STVs ranging from three to 50

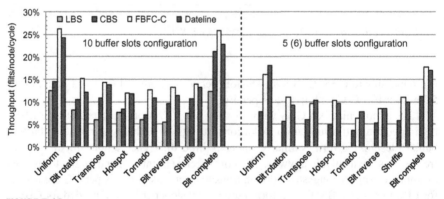

FIGURE 7.18

The performance for an 8 × 8 torus.

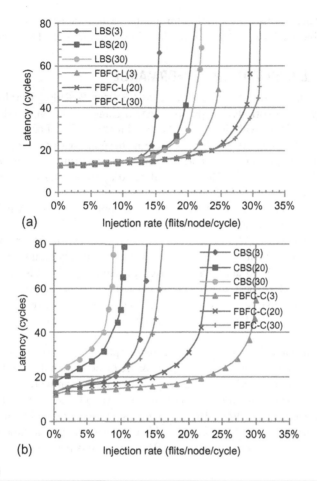

FIGURE 7.19

The performance with several STVs. (a) Localized schemes. (b) Critical schemes.

cycles. LBS and FBFC-L perform better with larger STVs until 30 cycles, and then their performance remains almost constant. We set the STVs in LBS and FBFC-L to 30 cycles.

CBS and FBFC-C can starve owing to the critical bubble stall. Also, FBFC-C has the same starvation as FBFC-L. FBFC-C uses two STVs, one for each type of starvation. We fix the STV in FBFC-C for the same starvation as FBFC-L to 30 cycles, and analyze the other type of starvation. As shown in Figure 7.19b, the smaller the STV, the higher the performance. The proactive transfer of the critical bubble does not prohibit injection; even with many false detections, there is no negative effect. In contrast, the performance drops if packets cannot be injected for a long time. With a 20-cycle STV, if one node suffers starvation, it will move the critical bubble after 20 cycles. Then it starts injecting packets. This lazy reaction not only increases the zero-load latency, but also limits the saturation throughput. Since the proactive transfer of the critical bubble needs two cycles, we set the critical bubble stall STV to three cycles.

7.7.6 **REAL APPLICATION PERFORMANCE**

Figure 7.20 shows the speedups relative to LBS for PARSEC workloads. FBFC supports higher network throughput, but system gains depend on workloads. FBFC benefits applications with heavy loads and bursty traffic. For blackscholes, fluidanimate, and swaptions, different designs perform similarly. Their computation phases have few barriers and their working sets fit into caches, creating light network loads. They are unaffected by techniques improving network throughput, such as FBFC.

Network optimizations affect the other seven applications. Both CBS and FBFC see gains. The largest speedup of FBFC over LBS is 22.7%, for canneal. Two factors bring the gains. First, these applications create bursty traffic and heavy loads. Second, the two VNs with hybrid-size packets have relatively high loads. Across the seven applications, VN0 and VN2, on average, have 70.8% loads, including read request, write-back request, read response, write-back ACK, and invalidation ACK. These relatively congested VNs emphasize FBFC's merit in delivering variable-size packets. Across all workloads, FBFC and CBS have average speedups of 13.0% and 7.5% over LBS.

Compared with synthetic traffic, the real application gains are lower. This is because the configured CMP places light pressure on network buffers. Other designs, such as concentration [19] or configuring fewer buffers, can increase the pressure. FBFC shows larger gains in these scenarios. For example, we also evaluate performance with five slots per VN. FBFC-C achieves an average speedup of 9.8% and a maximum speedup of 20.2% over CBS. Also, the application run time of FBFC-C with 5 slots per VN is similar to that of LBS with 10 slots per VN.

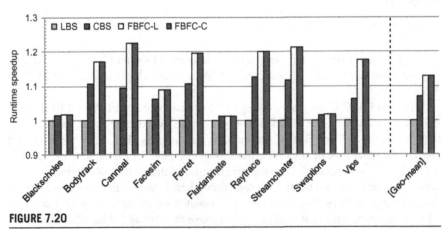

FIGURE 7.20

System speedups for PARSEC benchmarks.

7.7.7 **LARGE-SCALE SYSTEMS AND MESSAGE PASSING**

The advance of CMOS technology will result in hundreds or thousands of cores being integrated in a chip [9]. Some current many-core chips, including the 61-core Xeon Phi chip [28], codenamed Knights Corner, use the shared memory paradigm. Yet, cache coherence faces scalability challenges with more cores. Message passing is an alternative paradigm. For example, the 64-core TILE64 processor uses a message passing paradigm [58]. It remains an open problem to design an appropriate paradigm for large-scale systems [41]. We evaluate FBFC for both paradigms. As a case study, we use a 256-core platform organized as a 16×16 torus.

In large-scale systems, assuming uniform communication among all nodes is not reasonable [47], workload consolidation [37] and application mapping optimizations [14] increase traffic locality; we leverage an exponential locality traffic model [47], which exponentially distributes packet hop counts. For example, with distribution parameter $\lambda = 0.5$, the traffic average hop is $1/\lambda = 2$, and 95% traffic is within 6 hops, and 99% traffic is within 10 hops. We evaluate two distribution parameters, $\lambda = 0.5$ and $\lambda = 0.3$.

The packet length distribution of shared memory traffic is the same as the baseline configuration in Section 7.5; 80% of packets have one flit, and the others have five flits. All designs use 10 slots per VN. We assume that the packet length distribution of message passing traffic is similar to that of Blue Gene/L; packet sizes range from 32 to 256 bytes [4]. With a 16-byte flit width, packet lengths are uniformly distributed between two and 16 flits. All designs use 32 slots per VN.

Figure 7.21 shows the performance. The overall trends among different designs are similar to the trend for an 8×8 torus (Figure 7.18). LBS and CBS are limited by inefficient buffer utilization. LBS is further limited by its high injection buffer requirement. Dateline's limitation for packet forwarding restricts its performance. FBFC efficiently utilizes buffers, and yields the best performance.

The performance gaps between FBFC and the other bubble designs depend on the distribution parameter λ. The smaller the λ, the larger the average hops. Larger average hops emphasize efficient buffer utilization; thus, FBFC had greater performance gains. For shared memory traffic, FBFC-C performs 47.7% better than CBS with $\lambda = 0.5$, and this gain is 66.5% with $\lambda = 0.3$. Message passing traffic shows similar trends. FBFC-C performs 18.7% better than CBS with $\lambda = 0.5$, and the gain increases to 23.8% with $\lambda = 0.3$.

FBFC's gain for message passing traffic is lower than for shared memory traffic. With $\lambda = 0.5$, FBFC-C performs 105.2% better than LBS for shared memory traffic, while it is 68.8% for message passing traffic. The packet length distributions are different. For shared memory traffic, the average packet length is 1.8 flits, and LBS regards each packet as a five-flit packet. The average packet length of message passing traffic is nine flits, and LBS regards each packet as a 16-flit packet. The maximum packet length of shared memory traffic is approximately 2.8 times the average length, while it is two times the average length for message passing traffic. This causes the drop in FBFC-C's gain for message passing traffic.

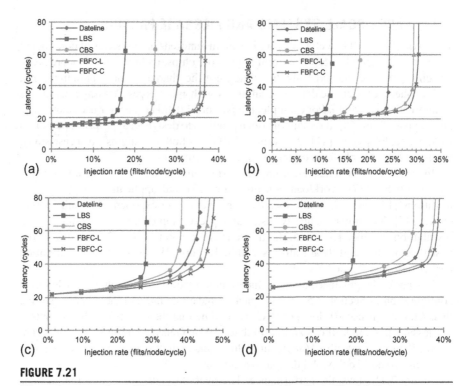

FIGURE 7.21

The performance for a 16×16 torus with exponential locality traffic. (a) Shared memory, $\lambda = 0.5$. (b) Shared memory, $\lambda = 0.3$. (c) Message passing, $\lambda = 0.5$. (d) Message passing, $\lambda = 0.3$.

7.8 OVERHEADS: POWER AND AREA

In this section we conduct power and area analysis of our designs. We also compare tori with meshes.

7.8.1 METHODOLOGY

We modify an NoC power and area model [5], which is integrated in BookSim [29]. We calculate both dynamic and static power. The dynamic power is formulated as $P = \alpha C V_{dd}^2 f$, where α is the switching activity, C is the capacitance, V_{dd} is the supply voltage, and f is the frequency. The switching activities of NoC components are obtained from BookSim. The capacitance, including gate and wire capacitances, is estimated on the basis of canonical modeling of component structures [5].

The static power is calculated as $P = I_{leak} V_{dd}$, where I_{leak} is the leakage current. The leakage current is estimated by taking account of both component structures and input states [5]. For example, the inserted repeater, composed of a pair of pMOS and nMOS devices, determines the wire leakage current. Since pMOS devices leak with high input and nMOS devices leak with low input, the repeater leaks in both high-input and low-input states. The wire leakage current is estimated as the average leakage current of a pMOS and an nMOS device [5].

The router area is estimated on the basis of detailed floor plans [5]. The wires are routed above other logic; the channel area includes only the repeater and flip-flop areas. The device and wire parameters are obtained from the International Technology Roadmap for Semiconductors report [52] for a 32-nm process, at 0.9 V and 70 °C. As a conservation assumption, all designs are assumed to operate at 1 GHz.

We assume a 128-bit flit width. The channel length is 1.5 mm; an 8×8 torus occupies approximately 150 mm^2. Repeaters are inserted to make signals traverse a channel in one cycle. The number and the size of the repeaters are chosen to minimize energy. VCs use SRAM buffers. We assume four VNs to avoid protocol-level deadlock. Allocators use the separable input-first structure [19]. We leverage the segmented crossbar [57] to allow a compact layout and reduce power dissipation. Packets are assumed to carry random payloads; two sequential flits cause half of the channel wires to switch.

7.8.2 POWER EFFICIENCY

Figure 7.22 shows the power consumption of an 8×8 torus with uniform random traffic. The 0% injection rate bars refer to static power. All designs use 10 slots per VN. We divide the NoC into the flit channel, crossbar, buffer, allocator, credit channel, and starvation channel. LBS and CBS cannot support an injection rate of more than 35%. All designs consume similar amounts of power with the same load. The flit channel, crossbar, and buffer together consume more than 93% of the overall power. These components are similar for all designs. The allocator consists of combinational logic, and it induces low power consumption.

The credit channel of bubble designs has three wires; 2 bits encode four VCs and one valid bit. Dateline's credit channel has four wires; 3 bits encode eight VCs and one valid bit. The starvation channel of LBS and FBFC-L has six wires. Three bits identify one node among eight nodes of a ring. Two bits encode four VNs, and one valid bit. CBS's starvation channel uses six wires. "*N2C*" and "*C2N*" signals need 2 bits, and both signals use 2 bits to encode four VNs. FBFC-C handles two types of starvation; its starvation channel has 12 wires. These credit channels and starvation channels are narrower than flit channels; they induce low power consumption.

To clarify the differences, Figure 7.22b shows the allocator and sideband channel power consumption. Starvation channels are not needed for dateline. Yet, their activities are low. For example, FBFC-L's starvation channels remains idle until the injection rate is 34%. Dateline uses large allocators. Also, dateline's credit channel has one more wire than bubble designs. Dateline consumes more power than bubble designs with injection rates of 20% and 40%.

Although the credit channels of bubble designs are narrower than starvation channels, two reasons cause credit channels to consume more dynamic power. First,

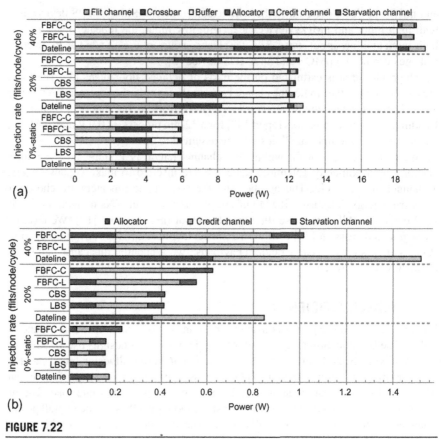

FIGURE 7.22

The power consumption of an 8×8 torus. (a) The overall power consumption. (b) The allocator and sideband channel power consumption.

starvation channels are not needed for injection/ejection ports. An 8×8 torus has 384 credit channels and 256 starvation channels. Second, credit channels' activities are higher. VCT routers return one credit for each packet, and wormhole routers return one credit for each flit. Credit channels of LBS and CBS consume less dynamic power than FBFC.

Figure 7.23a evaluates the PDP. Compared with LBS and CBS, FBFC reduces latencies for heavy loads, improving PDP. With injection rates of 20% and 30%, FBFC-L's PDP is 5.9% and 56.9% lower than that of LBS. FBFC-L's PDP is 27.6% lower than that of CBS with an injection rate of 30%. Dateline's power consumption and latency are similar to those of FBFC. Its PDP is similar to that of FBFC. We also evaluate other traffic patterns. The trends are similar to those for uniform random traffic. All designs consume similar amounts of power, and FBFC's latency optimization reduces the PDP. For example, with transpose, FBFC-L's PDP is 34.2%

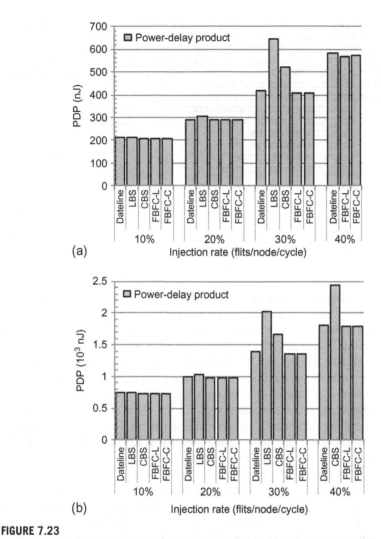

FIGURE 7.23

The PDP results. (a) 8 × 8 torus, uniform random. (b) 16 × 16 torus, λ = 0.3.

lower than that of LBS with an injection rate of 15%, and its PDP is 28.9% lower than that of CBS with an injection rate of 20%.

To show the impact of network scaling on power efficiency, Figure 7.23b gives the PDP on a 16 × 16 torus. The exponential locality traffic with $\lambda = 0.3$ in Section 7.7.7 is used; $\lambda = 0.5$ has a similar trend. Since FBFC optimizes latencies, it still offers power efficiency gains in larger NoCs. FBFC-L's PDP is 32.7% and 18.2% lower than that of LBS and CBS respectively for an injection rate of 30%, and its PDP is 26.6% lower than that of CBS for an injection rate of 40%.

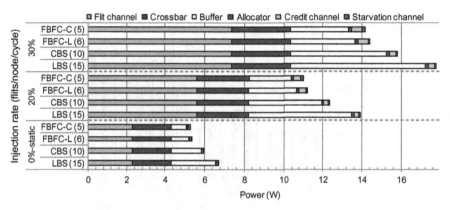

FIGURE 7.24

The power consumption results when bubble designs support similar saturation throughput in an 8 × 8 torus.

As shown in Section 7.7.3, with half as many buffers, FBFC performs the same as CBS in an 8 × 8 torus. With one third of the number of buffers, FBFC performs similarly to LBS. Figure 7.24 gives the power consumption of bubble designs when they performs similarly. FBFC-C and FBFC-L use five and six slots per VN. CBS and LBS use 10 and 15 slots per VN. FBFC-C's buffer static power consumption is 49.8% lower than that of CBS, which results in FBFC-C's network static power consumption being 11.5% lower than that of CBS. FBFC-C's overall network static power consumption is 21.3% lower than that of LBS. High loads increase dynamic power consumption. Yet, even with an injection rate of 20%, FBFC-C's power consumption is still 20.4% and 10.4% lower than that of LBS and CBS respectively. Since now all designs perform similarly, FBFC's PDP is lower as well.

In summary, with the same buffer size, all designs consume a similar amount of power. FBFC's starvation channels induce negligible power consumption. Since FBFC significantly outperforms existing bubble designs, it achieves a much lower PDP, in both 8 × 8 and 16 × 16 tori. When bubble designs perform similarly with different buffer sizes, FBFC consumes less power and offers PDP gains.

7.8.3 AREA

Figure 7.25 shows the area results. The areas of the flit channel and crossbar are similar for all designs. With the same buffer amount, the area differences among designs are mainly due to the allocator, credit channel, and starvation channel. Dateline's allocator is approximately two times larger than that of bubble designs; this causes dateline to consume the greatest area. With 10 slots per VN, dateline's overall area is 7.4% higher than that of FBFC-L. When bubble designs perform similarly, FBFC's area benefit is more significant. CBS's network area with 10 slots per VN is

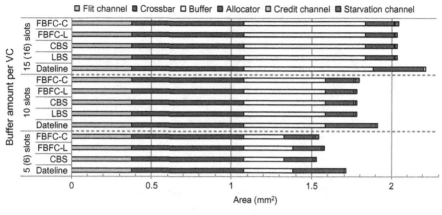

FIGURE 7.25

The area of an 8 × 8 torus.

15.6% higher than that of FBFC-C with 5 slots per VN. The overall area of LBS with 15 slots per VN is 31.9% higher than that of FBFC-C with 5 slots per VN.

7.8.4 COMPARISON WITH MESHES

We compare tori with meshes. The routing is DOR. The torus uses FBFC to avoid deadlock. On the basis of a floor-plan model [5], the mesh channel length is 33.3% shorter than that of the torus; it is 1.0 mm. Inserted repeaters make the channel delay be one cycle. Two meshes are evaluated. One uses a 128-bit channel width, which is the same as for the torus. Yet, its bisection bandwidth is half that of the torus. The other mesh uses a 256-bit width to achieve the same bisection bandwidth as the torus. All networks have the same packet size distribution. The 128-bit-width mesh uses 10 slots per VN. Its traffic has 80% five-flit packets and 20% SFPs. The 256-bit-width mesh uses five slots per VN. Its traffic has 80% three-flit packets and 20% SFPs.

Figure 7.26 gives the performance. Since the flit sizes of evaluated networks are different, the injection rate is measured in packets per node per cycle. The torus's wraparound channels reduce hop counts. For both patterns, the torus shows approximately 20% lower zero-load latencies than the mesh. With half of the bisection bandwidth, the saturation throughput of *mesh-128bits* is 24.2% and 37.0% lower than that of the torus for uniform random traffic and transpose traffic respectively. With the same bisection bandwidth, the saturation throughput of *mesh-256bits* is similar to that of the torus for uniform random traffic. The transpose traffic congests the mesh's center portion [43]; the saturation throughput of *mesh-256bits* is still 17.3% lower than that of the torus.

Figure 7.27 shows the power consumption and area. With the same channel width, the static power consumption of *mesh-128bits* is 30.5% lower than that of the torus; this is due to the optimization of buffers and flit channels. An 8 × 8 mesh has 288

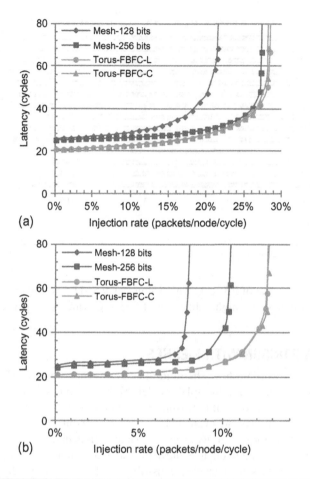

FIGURE 7.26

The performance comparison. (a) Uniform random. (b) Transpose.

ports with buffers, while an 8 × 8 torus has 320 ports. Two factors bring about power consumption reduction for mesh channels. First, an 8 × 8 mesh has 352 flit channels, while an 8 × 8 torus has 384 flit channels. Second, mesh channels are 33.3% shorter than torus ones. Thus, even 256-bit mesh channels consume less static power than torus ones. Yet, the 256-bit channel width quadratically increases crossbar power consumption. The overall static power consumption of *mesh-256bits* is 77.7% higher than that of the torus.

With high loads, mesh's center congestion increases dynamic power consumption; the benefit of *mesh-128bits* over FBFC-L decreases. With injection rates of 10% and 20%, its power consumption is 15.7% and 10.9% less than that of FBFC-L. With an injection rate of 20%, the channel power consumption of *mesh-256bits* is higher than that of FBFC-L. The PDP reflects power efficiency. With an injection rate of 10%, FBFC-L's PDP is 6.4% and 64.9% less than that of *mesh-128bits* and

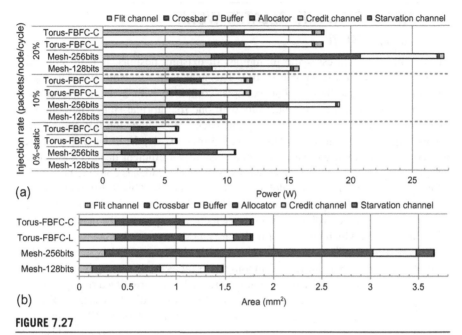

FIGURE 7.27

The power consumption and area comparison. (a) The power consumption with uniform random traffic. (b) The area.

mesh-256bits respectively. Its power efficiency increases with high loads. With an injection rate of 20%, FBFC-L's PDP is 33.7% less than that of *mesh-128bits*.

The network area of *mesh-128bits* is 17.4% less than that of FBFC-L. The *mesh-128bits* uses fewer buffers and channels than FBFC-L. Owing to the large crossbar of *mesh-256bits*, its network area is 107.2% greater than that of FBFC-L.

We also compare a 16×16 mesh with a 16×16 torus. The static power and area benefits of *mesh-128bits* decrease with larger networks. The mesh's benefit of using fewer flit channels and ports decreases. In a 64-node network, the torus has 9.1% more flit channels and 11.1% more ports than the mesh. These values are 4.3% (1536 vs. 1472) and 5.3% (1280 vs. 1216) in a 256-node network. The static power consumption of *mesh-128bits* in a 16×16 mesh is 24.3% less than that of a 16×16 torus, and its network area is 13.4% less. The PDP is more favorable to FBFC. FBFC-L's PDP is 24.5% and 57.9% less than that of *mesh-128bits* and *mesh-256bits* respectively at an injection rate of 15% for exponential locality traffic with $\lambda = 0.3$.

In summary, although with the same channel width, the mesh consumes less power and occupies less area than the torus, its performance is poor owing to limited bisection bandwidth. With FBFC applied, the torus is more power efficient than the mesh for the same channel width. With the same bisection bandwidth, the mesh consumes much more power than the torus. Applying FBFC on the torus is well scalable.

7.9 DISCUSSION AND RELATED WORK

7.9.1 DISCUSSION

FBFC efficiently addresses the limitations of existing designs. It is an important extension to packet-size bubble theory. The insight of *"leaving one slot empty"* enables other design choices. For example, combined with dynamic packet fragmentation [24], a packet can inject flits with one free normal slot. When one flit's injection will consume the critical slot, this packet stops injecting flits and changes the waiting flit into a head flit. This design allows VC depths to be shallower than the largest packet. On the basis of similar insight, an efficient deadlock avoidance design is proposed for wormhole networks [12]. Its basic idea is coloring buffer slots white, gray, or black to convey global buffer status locally. This is different from our design. FBFC uses local buffer status with hybrid flow control which combines VCT and wormhole flow control. Also, we mainly focus on improving the performance for coherence traffic, which consists of both long and short packets.

Similarly to dateline and packet-size bubble designs, FBFC is a general flow control. It can be adopted in various topologies as far as there is a ring in the network. For example, Immune [49] achieves fault tolerance by constructing a ring in arbitrary topologies for connectivity. The ring uses LBS to avoid deadlock. Instead, FBFC can be used; it will support higher performance with fewer buffers. MRR [1] leverages the ring of a rotary router [2] to support multicast. The ring uses LBS. FBFC can be used as well. By configuring a ring in the network, FBFC can support both unicast and multicast for streaming applications [3]. Also, FBFC can support fully adaptive routing [13, 50]. Bubble designs use one VC; there is head-of-line blocking. FBFC can combine with dynamically allocated multiqueue [46, 55] to mitigate this blocking and further improve buffer utilization.

7.9.2 RELATED WORK

The Ivy Bridge [21], Xeon Phi [28], and Cell [31] processors use ring networks. The ring is much simpler than the 2D or higher-dimensional torus, and it is easy to avoid deadlock through end-to-end backpressure or centralized control schemes [31, 33]. The ring networks of these chips [21, 28, 31] guarantee injected packets cannot be blocked, which is similar to bufferless networks. Bufferless designs generally do not consider deadlock as packets are always movable [24]. Our research is different. We focus on efficient deadlock avoidance designs for buffered networks, which support higher throughput than bufferless networks. In addition to dateline [19], LBS [10], and CBS [13], there are other designs. Priority arbitration is used for SFPs with single-cycle routers [33]. Prevention flow control combines priority arbitration with prevention slot cycling [30]; it has deadlock with variable-size packets. The turn model [26] allows only nonminimal routing in tori. A design allows deadlock formation, and then applies a recovery mechanism [53].

FBFC observes that most coherence packets are short. Several designs use this observation, including packet chaining [42], the NoX router [27], and whole packet forwarding [38]. Configuring more VNs, such as seven VNs in Alpha 21364 [44], can eliminate coexistence of variable-size packets. Yet, additional VNs have overheads; using minimum VNs is preferable. DASH [36], Origin 2000 [35], and Piranha [6] all apply protocol-level deadlock recovery to eliminate one VN; they utilize two VNs to implement three-hop directory protocols. These VNs all carry variable-size packets.

7.10 CHAPTER SUMMARY

Optimizing NoCs for coherence traffic improves the efficiency of many-core coherence protocols. We observed two properties of cache coherence traffic: short packets dominate the traffic, and short and long packets coexist in an NoC. We proposed an efficient deadlock avoidance theory, FBFC, for torus networks. It maintains one free *flit-size* buffer slot to avoid deadlock. Only one VC is needed, which achieves high frequency. Also, FBFC does not treat short packets as long ones; this yields high buffer utilization. With the same buffer size, FBFC significantly outperforms LBS and CBS, and is more power efficient as well. When bubble designs perform similarly, FBFC consumes much less power and area. With FBFC applied, the torus is more power efficient than the mesh.

REFERENCES

[1] P. Abad, V. Puente, J.-A. Gregorio, MRR: enabling fully adaptive multicast routing for CMP interconnection networks, in: Proceedings of the International Symposium on High-Performance Computer Architecture (HPCA), 2009, pp. 355–366. 248

[2] P. Abad, V. Puente, J. A. Gregorio, P. Prieto, Rotary router: an efficient architecture for CMP interconnection networks, in: Proceedings of the International Symposium on Computer Architecture (ISCA), 2007, pp. 116–125. 248

[3] A.H. Abdel-Gawad, M. Thottethodi, TransCom: transforming stream communication for load balance and efficiency in networks-on-chip, in: Proceedings of the International Symposium on Microarchitecture (MICRO), 2011, pp. 237–247. 248

[4] N.R. Adiga, M.A. Blumrich, D. Chen, P. Coteus, A. Gara, M.E. Giampapa, P. Heidelberger, S. Singh, B.D. Steinmacher-Burow, T. Takken, M. Tsao, P. Vranas, Blue Gene/L torus interconnection network, IBM J. Res. Dev. 49(2.3) (2005) 265–276. 216, 220, 239

[5] J. Balfour, W. Dally, Design tradeoffs for tiled CMP on-chip networks, in: Proceedings of the International Conference on Supercomputing (ICS), 2006, pp. 187–198. 240, 241, 245

[6] L.A. Barroso, K. Gharachorloo, R. McNamara, A. Nowatzyk, S. Qadeer, B. Sano, S. Smith, R. Stets, B. Verghese, Piranha: a scalable architecture based on single-chip multiprocessing, in: Proceedings of the International Symposium on Computer Architecture (ISCA), 2000, pp. 282–293. 216, 249

[7] D.U. Becker, W.J. Dally, Allocator implementations for network-on-chip routers, in: Proceedings of the Conference on High Performance Computing Networking, Storage and Analysis (SC), 2009, pp. 1–12. 227, 228

[8] C. Bienia, S. Kumar, J.P. Singh, K. Li, The PARSEC benchmark suite: characterization and architectural implications, in: Proceedings of the International Conference on Parallel Architectures and Compilation Techniques (PACT), 2008, pp. 72–81. 217, 228

[9] S. Borkar, Thousand core chips: a technology perspective, in: Proceedings of the Design Automation Conference (DAC), 2007, pp. 746–749. 239

[10] C. Carrion, R. Beivide, J. Gregorio, F. Vallejo, A flow control mechanism to avoid message deadlock in *k*-ary *n*-cube networks, in: Proceedings of the International Conference on High-Performance Computing (HiPC), 1997, pp. 322–329. 216, 218, 219, 224, 236, 248

[11] L. Chen, T. Pinkston, Personal communication, 2012. 225

[12] L. Chen, T.M. Pinkston, Worm-bubble flow control, in: Proceedings of the International Symposium on High-Performance Computer Architecture (HPCA), 2013, pp. 366–377. 248

[13] L. Chen, R. Wang, T.M. Pinkston, Critical bubble scheme: an efficient implementation of globally aware network flow control, in: Proceedings of the International Parallel & Distributed Processing Symposium (IPDPS), 2011, pp. 592–603. 216, 218, 219, 233, 236, 248

[14] C.-L. Chou, R. Marculescu, Run-time task allocation considering user behavior in embedded multiprocessor networks-on-chip, IEEE Trans. Comput. Aided Des. Integr. Circuits Syst. 29(1) (2010) 78–91. 239

[15] P. Conway, B. Hughes, The AMD Opteron northbridge architecture, IEEE Micro 27(2) (2007) 10–21. 218

[16] W. Dally, Virtual-channel flow control, IEEE Trans. Parallel Distrib. Syst. 3(2) (1992) 194–205. 216

[17] W. Dally, C. Seitz, Deadlock-free message routing in multiprocessor interconnection networks, IEEE Trans. Comput. C-36(5) (1987) 547–553. 218, 221

[18] W. Dally, B. Towles, Route packets, not wires: on-chip interconnection networks, in: Proceedings of the Design Automation Conference (DAC), 2001, pp. 684–689. 216

[19] W. Dally, B. Towles, Principles and Practices of Interconnection Networks, first ed., Morgan Kaufmann Publishers Inc., San Francisco, CA, 2003. 216, 218, 225, 226, 227, 228, 234, 238, 241, 248

[20] W.J. Dally, C.L. Seitz, The torus routing chip, Distrib. Comput. 1(4) (1986) 187–196. 216, 218, 221

[21] S. Damaraju, V. George, S. Jahagirdar, T. Khondker, R. Milstrey, S. Sarkar, S. Siers, I. Stolero, A. Subbiah, A 22 nm IA multi-CPU and GPU system-on-chip, in: Proceedings of the International Solid-State Circuits Conference Digest of Technical Papers (ISSCC), 2012, pp. 56–57. 216, 248

[22] J. Duato, A. Robles, F. Silla, R. Beivide, A comparison of router architectures for virtual cut-through and wormhole switching in a NOW environment, J. Parallel Distrib. Comput. 61(2) (2001) 224–253. 226

[23] N. Enright Jerger, L. Peh, On-Chip Networks, first ed., Morgan & Claypool, San Rafael, CA, 2009. 215, 225

[24] C. Fallin, C. Craik, O. Mutlu, CHIPPER: a low-complexity bufferless deflection router, in: Proceedings of the International Symposium on High-Performance Computer Architecture (HPCA), 2011, pp. 144–155. 216, 217, 248

[25] K. Gharachorloo, M. Sharma, S. Steely, S. Van Doren, Architecture and design of AlphaServer GS320, in: Proceedings of the International Conference on Architectural Support for Programming Languages and Operating Systems (ASPLOS), 2000, pp. 13–24. 216

[26] C. Glass, L. Ni, The turn model for adaptive routing, in: Proceedings of the International Symposium on Computer Architecture (ISCA), 1992, pp. 278–287. 248

[27] M. Hayenga, M. Lipasti, The NoX router, in: Proceedings of the International Symposium on Microarchitecture (MICRO), 2011, pp. 36–46. 228, 248

[28] Intel, Intel Xeon Phi coprocessor—datasheet, Technical Report, Intel, June 2013. 216, 239, 248

[29] N. Jiang, D.U. Becker, G. Michelogiannakis, J. Balfour, B. Towles, D. Shaw, J. Kim, W. Dally, A detailed and flexible cycle-accurate network-on-chip simulator, in: Proceedings of the International Symposium on Performance Analysis of Systems and Software (ISPASS), 2013, pp. 86–96. 227, 240

[30] A. Joshi, M. Mutyam, Prevention flow-control for low latency torus networks-on-chip, in: Proceedings of the International Symposium on Networks-on-Chip (NOCS), 2011, pp. 41–48. 248

[31] J.A. Kahle, M.N. Day, H.P. Hofstee, C.R. Johns, T.R. Maeurer, D. Shippy, Introduction to the cell multiprocessor, IBM J. Res. Dev. 49(4.5) (2005) 589–604. 216, 248

[32] P. Kermani, L. Kleinrock, Virtual cut-through: a new computer communication switching technique, Comput. Netw. 3(4) (1979) 267–286. 216

[33] J. Kim, H. Kim, Router microarchitecture and scalability of ring topology in on-chip networks, in: Proceedings of the International Workshop on Network on Chip Architectures (NoCArc), 2009, pp. 5–10. 248

[34] A. Kumar, P. Kundu, A. Singh, L.-S. Peh, N. Jha, A 4.6 Tbits/s 3.6 GHz single-cycle NoC router with a novel switch allocator in 65 nm CMOS, in: Proceedings of the International Conference on Computer Design (ICCD), 2007, pp. 63–70. 225

[35] J. Laudon, D. Lenoski, The SGI origin: a ccNUMA highly scalable server, in: Proceedings of the International Symposium on Computer Architecture (ISCA), 1997, pp. 241–251. 216, 249

[36] D. Lenoski, J. Laudon, K. Gharachorloo, A. Gupta, J. Hennessy, The directory-based cache coherence protocol for the DASH multiprocessor, in: Proceedings of the International Symposium on Computer Architecture (ISCA), 1990, pp. 148–159. 216, 249

[37] S. Ma, N. Enright Jerger, Z. Wang, DBAR: an efficient routing algorithm to support multiple concurrent applications in networks-on-chip, in: Proceedings of the International Symposium on Computer Architecture (ISCA), 2011, pp. 413–424. 239

[38] S. Ma, N. Enright Jerger, Z. Wang, Whole packet forwarding: efficient design of fully adaptive routing algorithms for networks-on-chip, in: Proceedings of the International Symposium on High-Performance Computer Architecture (HPCA), 2012, pp. 467–478. 217, 249

[39] S. Ma, Z. Wang, Z. Liu, N. Enright Jerger, Leaving one slot empty: flit bubble flow control for torus cache-coherent NoCs, IEEE Trans. Comput. Preprints (99) (2013), preprint. 215

[40] P.S. Magnusson, M. Christensson, J. Eskilson, D. Forsgren, G. Hallberg, J. Hogberg, F. Larsson, A. Moestedt, B. Werner, Simics: a full system simulation platform, Computer 35 (2002) 50–58. 228

[41] M.M.K. Martin, M.D. Hill, D.J. Sorin, Why on-chip cache coherence is here to stay, Commun. ACM 55(7) (2012) 78–89. 216, 239

[42] G. Michelogiannakis, N. Jiang, D. Becker, W. Dally, Packet chaining: efficient single-cycle allocation for on-chip networks, in: Proceedings of the International Symposium on Microarchitecture (MICRO), 2011, pp. 33–36. 228, 248

[43] A.K. Mishra, N. Vijaykrishnan, C.R. Das, A case for heterogeneous on-chip interconnects for CMPs, in: Proceedings of the International Symposium on Computer Architecture (ISCA), 2011, pp. 389–400. 245

[44] S.S. Mukherjee, P. Bannon, S. Lang, A. Spink, D. Webb, The Alpha 21364 network architecture, in: Proceedings of the Symposium on High Performance Interconnects (HOTI), 2001, pp. 113–118. 216, 249

[45] N. Neelakantam, C. Blundell, J. Devietti, M.M. Martin, C. Zilles, FeS2: a full-system execution-driven simulator for x86, in: Poster Presented at ASPLOS 2008, 2008. 218, 228

[46] C.A. Nicopoulos, D. Park, J. Kim, N. Vijaykrishnan, M.S. Yousif, C.R. Das, ViChaR: a dynamic virtual channel regulator for network-on-chip routers, in: Proceedings of the International Symposium on Microarchitecture (MICRO), 2006, pp. 333–346. 216, 217, 248

[47] G.P. Nychis, C. Fallin, T. Moscibroda, O. Mutlu, S. Seshan, On-chip networks from a networking perspective: congestion and scalability in many-core interconnects, in: Proceedings of the SIGCOMM Conference on Applications, Technologies, Architectures, and Protocols for Computer Communication (SIGCOMM), 2012, pp. 407–418. 239

[48] L.-S. Peh, W. Dally, A delay model and speculative architecture for pipelined routers, in: Proceedings of the International Symposium on High-Performance Computer Architecture (HPCA), 2001, pp. 255–266. 227

[49] V. Puente, J.A. Gregorio, F. Vallejo, R. Beivide, Immunet: dependable routing for interconnection networks with arbitrary topology, IEEE Trans. Comput. 57(12) (2008) 1676–1689. 248

[50] V. Puente, C. Izu, R. Beivide, J.A. Gregorio, F. Vallejo, J. Prellezo, The adaptive bubble router, J. Parallel Distrib. Comput. 61(9) (2001) 1180–1208. 219, 248

[51] S.L. Scott, G.M. Thorson, The Cray T3E network: adaptive routing in a high performance 3D torus, in: Proceedings of the the Symposium on High Performance Interconnects (HOTI), 1996, pp. 1–10. 216, 227

[52] Semiconductor Industry Association, International Technology Roadmap for Semiconductors, 2010 ed., http://www.itrs.net, 2010. 241

[53] M. Shin, J. Kim, Leveraging torus topology with deadlock recovery for cost-efficient on-chip network, in: Proceedings of the International Conference on Computer Design (ICCD), 2011, pp. 25–30. 216, 248

[54] A. Singh, W.J. Dally, A.K. Gupta, B. Towles, GOAL: a load-balanced adaptive routing algorithm for torus networks, in: Proceedings of the International Symposium on Computer Architecture (ISCA), 2003, pp. 194–205. 216

[55] Y. Tamir, G. Frazier, High-performance multiqueue buffers for VLSI communication switches, in: Proceedings of the International Symposium on Computer Architecture (ISCA), 1988, pp. 343–354. 248

[56] A. Vaidya, A. Sivasubramaniam, C. Das, Impact of virtual channels and adaptive routing on application performance, IEEE Trans. Parallel Distrib. Syst. 12(2) (2001) 223–237. 230

[57] H. Wang, L. Peh, S. Malik, Power-driven design of router microarchitectures in on-chip networks, in: Proceedings of the International Symposium on Microarchitecture (MICRO), 2003, pp. 105–116. 241

[58] D. Wentzlaff, P. Griffin, H. Hoffmann, L. Bao, B. Edwards, C. Ramey, M. Mattina, C.-C. Miao, J.F. Brown III, A. Agarwa, On-chip interconnection architecture of the TILE processor, IEEE Micro 27(5) (2007) 15–31. 239

Programming paradigms

Designing efficient parallel programming paradigms is one of the most critical challenges for the development of many-core processors. The efficiency of programming paradigm largely depends on the communication scheme as the communication is the basis of core cooperation. There are generally two types of paradigms, the shared memory and the message passing paradigms. The shared memory paradigm uses cache coherence protocols to maintain a coherent memory view; cache coherence protocols determine traffic characteristics. Without hardware support, coherent collective communications, including multicast and reduction ones, easily become system bottlenecks. For message passing paradigms, conventional software implementations of message passing interface (MPI) functions cause large latencies. Since most functions are built from a few primitives, hardware implementations of primitives can efficiently scale up performance. In addition, the MPI protocol leverages two communication modes, the buffered and the synchronous modes; the buffered one is efficient with large *receiving buffers, while the synchronous one is appropriate with limited buffers. Adaptively adjusting the communication mode on the basis of buffer status can provide robust and high performance. On the basis of the aforementioned analysis, this part delves into the co-design of network-on-chip (NoC) and high-level programming paradigms in three chapters.*

Chapter 8 studies hardware support for collective communications in cache coherence protocols. Support for multicast communication in NoCs has achieved substantial throughput gains and power savings. This chapter explores support for reduction communications. As a case study, we focus on acknowledgment messages (ACK) that must be collected in a directory protocol before a cache line may be upgraded to or installed in the modified state. This chapter makes two primary contributions, an efficient framework to support the reduction of ACK packets and a novel balanced, adaptive multicast (BAM) routing algorithm. The message combination framework complements several multicast algorithms. By combining ACKs during transmission, this framework not only reduces packet latency for low-to-medium network loads, but also improves the network saturation throughput with little overhead. The balanced buffer resource configuration of BAM results in additional saturation throughput improvements.

Chapter 9 presents a NoC design that optimizes the well-known parallel programming model, MPI, to boost applications by exploiting hardware features available in NoC-based many-core architectures. Conventional MPI functions are normally implemented in software because of their enormity and complexity, resulting in large communication latencies. We propose a novel hardware implementation of basic MPI primitives. The premise is that all other MPI functions can be efficiently built from these three MPI primitives. The design includes two important features: the customized NoC design incorporating virtual buses into NoCs and the optimized MPI unit efficiently executing MPI-related transactions. The proposed designs effectively boost the performance of MPI communication functions.

Chapter 10 explores designing MPI communication protocols over NoCs. We advocate a hardware-supported communication mechanism using a protocol-adaptive approach to adjust to varying NoC configurations (e.g. number of buffers) and workload behavior (e.g. number of messages). This chapter proposes the adaptive communication mechanism (ADCM), a hybrid protocol that involves behavior similar to buffered communication when sufficient buffer is available in the receiver and that similar to a synchronous protocol when buffers in the receiver are limited. ADCM adapts dynamically by deciding on the communication protocol on a per-request basis using a local estimate of recent buffer utilization. ADCM attempts to combine the advantages of both buffered and synchronous communication modes to achieve enhanced throughput and performance. The proposed communication mechanism can be effectively used in future NoC designs.

Supporting cache-coherent collective communications†

†This research was first presented at the 18th IEEE International Symposium on High Performance Computer Architecture (HPCA-2012) [29].

Networks-on-Chip. http://dx.doi.org/10.1016/B978-0-12-800979-6.00008-1
Copyright © 2015 China Machine Press/Beijing Huazhang Graphics & Information Co., Ltd.
Published by Elsevier Inc. All rights reserved.

8.1 INTRODUCTION

Efficient and scalable on-chip communication will be required to realize the performance potential of many-core architectures. To harness this performance, it is imperative that networks-on-chip (NoCs) be designed to efficiently handle a variety of communication primitives. Collective communication lies on the critical path for many applications; the criticality of such communication is evident in the dedicated collective and barrier networks employed in several supercomputers, such as the NYU Ultracomputer [15], CM-5 [25], Cray T3D [2], Blue Gene/L [3], and TianHe-1A [45]. Likewise, many-core architectures will benefit from hardware support for collective communications but may not be able to afford separate, dedicated networks owing to rigid power and area budgets [36]; this chapter explores integrating collective communication support directly into the existing NoC.

Various parallel applications and programming paradigms require collective communication such as broadcast, multicast, and reduction. For example, a directory-based coherence protocol relies heavily on multicasts to invalidate shared data spread across multiple caches [19] and token coherence uses multicasts to collect tokens [31]. Reductions and multicasts are used for barrier synchronization [37, 44]. These collective communications can have a significant effect on many-core system performance. Without any special hardware mechanisms, even if 1% of injected packets are multicast, there is a sharp drop in saturation throughput [12]. Recent work proposes efficient multicast routing support to improve NoC performance [1, 12, 21, 27, 39–41, 43].

Often, a multicast will trigger an operation, such as invalidating a cache line [19] or counting available tokens [31]. To notify the source of the completion of these operations, the multicast destination nodes send out responses. The resulting many-to-one communication operation is called a reduction [9]. Figure 8.1 shows that a cache line invalidation message triggers on average 7.44 acknowledgment messages (ACKs) for the PARSEC benchmarks [4] in a 16-core system.[1]

Prior NoC multicast proposals [1, 12, 21, 27, 39–41, 43] implicitly assumed several unicast packets will deliver these messages to a single destination; this can lead to redundant network traversal operations and create transient hotspots inside the network. To provide high performance, scalable NoCs should handle traffic in an intelligent fashion by eliminating these redundant messages.

Furthermore, the multicast-reduction transaction cannot be completed until all responses have been received [5, 10, 19, 23, 26, 31]. As a result, the transmission of reduction messages lies on the critical path of a multicast-reduction transaction. These multicast-reduction operations are often associated with stores; for out-of-order cores, stores do not lie on the critical path. However, these stores can delay subsequent loads to hotly contended cache lines. For chip multiprocessors (CMPs) that employ simple, in-order cores, stores will lie on the critical path and can significantly impact performance. Figure 8.2 shows the completion latency of multicast-reduction

---

[1] See Section 8.5 for the detailed experimental configuration.

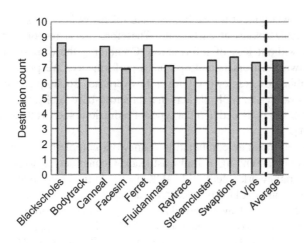

FIGURE 8.1

Average destinations per multicast for PARSEC benchmarks.

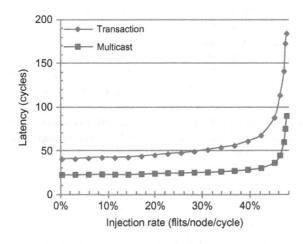

FIGURE 8.2

The latency of multicast-reduction transactions: for "multicast," the latency of the last-arriving multicast replica; for "transaction," the latency of the last-arriving ACK. The network is routed by BAM+NonCom as described in Section 8.5.

transactions in a 4 × 4 mesh running uniform random traffic. The transmission of reduction packets accounts for approximately 40% of the total transaction latency. We propose a novel packet reduction mechanism to improve performance of the full multicast-reduction transaction.

A noteworthy property of coherence-based reduction messages is that they carry similar information in a simple format. For example, invalidation ACKs only carry the acknowledgment for each node, and token count replies merely carry the count of available tokens at each node. Therefore, these response messages can be combined without loss of information. Combining these messages eliminates redundant network traversals and optimizes performance. We propose an efficient message combination framework with little overhead. To simplify the discussion, we focus on invalidation ACK packets in a directory-based coherence protocol. Our design can be easily extended to other types of reductions such as those used in token coherence [31].

Our proposed message combination framework complements several multicast routing algorithms. Sending a multicast packet constructs a logical tree in the network. The framework steers each ACK packet to traverse the same logical tree back to the root (source) of the multicast. In each router, a small message combination table (MCT) records total and received ACK counts for active multicast transactions. When an ACK packet arrives at the router, the MCT is checked. If the router has not received all expected ACKs, the table is updated and the incoming ACK is discarded. If the router has received all expected ACKs, the incoming ACK packet is updated and forwarded to the next node in the logical tree. Dropping in-flight ACK packets reduces network load and power consumption.

Our goal is to improve overall network performance in the presence of both unicast and multicast-reduction traffic. The recently proposed recursive partitioning multicast (RPM) routing algorithm utilizes two virtual networks (VNs) to avoid deadlock for multicasts [41]. However, the division of these two VNs results in unbalanced buffer resources between vertical and horizontal dimensions, which negatively affects performance. Therefore, we propose a novel multicast routing algorithm, *balanced, adaptive multicast* (BAM), which does not need two VNs to avoid multicast deadlock. BAM balances the buffer resources between different dimensions, and achieves efficient bandwidth utilization by computing an output port on the basis of all the multicast destination positions.

To summarize, our main contributions are as follows:

- An efficient message combination framework that reduces latency by 14.1% and the energy-delay product (EDP) by 20-40% for low-to-medium network loads and improves saturation throughput by 9.6%.
- A novel multicast routing algorithm which balances buffer resources across different dimensions and improves network throughput by an additional 13.8%.

8.2 MESSAGE COMBINATION FRAMEWORK

In this section, we describe the proposed message combination framework. We use a multicast-reduction example to illustrate the framework. One multicast packet with destinations 0, 7, and 15 is injected by node 9. A logical multicast tree [13] is built as shown in Figure 8.3a; gray nodes indicate destinations, while white nodes are branches that are only traversed by the packet. Each multicast destination

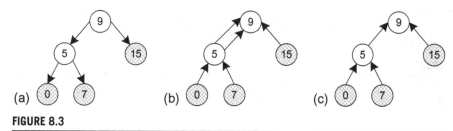

FIGURE 8.3

The message combination framework overview. (a) Logical multicast tree. (b) Logical ACK tree (without combination). (c) Logical ACK tree (with combination).

responds with an ACK message to the root node, node 9. Without combination, the ACKs are transmitted as unicast packets back to node 9 (Figure 8.3b).[2] These ACK packets traverse some common channels; merging them can reduce the network load. Figure 8.3c shows the logical ACK tree with message combination, which is the same as the logical multicast tree except with the opposite transmission direction. In this example, routers 5 and 9 serve as fork routers which are responsible for gathering and forwarding ACKs. Next, we address two important issues associated with our message combination framework: ensuring that ACK packets traverse the same logical tree as the multicast packet and ensuring that the fork routers are aware of the expected total and currently received ACKs.

In an $n \times n$ mesh network, a $\log(n \times n)$-bit field in the multicast header is reserved to identify the router where the last replication happened (pre_rep_router). This field is initially set to be the source node and is updated when a multicast packet replicates during transmission. A 3-bit ID field is used to differentiate multicast packets injected by the same source node. This field is incremented when a new multicast packet is injected. The src field encodes the source node.

A small MCT is added to each router. A multicast allocates an MCT entry upon replication. This entry records the identity of the router where the last multicast replication occurred and the total expected ACK count. The transmission of a multicast packet establishes a logical tree. Each branch in the logical tree has an MCT entry pointing to the previous fork router.

Each multicast destination responds with an ACK packet. A $\log(n \times n)$-bit field (cur_dest) in the ACK header serves to identify the intermediate destination. Its value is set to the pre_rep_router field in the triggering multicast packet. Each ACK packet has two fields named multicast_src and multicast_ID, which correspond to src and ID of the triggering multicast packet respectively. An additional $\log(n \times n)$-bit field (ACK_count) is used to record the carried ACK response count of the combined packet.

When an ACK packet arrives at its current destination, it accesses the MCT. If the router has not yet received all expected ACKs, the incoming packet is discarded and

---

[2]We assume the ACKs sent out by nodes 0 and 7 both traverse node 5.

V	src	ID	pre_rep_router	incoming_port	expected_count	cur_ACK_count
1 bit	4 bits	3 bits	4 bits	3 bits	4 bits	4 bits

FIGURE 8.4

The MCT entry format. The port encoding is as follows: east, 0; west, 1; south, 2; north, 3; local, 4. It is assumed there are 16 nodes.

the entry's received ACK count is incremented. If the router has received all expected ACKs, the incoming ACK packet updates its cur_dest field to be the next replication router. It will be routed to the fork router at the next level; thus, ACK packets traverse the same logical tree as multicast in the opposite direction.

8.2.1 MCT FORMAT

Figure 8.4 illustrates the format of an MCT entry. The V field is the valid bit for the entry. The src, ID, and pre_rep_router fields are the same as the corresponding fields in the multicast packet initializing this entry. The MCT is a content-addressable memory (CAM); the src and ID fields work together as the tag. The incoming_port field records the incoming port of the multicast packet. The expected_count field indicates the total expected ACK count, which is equal to the number of destinations at this branch of the multicast tree. The value of the cur_ACK_count field tracks the current received ACK count. As we will show later, recording the total expected ACK count instead of simply counting the number of direct successors is needed for handling full MCTs.

8.2.2 MESSAGE COMBINATION EXAMPLE

Figure 8.5a gives a multicast example within a 4 × 4 mesh network which is the same as that in Figure 8.3a. Figure 8.5b shows the multicast header values. Although our framework is independent of the multicast packet format, we assume bit string encoding [6] for the destination addresses in the destinations field of the header for clarity. M_a is the injected packet; its destinations field contains the three destinations. M_a replicates into two packets, M_b and M_c, at router 9. An MCT entry is created. The src, ID, and pre_rep_router fields of this entry are fetched from M_a. The incoming_port field is set to 4 to indicate that M_a comes from the local input port. The expected_count field is set to the total destination count of M_a: 3. The cur_ACK_count field is set to 0. At router 5, M_b replicates into two packets: M_d and M_e. An MCT entry is created with an expected_count of 2. The pre_rep_router fields of both M_d and M_e are updated to 5 since the last replication occurred at router 5.

After a destination node receives a multicast, it responds with an ACK packet. Figure 8.6a shows the transmission of ACK packets corresponding to the multicast shown in Figure 8.5a. Figure 8.6b gives the ACK header values.

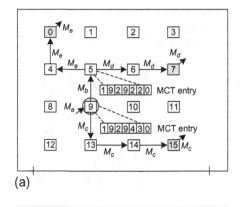

(a)

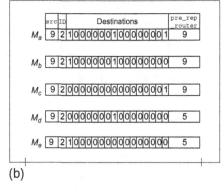

(b)

FIGURE 8.5

A multicast packet transmission example. (a) Transmission procedure. (b) Header value.

The A_c, A_d, and A_e packets are triggered by the M_c, M_d, and M_e multicast packets respectively. The `multicast_src`, `multicast_ID`, and `cur_dest` fields of the ACK packets are equal to the `src`, `ID`, and `pre_rep_router` fields, respectively, of the triggering multicast packet. The `ACK_count` fields of these three ACK packets are set to 1 since they all carry an ACK response from only one node.

The `cur_dest` field defines the current destination of the ACK packet. As shown in Figure 8.6a, A_e can be routed along a different path than M_e to reach its intermediate destination at router 5. ACK packets need only follow the same *logical* tree as the multicast packet, giving our design significant flexibility. A similar scenario can be seen for A_c. A_d and A_e both set their `cur_dest` fields as router 5; at router 5, they will be merged. Analysis shows that the possibility of multiple simultaneous MCT accesses is quite low (0.1% or less) as ACK packets will experience different congestion. A small arbiter is used to serialize concurrent accesses. Assuming A_d arrives earlier than A_e, its `multicast_src` and `multicast_ID` are used together as the tag to search the MCT. The sum of the `cur_ACK_count` field of the matched entry

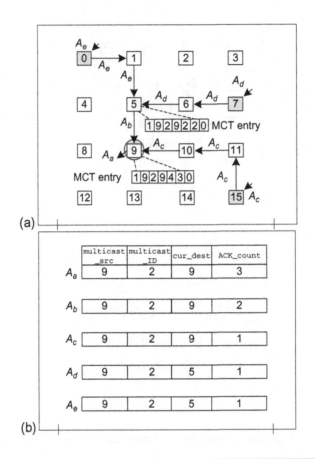

(a)

(b)

FIGURE 8.6

An ACK packet transmission example. (a) Transmission procedure. (b) Header value.

and the carried ACK_count of A_d is 1, which is smaller than the expected_count of 2 in that entry. Therefore, A_d is discarded and the cur_ACK_count field is incremented by 1.

When A_e arrives at router 5, it accesses the MCT. Since router 5 has received all expected ACKs, A_e will remain in the network. Its cur_dest field is updated to the pre_rep_router field of the matched entry and its ACK_count field is updated to 2 since it now carries the ACK responses of nodes 0 and 7 (see A_b). A_b uses the incoming_port field of the matched entry as the output port. The combined ACK packet is required to use the same multicast path for one hop to avoid an additional routing computation (RC) stage, which would add an additional cycle of latency. Now that router 5 has received all expected ACKs, the corresponding MCT entry is freed. Finally, node 9 receives A_b and A_c and combines them into A_a. The multicast-reduction transaction is complete.

8.2.3 INSUFFICIENT MCT ENTRIES

So far, we have assumed that there is always an available MCT entry when a multicast packet replicates. However, since the table size is finite, we must be able to handle a full MCT. If there are no free MCT entries, the replicated multicast packet will not update its `pre_rep_router` field. In the previous example, if there is no available MCT entry at router 5 when M_b replicates, M_d and M_e will keep their `pre_rep_router` field as 9. When A_d and A_e are injected, their `cur_dest` fields are set to 9; they will combine in router 9 instead of router 5. Both A_d and A_e must travel to router 9. In this case, router 9 will receive two ACK packets for the north-bound replication branch; this is why we record the expected total count of ACKs in the MCT instead of recording the number of direct successors in the logic multicast tree. In our design, insufficient MCT entries may affect performance, but do not pose any correctness issues. We evaluate the effect of the MCT size in Section 8.5.3.

8.3 BAM ROUTING

In this section, we describe our BAM routing algorithm. To achieve efficient bandwidth utilization, a multicast routing algorithm must compute the output port on the basis of all destination positions in a network [41]. A simple and efficient routing algorithm, RPM, was recently proposed to deliver high performance [41]. As shown in Figure 8.7, RPM partitions the network into at most eight parts on the basis of the current position, and applies several priority rules to avoid redundant replication for destinations located in different parts; the goal is to deliver a multicast packet along a common path as far as possible, then replicate and forward each copy on a different channel bound for a unique destination subset.

We observe that although RPM provides efficient bandwidth utilization, it suffers from unbalanced buffer resources between different dimensions, which negatively

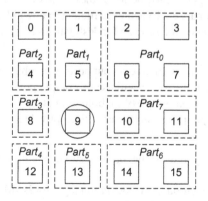

FIGURE 8.7

Network partitioning. The current position of the packet is at router 9.

affects network performance. To avoid deadlock for multicast routing, RPM divides the physical network into two VNs: VN0 is for upward packets and VN1 is for downward ones. The horizontal virtual channel (VC) buffers must be split into two disjoint subsets for the two VNs, while the vertical ones can be exclusively used by one VN [9, 41]. When a packet is routed in each VN, there are $2x$ more available vertical buffers than horizontal ones. This unbalanced buffer configuration negatively affects both unicast and multicast routing, since the more limited horizontal VCs become a performance bottleneck. Configuring different VC counts for different dimensions may mitigate this effect. However, it requires different control logic for each input port as the size of the arbiters in VCs and switch allocators is related to the VC count; a heterogeneous router requires extra design effort [33]. Also, the critical path may increase since it is determined by the largest arbiter. Therefore, we assume a homogeneous NoC router architecture in this chapter.

On the basis of these observations, we propose a novel adaptive multicast routing algorithm: BAM. The deadlock freedom of BAM is achieved by utilizing Duato's unicast deadlock-avoidance theory [8] for multicast packets, rather than leveraging multiple VNs. The multicast packets in NoCs are generally short as they carry only control packets for the coherence protocol; these are most likely single-flit packets [12], making the routing of each multicast branch independent. Thus, Duato's unicast theory can be applied to multicasts by regarding the routing of each multicast branch as an independent unicast. In Duato's theory, VCs are classified into escape and adaptive VCs. When a packet resides in an adaptive VC, it can be forwarded to any permissible output port. This property enables BAM to select the best output port on the basis of all destination positions. An additional advantage is that this design is compatible with an adaptive unicast routing algorithm.

Figure 8.7 shows the partitioning of destinations into eight parts. For $Part_1$, $Part_3$, $Part_5$, or $Part_7$, there is only one admissible output port. For $Part_0$, $Part_2$, $Part_4$, or $Part_6$, there are two alternative output ports. If a multicast packet has some destinations located in $Part_1$, $Part_3$, $Part_5$, and $Part_7$, the corresponding north, west, south, and east output ports must be used; these ports are called *obligatory* output ports for this multicast packet. To achieve efficient bandwidth utilization, we design a heuristic output port selection scheme for destinations located in $Part_0$, $Part_2$, $Part_4$, and $Part_6$:

(1) if only one of the two alternative output ports is an obligatory output port for the multicast packet, the router will use this output port to reach the destination;
(2) if the two alternative output ports are both obligatory output ports or are both not obligatory output ports, the router will adaptively select the one with less congestion.

This scheme maximally reuses the obligatory output ports to efficiently utilize bandwidth.

Figure 8.8 shows the output port calculation logic for BAM. The 1-bit P_i indicates whether there is a destination in $Part_i$. Take $Part_0$ as an example: N_{p_0} and E_{p_0} indicate that the router uses the north or east port to reach the destinations located in $Part_0$. N_{ne}

$$N_{p_0} = P_1 \cdot \overline{P_7} + P_1 \cdot P_7 \cdot N_{ne} + \overline{P_1} \cdot \overline{P_7} \cdot N_{ne} \qquad N = P_1 + N_{p_0} + N_{p_2}$$
$$E_{p_0} = \overline{P_1} \cdot P_7 + P_1 \cdot P_7 \cdot E_{ne} + \overline{P_1} \cdot \overline{P_7} \cdot E_{ne} \qquad W = P_3 + W_{p_2} + W_{p_4}$$
$$N_{p_2} = P_1 \cdot \overline{P_3} + P_1 \cdot P_3 \cdot N_{nw} + \overline{P_1} \cdot \overline{P_3} \cdot N_{nw} \qquad E = P_7 + E_{p_0} + E_{p_6}$$
$$W_{p_2} = \overline{P_1} \cdot P_3 + P_1 \cdot P_3 \cdot W_{nw} + \overline{P_1} \cdot \overline{P_7} \cdot W_{nw} \qquad S = P_5 + S_{p_4} + S_{p_6}$$
$$S_{p_4} = P_5 \cdot \overline{P_3} + P_5 \cdot P_3 \cdot S_{sw} + \overline{P_5} \cdot \overline{P_3} \cdot S_{sw}$$
$$W_{p_4} = \overline{P_5} \cdot P_3 + P_5 \cdot P_3 \cdot W_{sw} + \overline{P_5} \cdot \overline{P_3} \cdot W_{sw} \qquad Escape_n = \overline{N_{p_0}} \cdot \overline{N_{p_2}} \cdot P_1$$
$$S_{p_6} = P_5 \cdot \overline{P_7} + P_5 \cdot P_7 \cdot S_{se} + \overline{P_5} \cdot \overline{P_7} \cdot S_{se} \qquad Escape_w = W$$
$$W_{p_6} = \overline{P_5} \cdot P_7 + P_5 \cdot P_7 \cdot E_{se} + \overline{P_5} \cdot \overline{P_7} \cdot E_{se} \qquad Escape_s = \overline{S_{p_4}} \cdot \overline{S_{p_6}} \cdot P_5$$
$$Escape_e = E$$

FIGURE 8.8

The BAM routing logic.

and E_{ne} signals indicate whether the north or east output has less relative congestion in the northeast quadrant. These signals are provided by the RC module.

$Escape_n$, $Escape_w$, $Escape_s$, and $Escape_e$ indicate whether the multicast packet can use the escape VC for the north, west, south, and east output ports respectively. If a multicast packet uses the north output port to reach nodes in Part$_0$ or Part$_2$, it is not allowed to use the north escape VC since this packet will make a turn forbidden by dimensional order routing (DOR). A similar rule is applied to the south escape VC. The east and west escape VCs are always available for routing. If a multicast packet resides in an escape VC, it will replicate according to DOR, similarly to that in virtual circuit tree multicasting [12]. Once a multicast packet enters an escape VC, it can be forwarded to the destinations using only escape VCs; there is no deadlock among escape VCs. Any multicast packet residing in an adaptive VC has an opportunity to use an escape VC. This design is deadlock-free [8]. Compared with RPM, BAM does not need to partition the physical network into two VNs to avoid multicast deadlock; it achieves balanced buffer resources across vertical and horizontal dimensions. Moreover, BAM achieves efficient bandwidth utilization as well.

8.4 ROUTER PIPELINE AND MICROARCHITECTURE

Our baseline is a speculative VC router [7, 11, 38]. Lookahead signals transmit the unicast routing information one cycle ahead of the flit traversal to overlap the RC and link traversal (LT) stages [17, 24]. We use a technique to preselect the preferred output for adaptive routing; the optimal output port for each quadrant is selected one cycle ahead on the basis of network status [16, 22, 28]. The pipeline for unicast packets is two cycles plus one cycle for LT, as shown in Figure 8.9a.

Including multicast routing information in the lookahead signals requires too many bits; therefore, we assume a three-cycle router pipeline for multicasts, as shown in Figure 8.9b. A multicast packet replicates inside the router if multiple output ports are needed to reach the multicast destinations. We use asynchronous replication to eliminate lock-step traversal among several branches; the multicast

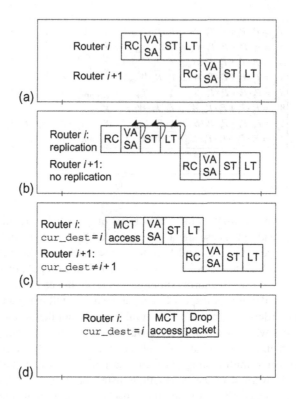

FIGURE 8.9

The router pipeline. (a) Normal unicast packet. (b) Multicast packet. (c) Updated and forwarded ACK. (d) Dropped ACK.

packet is handled as multiple independent unicast packets in the VC allocation (VA) and switch allocation (SA) stage, except that a flit is not removed from the input VC until all requested output ports are satisfied [21, 41].

ACK packets are handled differently from other unicast packets. When an ACK packet arrives, its `cur_dest` field is checked. If this field does not match the current router, the ACK packet is handled like a normal unicast packet (Figure 8.9a). If they match, the ACK packet accesses the MCT instead of performing the RC. The MCT access overlaps with the RC stage. As we show in Section 8.5.3, this operation can fit within a single pipeline stage; it does not add additional latency to the critical path. Figure 8.9c and 8.9d illustrates the ACK packet pipeline.

Figure 8.10 illustrates the proposed router microarchitecture. If a multicast packet needs multiple output ports after the RC, an entry is allocated in the MCT. This operation overlaps with the VA/SA operations and does not add delay to the critical path. The `Port Preselection` module provides eight signals indicating the optimal output port for each quadrant [16, 22, 28]. These signals are used by both unicasts and multicasts to avoid network congestion.

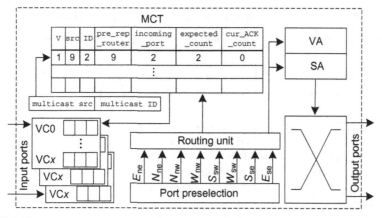

FIGURE 8.10

The router microarchitecture.

8.5 EVALUATION

We evaluate our message combination framework with RPM and BAM using synthetic traffic and real application workloads. We modify the cycle-accurate BookSim simulator [20] to model the router pipelines and microarchitecture described in Section 8.4. For synthetic traffic, we configure two VNs to avoid protocol-level deadlock [7]: one for multicasts and one for ACKs. RPM further divides the multicast VN into two sub-VNs: one for upward packets and one for downward ones. BAM does not need to sub-divide the multicast VN. Normal unicast packets can use any VN. However, once a packet has been injected into the network, its VN is fixed and cannot change during transmission. Unicast packets are routed by an adaptive routing algorithm. For BAM's multicast, BAM's ACK, and RPM's ACK VNs, the algorithm is designed on the basis of Duato's theory [8] and uses one VC as the escape VC. The two sub-VNs of RPM's multicast VN enable adaptive routing without requiring escape VCs. We use a local selection strategy for adaptive routing: when there are two available output ports, the selection strategy chooses the one with more free buffers.

Multicasts and ACKs are single-flit packets, while the normal unicast packet lengths are bimodally distributed; the unicast packets consist of five-flit packets (50%) and one-flit packets (50%). We use several synthetic unicast traffic patterns [7], including uniform random, transpose, bit rotation, and hotspot, to stress the network for detailed insight. We control the percentage of multicast packets relative to whole injected packets. For multicasts, the destination counts and positions are uniformly distributed. A cache's ACK packet response delay is uniformly distributed between one and four cycles. We assume a 64-entry MCT; in Section 8.5.3, we explore the impact of MCT size. Table 8.1 summarizes the baseline configuration and variations used in the sensitivity studies.

Table 8.1 The Baseline Simulation Configuration and Variations

Characteristic	Baseline	Variations
Topology (mesh)	4 × 4	8 × 8, 16 × 16
VC configuration	4 flits per VC, 8 VCs per port	4 and 6 VCs per port
Packet length (flits)	Normal, 1 and 5 (bimodal); ACK, 1; multicast, 1	–
Unicast traffic	Uniform random, transpose bit rotation, hotspot	–
Multicast ratio	10%	5%, 15%, 20%
Multicast destination count	2-10 (uniformly distributed)	2-4, 4-14, 10-14, 15
ACK response cycles	1-4 (uniformly distributed)	–
MCT entries	64	0, 1, 4, 16
Warm-up and total	10,000 and 100,000 cycles	–

Table 8.2 The Full System Simulation Configuration

Parameter	Value
No. of cores	16
L1 cache (D and I)	Private, 4-way, 32 kB each
L2 cache	Private, 8-way, 512 kB each
Cache coherence	MOESI distributed directory
Topology	4 × 4 mesh

L1, first level; L2, second level.

To measure full-system performance, we leverage two existing simulation frameworks: FeS2 [35] for x86 simulation and BookSim for NoC simulation. FeS2 is a timing-first, multiprocessor x86 simulator, implemented as a module for Virtutech Simics [30]. We run the PARSEC benchmarks [4] with 16 threads on a 16-core chip multiprocessor consisting of an Intel Pentium 4-like CPU. We assume cores optimized for clock frequency; they are clocked five times faster than the network. We use a distributed, directory-based MOESI coherence protocol that needs four VNs for protocol-level deadlock freedom. The cache line invalidation packets (multicasts) are routed in VN1, while the ACK packets are routed in VN2. The number of VCs per VN, VC depth, and MCT size are the same as the baseline (Table 8.1). Cache lines are 64 bytes wide and the network flit width is 16 bytes. All benchmarks use the *simsmall* input sets to reduce simulation time. The total run time is used as the metric for full-system performance. Table 8.2 gives the system configuration.

8.5.1 PERFORMANCE

We evaluate four scenarios: RPM without message combination (RPM + NonCom), RPM with message combination (RPM + Com), BAM without message combination (BAM + NonCom), and BAM with message combination (BAM + Com).

8.5.1.1 Overall network performance

Figure 8.11 illustrates the overall network performance. For both RPM + Com and BAM + Com there are performance improvements compared with RPM + NonCom and BAM+NonCom respectively; not only are the network latencies reduced, but the saturation throughputs[3] are also improved. Detailed analysis reveals that the combination framework reduces the average channel traversal count of ACK packets from 4.7 to 2.5, reducing the network operations by approximately 45% for ACKs.

The network latency for RPM + Com is reduced by 10-20% under low-to-medium injection rates[4] compared with RPM + NonCom (average 14.1%); larger

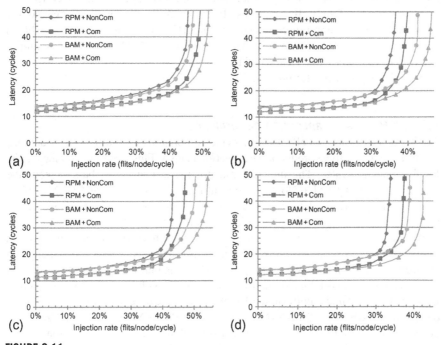

FIGURE 8.11

The overall network performance (10% multicast, average of six destinations). (a) Uniform random. (b) Transpose. (c) Bit rotation. (d) Hotspot.

[3]The saturation point is measured as the injection rate at which the average latency is three times the zero-load latency.
[4]The low-to-medium injection rate is the injection rate at which the average latency is less than two times the zero-load latency.

improvements are seen at high injection rates. Similar latency reductions are seen for BAM + Com versus BAM + NonCom. Packet dropping shortens the average hop distance, resulting in latency reductions at both low and high network loads. The mitigation of ejection-side congestion is also beneficial to latency reduction. Saturation throughput improvements resulting from the framework (RPM + Com vs. RPM + NonCom) range from 8.5% to 11.2% (average 9.6%). There are similar throughput improvements for BAM + Com over BAM + NonCom. Discarding in-flight ACKs reduces the network load, which is helpful to improve the saturation throughput.

Across the four patterns, both BAM + NonCom and BAM + Com improve the saturation throughput over RPM + NonCom and RPM + Com respectively. For transpose, bit rotation and hotspot patterns, although BAM + NonCom has larger latencies than RPM + Com under low loads, its saturation throughputs are higher. The balanced buffer configuration between vertical and horizontal dimensions helps BAM to improve the saturation throughput. BAM + Com improves the saturation throughput by 14.2%, 27.6%, 26.4%, and 25.1% (average 23.4%) over RPM + NonCom for the four traffic patterns. Both the ACK packet dropping and the balanced buffer configuration contribute to this performance improvement. As shown in Figure 8.11, the trend between BAM + NonCom and RPM + NonCom is similar to the trend between BAM + Com and RPM + Com; thus, we omit BAM + NonCom in the following sections for brevity.

8.5.1.2 Multicast transaction performance

To clearly understand the effects of message combination on multicast transactions, we measure the multicast-reduction transaction latency. Figure 8.12 shows the results for five injection rates. The injection rate for the last group of bars exceeds the saturation point of RPM + NonCom. For all rates, RPM + Com and BAM + Com have lower transaction latencies than RPM + NonCom; dropping ACK packets reduces network congestion and accelerates multicast-reduction transactions. ACK packet acceleration contributes more to the latency reduction than multicasts. A multicast needs to send out multiple replicas; releasing its current VC depends on the worst congestion each replica may face. Thus, the multicast is not as sensitive as the ACK to the network load reduction. For example, with low-to-medium injection rates (0.30 or less) under uniform random traffic (Figure 8.12a), the average multicast delay is reduced by 9.5% for BAM + Com versus RPM + NonCom. Yet, ACK packet delay is reduced by 17.6%. These two factors result in an average transaction latency reduction of 13.4%. The transaction acceleration increases with higher network load. Merging ACKs reduces the number of packets the source needs to wait for to finish a transaction. Waiting for only one ACK instead of multiple ACKs can improve performance since multiple packets may encounter significantly more congestion than a single packet, especially under high network load. For uniform random traffic with a high injection rate (0.40), RPM + Com and BAM + Com reduce the transaction latency by 46.2% and 57.6% respectively compared with RPM + NonCom.

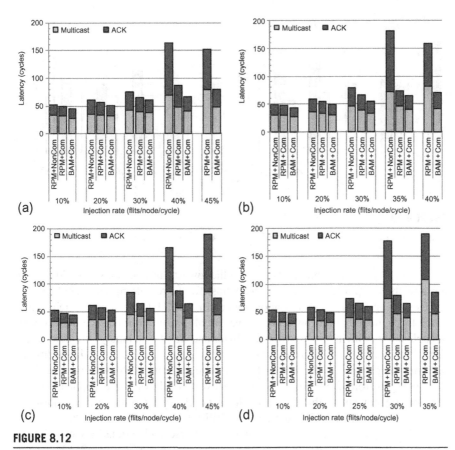

FIGURE 8.12

The multicast-reduction transaction latency (10% multicast, average of six destinations).
(a) Uniform random. (b) Transpose. (c) Bit rotation. (d) Hotspot.

The message combination framework accelerates the total transaction by a factor of almost 2. As the injection rate increases, BAM + Com outperforms RPM + Com by a significant margin.

8.5.1.3 Real application performance

Figure 8.13 shows the speedups over RPM + NonCom for the PARSEC benchmarks. Although the message combination framework mainly optimizes the performance for one of the four VNs (the ACK VN) used in full-system simulation, collective communication often lies on the critical path for applications. Also, dropping ACK packets in one VN reduces SA contention and improves the performance of other VNs. These factors result in RPM + Com achieving speedups over RPM + NonCom ranging from 5.3% to 8.3% for all applications. The efficiency of message combination depends on the multicast destination count. The multicast destination counts of

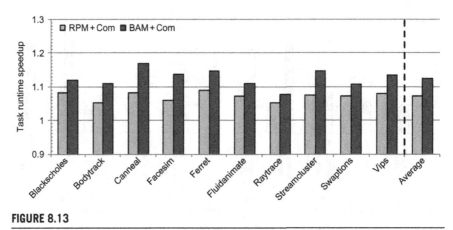

FIGURE 8.13

System speedups against RPM + NonCom for PARSEC workloads.

bodytrack, facesim, and raytrace are the lowest; the speedups of RPM + Com for these three applications are lower than those for the remaining seven applications.

The balanced buffer configuration utilized in BAM + Com supports higher saturation throughput than the unbalanced one. BAM + Com improves the performance of applications which have high network loads and significant bursty communication. For blackscholes, fluidanimate, raytrace, and swaptions, the additional performance gain due to the balanced buffer configuration (BAM + Com vs. RPM + Com) ranges from 2.3% to 3.7%. The network loads of these applications are low and do not stress the network. However, BAM + Com achieves additional speedups ranging from 5.5% to 8.6% over RPM + Com for bodytrack, canneal, facesim, ferret, streamcluster, and vips. These applications have more bursty communication and higher injection rates. For all ten applications, BAM + Com achieves an average speedup of 12.7% over RPM + NonCom. The maximal speedup is 16.8%, for canneal.

8.5.2 COMPARING MULTICAST VN CONFIGURATIONS

In this section, we delve into the effect of the unbalanced buffer configuration used in the multicast VN of RPM on both unicast and multicast packet routing. Since the ACK VNs of RPM and BAM are the same, we assume the network has only one VN: the multicast VN. Four VCs are configured in this VN, and RPM further divides this VN into two sub-VNs: the horizontal VC count of each sub-VN is two, while the vertical count is four.

8.5.2.1 Unicast performance

Figure 8.14 compares the performance of BAM's and RPM's multicast VN configuration using only unicast packets. To extensively understand the effect of the unbalanced buffer configuration, we evaluate the performance of *XY*, *YX*, and a

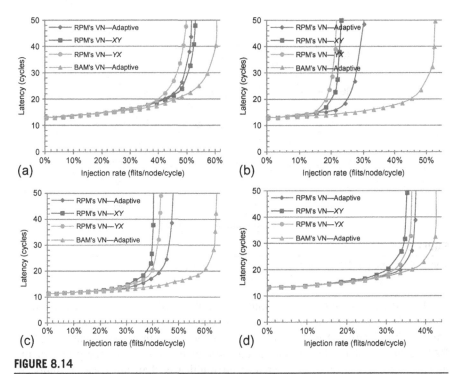

FIGURE 8.14

The unicast traffic performance for RPM's and BAM's multicast VN. (a) Uniform random. (b) Transpose. (c) Bit rotation. (d) Hotspot.

locally adaptive routing algorithm (Adaptive) in RPM's multicast VN. *XY* efficiently distributes uniform random traffic and achieves the highest performance for this pattern. For the other three patterns, Adaptive has the highest performance; adaptively choosing the output port mitigates the negative effect of unbalanced buffer resources. Therefore, this work uses locally adaptive routing for unicast packets.

Although Adaptive has better performance than *XY* and *YX*, its performance is still limited by the unbalanced buffer resources used in each of RPM's multicast sub-VNs; the horizontal dimension has half the buffer resources of the vertical one. However, in BAM's VN, the numbers of buffers of different dimensions are equal. Adaptive routing in BAM's VN shows substantial performance improvement over Adaptive routing in RPM's VN. Transpose has the largest performance gain, with a 73.2% saturation point improvement. Across these four traffic patterns, BAM's VN-Adaptive achieves an average saturation throughput improvement of 35.3% over RPM's VN-Adaptive.

8.5.2.2 Multicast performance

Figure 8.15 shows the performance using 100% multicast packets. The Adaptive curve shows the performance of the adaptive multicast routing algorithm without

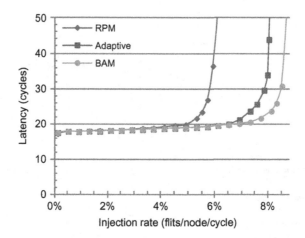

FIGURE 8.15

The performance of 100% multicast traffic (average of six destinations).

our heuristic replication scheme; multicast replicas adaptively choose the output ports without considering the reuse of obligatory ports. This negatively affects bandwidth utilization. BAM has 8.7% higher saturation throughput than Adaptive. BAM achieves 47.1% higher saturation throughput over RPM. The effect of the unbalanced buffer resources is greater on multicasts than unicasts. A multicast packet is removed from its current VC only after all its replicas have been sent out; the horizontal VC bottleneck affects multicast performance more strongly than unicast performance. The average switch traversal counts are 8.6, 8.8, and 8.4 for RPM, Adaptive, and BAM respectively, which further demonstrates that our applied heuristic replication scheme achieves efficient bandwidth utilization.

8.5.3 MCT SIZE

As described in Section 8.2.3, the MCT size affects network performance but not correctness. Too few MCT entries will hamper ACK packet combination and force the ACK packets to traverse more hops than with combination. To determine the appropriate size, we simulate an infinite MCT using the baseline configuration (Table 8.1) and uniform random traffic. Multicast packets are routed using BAM. Figure 8.16a presents the maximum and average concurrently valid MCT entries. For low-to-medium injection rates (less than 0.39), the maximum number of concurrently valid entries is less than 10 and the average number is less than 1.5. Even when the network is at saturation (injection rate of 0.52), the maximum number of concurrently valid entries is 49 and the average is 10.15. These experimental results indicate that a small MCT can provide good performance.

Figure 8.16b shows the performance for different table sizes. The 0-entry curve corresponds to no ACK combination; all ACK packets are injected into the network

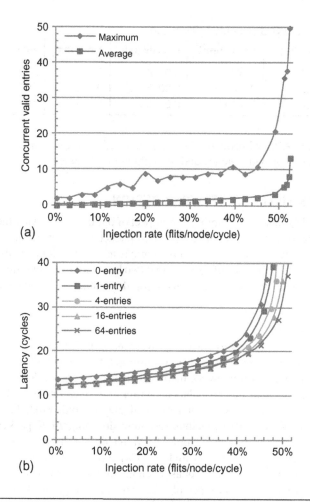

FIGURE 8.16

The evaluation of MCT sizes (10% multicast, average of six destinations). (a) Concurrent valid entries. (b) Performance.

with their destination set to the source node of the triggering multicast packet. Even with only one entry per router, ACK combination reduces the average network latency by 10%. More entries reduce the latency further, especially for high injection rates. Saturation throughput improvements range from 3.3% to 12.1% for one to 64 entries.

Cacti [34] is used to calculate the power consumption, area, and access latency for the MCT in a 32-nm technology process. Table 8.3 shows the results. If we assume a clock frequency of 1 GHz, a 64-entry table can be accessed in one cycle. This size provides good performance for nearly all injection rates for various traffic patterns. In the full-system evaluation, we observe that the maximal number of concurrently valid

Table 8.3 The MCT Overhead

Entries	Area (mm^2)	Energy (nJ)	Time (ns)	Bytes
16	0.0011	0.0008	0.138	48
32	0.0017	0.0013	0.146	96
64	0.0031	0.0026	0.153	192

entries for the PARSEC benchmarks is less than 25. For area-constrained designs, fewer entries still provide latency and throughput improvements.

8.5.4 SENSITIVITY TO NETWORK DESIGN

To understand further the scalability and impact of our design, we vary the VC count, multicast ratio, destination count per multicast packet, and network size. Figure 8.17 presents the average performance improvement across the four synthetic unicast traffic patterns. In each group of bars, the first two bars show the saturation throughput improvement, and the second two bars show the latency reduction under low-to-medium loads.

8.5.4.1 VC count

Figure 8.17a shows the performance improvement with eight, six, and four VCs per physical channel. One interesting trend is observed: for smaller VC counts, the gain due to the combination framework increases (RPM + Com vs. RPM + NonCom), while the improvement due to balanced buffer resources declines (BAM + Com vs. RPM + Com). The reasons for this trend are twofold. First, fewer VCs per port makes the VCs a more precious resource; dropping ACK packets improves the reuse of this resource. For example, RPM + Com has 14.8% higher saturation throughput than RPM + NonCom with four VCs per physical channel. As the number of VCs increases, this resource is not as precious; its effect on performance declines. However, even with eight VCs per physical channel, dropping ACK packets still improves the saturation throughput by 9.6%.

Second, BAM + Com uses escape VCs to avoid deadlock. The horizontal escape VC can always be used, while the vertical one can be used only by DOR; there is some imbalance in the utilization of escape VCs, and this imbalance increases with fewer VCs. However, the situation is worse for RPM + Com. In RPM's multicast VN configuration, the vertical dimension always has twice as many VCs as the horizontal one. Even with four VCs per port, the saturation point improvement of BAM + Com is still larger than RPM + Com's improvement by about 9.0%. With fewer VCs, RPM + Com's latency reduction increases; it achieves an 18.5% reduction with four VCs per port. BAM + Com further reduces latency owing to the balanced buffer configuration among different dimensions. The latency difference between BAM + Com and RPM + Com is not as significant as the saturation throughput improvement. Adaptive routing mainly accelerates packet transmission at high injection rates by avoiding network congestion.

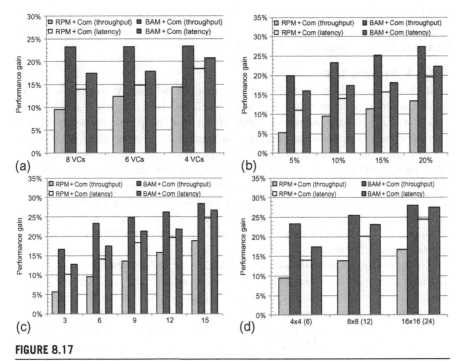

FIGURE 8.17

Performance gains of RPM + Com and BAM + Com over RPM + NonCom for sensitivity studies. (a) VC counts. (b) Multicast ratios. (c) Multicast destinations. (d) Network sizes.

8.5.4.2 Multicast ratio

Figure 8.17b presents the performance improvement for several multicast ratios: 5%, 10%, 15%, and 20%. Increasing the multicast portion leads to greater throughput improvements owing to message combination. The improvement contributed by the balanced buffer configuration remains almost constant (BAM + Com vs. RPM + Com). A higher multicast packet ratio triggers more ACK packets; ACK combination has more opportunity to reduce network load. The framework becomes more effective. BAM + Com achieves a saturation throughput improvement of 27.2% with a multicast ratio of 20%. A higher multicast ratio also results in larger latency reductions. The VC count per physical channel is kept constant (eight VCs per port) in this experiment, so the gap between BAM + Com and RPM + Com remains almost the same.

8.5.4.3 Destinations per multicast

Figure 8.17c illustrates the performance gain for different average numbers of multicast destinations: 3, 6, 9, 12, and 15 (broadcast). The trend is similar to varying the multicast ratio. Although the multicast ratio remains constant (10%), more destinations per multicast trigger more ACK packets. The framework combines more of these ACKs during transmission. As the destination count varies from 3 to 15,

BAM + Com improves the saturation throughput in the range from 17% to 28%. RPM + Com reduces the latency in the range from 10% to 25%; BAM + Com's latency reduction ranges from 13% to 27%.

8.5.4.4 Network size

Figure 8.17d shows the performance improvement for different network sizes: 4 × 4, 8 × 8, and 16 × 16 mesh networks.[5] Since 8 × 8 and 16 × 16 networks have more nodes, we increase the average number of destinations per multicast to 12 and 24 respectively. The message combination framework is more efficient at larger network sizes since packets traverse more hops on average. As a result, combining ACK packets eliminates more network operations. For the 16 × 16 network, the message combination framework improves the saturation point by 16.8%. Similarly, larger network sizes show greater latency reductions. RPM + Com and BAM + Com achieve latency reductions of 25% and 27% respectively for a 16 × 16 mesh. The efficiency of the balanced buffer configuration utilized by BAM + Com remains constant with different network sizes.

Throughout the sensitivity studies, the MCT size is 64 entries. One may think that with more multicast destinations, a higher multicast ratio, or a larger network size that more entries will be required. Yet, analysis reveals changing these aspects reduces the injection rate at which the network becomes saturated. Thus, the maximal concurrently active multicast transactions supported by the network is reduced. As a result, a 64-entry MCT is able to achieve high performance for these different network design points.

8.6 POWER ANALYSIS

An NoC power model [32] is leveraged to determine overall network power consumption; network power consumption is contributed by three main components: channels, input buffers, and router control logic. The activity of these components is obtained from BookSim. Leakage power is included for buffers and channels. The power consumption of an MCT access is also integrated into the power model. We assume a 128-bit flit width. The technology process is 32 nm and the network clock is 1 GHz on the basis of a conservative assumption of router frequency. Figure 8.18a and 8.18b shows the power consumption using transpose traffic for unicast messages. We measure the MCT access power and the static and dynamic network power for two multicast ratios: 10% and 20%.

The MCT access power comprises only a small portion of the total power. The reason is twofold. The MCT is very small; each entry is only 3 bytes. With 64 entries, the size of the MCT is 192 bytes, which is only 7.5% of the size of flit buffers. Second and more importantly, the MCT access activity is very low. Even when the network is

---

[5]Bit string encoding is used in all experiments to encode the destination set; this method is impractical in large networks. However, further exploration of this issue is orthogonal to the message combination framework.

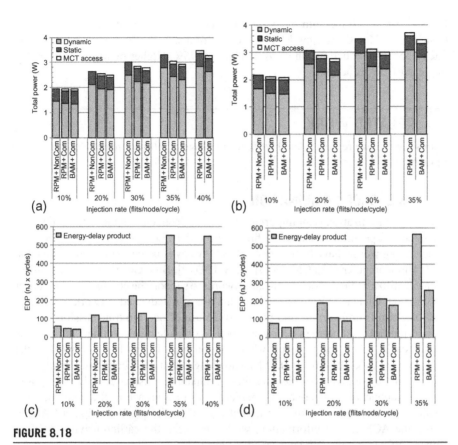

FIGURE 8.18

Power consumption and EDP results (average of six destinations). (a) Power (10% multicast). (b) Power (20% multicast). (c) EDP (10% multicast). (d) EDP (20% multicast).

saturated, only 7.2% of all cycles have an MCT access; if the network is not saturated, the activity is even lower.

RPM + Com reduces power consumption compared with RPM + NonCom for all injection rates owing to the dropping of ACK packets. BAM + Com further reduces the power consumption owing to more balanced buffer utilization among different dimensions. As the injection rate increases, the reduction in power consumption becomes more obvious. For an injection rate of 0.35 with a multicast ratio of 10%, RPM + Com and BAM + Com achieve power reductions of 7.6% and 10.8% respectively over RPM + NonCom. With larger multicast ratios, the combination framework is able to reduce power consumption more for both RPM + Com and BAM + Com.

The EDP [14] of the whole network further highlights the energy efficiency of our design, as shown in Figure 8.18c and 8.18d. Dropping ACK packets during transmission not only results in fewer network operations, but also reduces network latency. For low-to-medium injection rates with a multicast ratio of 10%, RPM + Com and BAM + Com show EDP reductions of about 20-40%. At a high injection rate (0.35), this reduction can be as much as 60-75%. Higher multicast ratios result in greater EDP reduction.

8.7 RELATED WORK

In this section, we review related work for message combination and NoC multicast routing.

8.7.1 MESSAGE COMBINATION

Barrier synchronization is an important class of collective communication, in which a reduction operation is executed first, followed by a broadcast. Gathering and broadcasting worms have been proposed [37]. Many supercomputers, including the NYU Ultracomputer [15], CM-5 [25], Cray T3D [2], Blue Gene/L [3], and TianHe-1A [45], utilize dedicated or optimized networks to support the combination of barrier information. Oh *et al.* [36] observed that the use of a dedicated network in a many-core platform is unfavorable and proposed the use of on-chip transmission lines to support multiple fast barriers. Our work focuses on collective communication that is used in cache coherence protocols where the multicast is sent first, followed by the collection of ACKs. Bolotin *et al.* [5] acknowledged that ACK combination might be useful, but did not give a detailed design or evaluation. Krishna *et al.* [23] proposed efficient support for collective communications in coherence traffic. Our message combination mechanism is quite different from their design. We utilize a CAM to record the ACK arrival information, while they keep the earlier-arriving ACK in network VCs to merge with later-arriving ones [23].

8.7.2 NoC MULTICAST ROUTING

Recent work explores various NoC multicast routing algorithms. Lu *et al.* [27] used path-based multicast routing, which requires path setup and ACKs, resulting in a long latency overhead. Tree-based multicast mechanisms in NoCs avoid this latency overhead. Virtual circuit tree multicasting [12] is based on the concept of virtual multicast trees. The bLBDR design [39] uses broadcasting in a small region to implement multicasting. RPM [41] focuses on achieving bandwidth-efficient multicasting. On the basis of the isolation mechanism proposed in bLBDR, Wang *et al.* [43] extended RPM for irregular regions. MRR [1] is an adaptive multicast routing algorithm based on the rotary router. Whirl [23] provides efficient support for broadcasts and dense multicasts. The deadlock avoidance mechanisms of Whirl and our proposed BAM are similar; both are based on Duato's theory [8]. Both BAM and RPM use bit string encoding for multicast packet destinations; this method is not scalable for large networks. Some compression methods [42] can improve the scalability of this encoding scheme. In addition, coarse bit vectors [18], similar to what has been proposed for directories, are another possible approach to reduce the size of the destination set encodings. This type of encoding will increase the number of destinations per multicast and receive greater benefits from our proposal. Delving into this issue is left as future work.

8.8 CHAPTER SUMMARY

Scalable NoCs must handle traffic in an intelligent fashion; to improve performance and reduce power they must eliminate unnecessary or redundant messages. The proposed message combination framework does just that by combining in-flight ACK responses to multicast requests. A small 64-entry CAM is added to each router to coordinate combination. In addition to the framework, we propose a novel multicast routing algorithm that balances buffer resources between different dimensions to improve performance. Simulation results show that our message combination framework not only reduces latency by 14.1% for low-to-medium network loads, but also improves the saturation point by 9.6%. The framework is more efficient for fewer VCs, larger network size, a higher multicast ratio, or more destinations per multicast. The balanced buffer configuration achieves an additional 13.8% saturation throughput improvement. Our proposed message combination framework can be easily extended to support the reduction operations in token coherence and other parallel architectures.

REFERENCES

[1] P. Abad, V. Puente, J.-A. Gregorio, MRR: enabling fully adaptive multicast routing for CMP interconnection networks, in: Proceedings of the International Symposium on High-Performance Computer Architecture (HPCA), 2009, pp. 355–366. 256, 280
[2] D. Adams, CRAY T3D system architecture overview, Technical Report, Cray Research Inc., 1993. 256, 280
[3] N. Adiga, G. Almasi, G. Almasi, et al., An overview of the Blue Gene/L supercomputer, in: Proceedings of the ACM/IEEE Conference on Supercomputing (SC), 2002, pp. 1–22. 256, 280
[4] C. Bienia, S. Kumar, J.P. Singh, K. Li, The PARSEC benchmark suite: characterization and architectural implications, in: Proceedings of the International Conference on Parallel Architectures and Compilation Techniques (PACT), 2008, pp. 72–81. 256, 268
[5] E. Bolotin, Z. Guz, I. Cidon, R. Ginosar, A. Kolodny, The power of priority: NoC based distributed cache coherency, in: Proceedings of the International Symposium on Networks-on-Chip (NOCS), 2007, pp. 117–126. 256, 280
[6] C.-M. Chiang, L.M. Ni, Multi-address encoding for multicast, in: Proceedings of the International Workshop on Parallel Computer Routing and Communication (PCRCW), 1994, pp. 146–160. 260
[7] W. Dally, B. Towles, Principles and Practices of Interconnection Networks, first ed., Morgan Kaufmann Publishers Inc., San Francisco, CA, 2003. 265, 267

[8] J. Duato, A new theory of deadlock-free adaptive routing in wormhole networks, IEEE Trans. Parallel Distrib. Syst. 4(12) (1993) 1320–1331. 264, 265, 267, 280

[9] J. Duato, S. Yalamanchili, L. Ni, Interconnection Networks: An Engineering Approach, first ed., Morgan Kaufmann Publishers Inc., San Francisco, CA, 2003. 256, 264

[10] N. Enright Jerger, SigNet: network-on-chip filtering for coarse vector directories, in: Proceedings of the Design, Automation & Test in Europe Conference & Exhibition (DATE), 2010, pp. 1378–1383. 256

[11] N. Enright Jerger, L. Peh, On-Chip Networks, first ed., Morgan & Claypool, San Rafael, CA, 2009. 265

[12] N. Enright Jerger, L.-S. Peh, M. Lipasti, Virtual circuit tree multicasting: a case for on-chip hardware multicast support, in: Proceedings of the International Symposium on Computer Architecture (ISCA), 2008, pp. 229–240. 256, 264, 265, 280

[13] N.D. Enright Jerger, L.-S. Peh, M.H. Lipasti, Virtual tree coherence: leveraging regions and in-network multicast trees for scalable cache coherence, in: Proceedings of the International Symposium on Microarchitecture (MICRO), 2008, pp. 35–46. 258

[14] R. Gonzalez, M. Horowitz, Energy dissipation in general purpose microprocessors, IEEE J. Solid-State Circuits 31(9) (1996) 1277–1284. 279

[15] A. Gottlieb, R. Grishman, C.P. Kruskal, K.P. McAuliffe, L. Rudolph, M. Snir, The NYU ultracomputer—designing a MIMD, shared-memory parallel machine, in: Proceedings of the International Symposium on Computer Architecture (ISCA), 1982, pp. 27–42. 256, 280

[16] P. Gratz, B. Grot, S. Keckler, Regional congestion awareness for load balance in networks-on-chip, in: Proceedings of the International Symposium on High-Performance Computer Architecture (HPCA), 2008, pp. 203–214. 265, 266

[17] P. Gratz, K. Sankaralingam, H. Hanson, P. Shivakumar, R. McDonald, S. Keckler, D. Burger, Implementation and evaluation of a dynamically routed processor operand network, in: Proceedings of the International Symposium on Networks-on-Chip (NOCS), 2007, pp. 7–17. 265

[18] A. Gupta, W.-D. Weber, T. Mowry, Reducing memory and traffic requirements for scalable directory-based cache coherence schemes, in: Proc of the International Conference on Parallel Processing (ICPP), 1990, pp. 312–321. 280

[19] J.L. Hennessy, D.A. Patterson, Computer Architecture: A Quantitative Approach, third ed., Morgan Kaufmann Publishers Inc., San Francisco, CA, 2003. 256

[20] N. Jiang, D.U. Becker, G. Michelogiannakis, J. Balfour, B. Towles, D. Shaw, J. Kim, W. Dally, A detailed and flexible cycle-accurate network-on-chip simulator, in: Proceedings of the International Symposium on Performance Analysis of Systems and Software (ISPASS), 2013, pp. 86–96. 267

[21] Y.H. Kang, J. Sondeen, J. Draper, Multicast routing with dynamic packet fragmentation, in: Proceedings of the Great Lakes Symposium on VLSI (GLSVLSI), 2009, pp. 113–116. 256, 266

[22] J. Kim, D. Park, T. Theocharides, N. Vijaykrishnan, C.R. Das, A low latency router supporting adaptivity for on-chip interconnects, in: Proceedings of the Design Automation Conference (DAC), 2005, pp. 559–564. 265, 266

[23] T. Krishna, L.-S. Peh, B.M. Beckmann, S.K. Reinhardt, Towards the ideal on-chip fabric for 1-to-many and many-to-1 communication, in: Proceedings of the International Symposium on Microarchitecture (MICRO), 2011, pp. 71–82. 256, 280

[24] A. Kumar, P. Kundu, A. Singh, L.-S. Peh, N. Jha, A 4.6 Tbits/s 3.6 GHz single-cycle NoC router with a novel switch allocator in 65 nm CMOS, in: Proceedings of the International Conference on Computer Design (ICCD), 2007, pp. 63–70. 265

[25] C.E. Leiserson, Z.S. Abuhamdeh, D.C. Douglas, C.R. Feynman, M.N. Ganmukhi, J.V. Hill, D. Hillis, B.C. Kuszmaul, M.A. St Pierre, D.S. Wells, et al., The network architecture of the connection machine CM–5, J. Parallel Distrib. Comput. 33 (1996) 272–285. 256, 280

[26] D. Lenoski, J. Laudon, K. Gharachorloo, A. Gupta, J. Hennessy, The directory-based cache coherence protocol for the DASH multiprocessor, in: Proceedings of the International Symposium on Computer Architecture (ISCA), 1990, pp. 148–159. 256

[27] Z. Lu, B. Yin, A. Jantsch, Connection-oriented multicasting in wormhole-switched networks on chip, in: Proceedings of the International Symposium on Emerging VLSI Technologies and Architectures (ISVLSI), 2006, pp. 205–210. 256, 280

[28] S. Ma, N. Enright Jerger, Z. Wang, DBAR: an efficient routing algorithm to support multiple concurrent applications in networks-on-chip, in: Proceedings of the International Symposium on Computer Architecture (ISCA), 2011, pp. 413–424. 265, 266

[29] S. Ma, N. Enright Jerger, Z. Wang, Supporting efficient collective communication in NoCs, in: Proceedings of the International Symposium on High-Performance Computer Architecture (HPCA), 2012, pp. 165–176. 255

[30] P.S. Magnusson, M. Christensson, J. Eskilson, D. Forsgren, G. Hallberg, J. Hogberg, F. Larsson, A. Moestedt, B. Werner, Simics: a full system simulation platform, Computer 35 (2002) 50–58. 268

[31] M. Martin, M. Hill, D. Wood, Token coherence: decoupling performance and correctness, in: Proceedings of the International Symposium on Computer Architecture (ISCA), 2003, pp. 182–193. 256, 258

[32] G. Michelogiannakis, D. Sanchez, W.J. Dally, C. Kozyrakis, Evaluating bufferless flow control for on-chip networks, in: Proceedings of the International Symposium on Networks-on-Chip (NOCS), 2010, pp. 9–16. 278

[33] A.K. Mishra, N. Vijaykrishnan, C.R. Das, A case for heterogeneous on-chip interconnects for CMPs, in: Proceedings of the International Symposium on Computer Architecture (ISCA), 2011, pp. 389–400. 264

[34] N. Muralimanohar, R. Balasubramonian, N. Jouppi, CACTI 6.0: a tool to model large caches, Technical Report HPL-2009-85, HP Laboratories, April 2009. 275

[35] N. Neelakantam, C. Blundell, J. Devietti, M.M. Martin, C. Zilles, FeS2: a full-system execution-driven simulator for x86, in: Poster presented at ASPLOS 2008, 2008. 268

[36] J. Oh, M. Prvulovic, A. Zajic, TLSync: support for multiple fast barriers using on-chip transmission lines, in: Proceedings of the International Symposium on Computer Architecture (ISCA), 2011, pp. 105–115. 256, 280

[37] D. Panda, Fast barrier synchronization in wormhole k-ary n-cube networks with multidestination worms, in: Proceedings of the International Symposium on High-Performance Computer Architecture (HPCA), 1995, pp. 200–209. 256, 280

[38] L.-S. Peh, W. Dally, A delay model and speculative architecture for pipelined routers, in: Proceedings of the International Symposium on High-Performance Computer Architecture (HPCA), 2001, pp. 255–266. 265

[39] S. Rodrigo, J. Flich, J. Duato, M. Hummel, Efficient unicast and multicast support for CMPs, in: Proceedings of the International Symposium on Microarchitecture (MICRO), 2008, pp. 364–375. 256, 280

[40] F. Samman, T. Hollstein, M. Glesner, New theory for deadlock-free multicast routing in wormhole-switched virtual-channelless networks-on-chip, IEEE Trans. Parallel Distrib. Syst. 22(4) (2011) 544–557.

[41] L. Wang, Y. Jin, H. Kim, E.J. Kim, Recursive partitioning multicast: a bandwidth-efficient routing for networks-on-chip, in: Proceedings of the International Symposium on Networks-on-Chip (NOCS), 2009, pp. 64–73. 256, 258, 263, 264, 266, 280

[42] L. Wang, P. Kumar, R. Boyapati, K.H. Yum, E.J. Kim, Efficient lookahead routing and header compression for multicasting in networks-on-chip, in: Proceedings of ACM/IEEE Symposium on Architectures for Networking and Communications Systems (ANCS), 2010, pp. 1–10. 280

[43] X. Wang, M. Yang, Y. Jiang, P. Liu, On an efficient NoC multicasting scheme in support of multiple applications running on irregular sub-networks, Microprocess. Microsyst. 35 (2011) 119–129. 256, 280

[44] H. Xu, P. McKinley, L. Ni, Efficient implementation of barrier synchronization in wormhole-routed hypercube multicomputers, in: Proceedings of the International Conference on Distributed Computing Systems (ICDCS), 1992, pp. 118–125. 256

[45] X. Yang, X. Liao, K. Lu, Q. Hu, J. Song, J. Su, The TianHe-1A supercomputer: its hardware and software, J. Comput. Sci. Technol. 26(3) (2011) 344–351. 256, 280

Network-on-chip customizations for message passing interface primitives[†]

9

[†]Part of this research was first presented at the IEEE 23rd International Conference on Application-Specific Systems, Architectures and Processors (ASAP-2012) [23].

Networks-on-Chip. http://dx.doi.org/10.1016/B978-0-12-800979-6.00009-3

9.1 INTRODUCTION

To enable continuous exponential performance scaling for parallel applications, multicore designs have become the dominant organization forms for future high-performance microprocessors. The availability of massive on-chip transistors to hardware architects gives rise to the expectation that the exponential growth in the number of cores on a single processor will soon facilitate the integration of hundreds of cores into mainstream computers [3].

Traditional shared bus interconnects are incapable of sustaining the multicore architecture with the appropriate degree of scalability and required bandwidth to facilitate communication among a large number of cores. Moreover, full crossbars become impractical with the growing number of cores. Networks-on-chip (NoCs) are becoming the most viable solution to mitigate this problem. The NoCs are conceived to be more cost-effective than the bus in terms of traffic scalability, area, and power in large-scale systems. Such networks are ideal for component reuse, design modularity, plug-and-play, and scalability while avoiding issues with global wire delays. Recent proposals, such as the 64-core TILE64 processor from Tilera [51], Intel's 80-core Teraflops chip [22], Arteris's NoC interconnect IPs [2], and NXP-Philips' AEtheral NoC [18], have successfully demonstrated the potential effectiveness of NoC designs.

However, most existing general-purpose NoC designs do not support the high-level programming model well. This condition compromises performance and efficiency when programs are mapped onto NoC-based hardware architectures. That is, most current programming model optimizations aiming to address these problems maintain a firm abstraction of the interconnection network fabric as a communication medium: protocol optimizations comprise end-to-end messages between requestor and answer nodes, whereas network optimizations separately aim at reducing the communication latency and improving the throughput for data messages. Reducing or even eliminating the gap between the multicore programming models with the underlying NoC-based hardware is a demanding task. Thus, a key challenge in multicore research is the provision of efficient support for parallel programming models to boost applications by exploiting all hardware features available in NoC-based multicore architectures.

A large body of research has recently focused on integrating the message passing interface (MPI) standard into multicore architectures. The MPI is a standard, public-domain, platform-independent communications library for message-passing programming. Numerous applications have now been ported to or developed for the MPI model. The performance optimization of such models is a necessity for multicore architectures. These research efforts include the systems-on chip (SoC) MPI library implemented on a Xilinx Virtex field-programmable gate array (FPGA) [30], the rMPI targeting embedded systems using MIT's Raw processor [39], the TMD-MPI (a lightweight subset implementation of the MPI standard) focusing on the parallel programming of multicore multi-FPGA systems [42], the MPI communication layer for reconfigurable cluster-on-chip architecture [52], and the lightweight MPI (LMPI) for embedded heterogeneous systems [1]. These studies have successfully demonstrated the effectiveness of adapting MPI into

NoC-based multicore processors. However, given the enormity and complexity of MPI, the aforementioned solutions are implemented in software and do not consider the refinement of NoC designs, thus resulting in large communication latencies.

The software overhead has been found to contribute a large percentage of message latency, particularly for small messages and collective communication messages. The software overhead problem will worsen when high-speed parallel communication channels are used to transmit messages. To accelerate the software processing time, the hardware support features of NoC designs require further exploration. The subsequent work on the TMD-MPI approach has considered the NoC designs (TMD-MPE, message passing engine) for low-overhead transmission [38, 44]. However, TMD-MPE is implemented on an FPGA network, which is designed without consideration of the underlying network infrastructure; this has performance drawbacks especially for small and collective messages. Thus, the network infrastructure and hardware design could be further optimized for small or collective messages in NoC-based multicore processors.

In this chapter, we present the hardware acceleration architecture for implementing basic MPI functions in NoC-based multicore processors. To minimize the processing delay overhead of MPI functions, we optimize the communication architecture in two aspects: the underlying NoC design and the optimized MPI unit (MU). First, the current multihop feature and inefficient multicast (one-to-many) or broadcast (one-to-all) support have degraded the performance of MPI communications. We thus designed a specialized NoC to decrease this transmission delay. This special NoC incorporates a virtual bus (VB) into NoCs, where the conventional NoC point-to-point links can be dynamically used as bus transaction links for VB requests to achieve low latency while sustaining high throughput for both unicast (one-to-one) and multicast (one-to-many) communications at low cost. Second, an MPI processing unit (MPU) is introduced for the direct execution of MPI functions. Transferring the MPI functionality into a hardware block (the MU) aims at improving the overall performance of the communication system by reducing the latency, increasing the message throughput, and relieving the processor from handling the message passing protocol. We implement a basic subset of MPI functions that could improve the performance of both point-to-point and collective communications. To show the effectiveness of the proposed design, we performed the evaluation by integrating our work into the conventional communication architecture.

9.2 BACKGROUND

There are a large number of research works on the software optimization and hardware support for improving the performance of MPI applications. For multiprocessor systems, the MPI optimization is an extensively investigated domain [11, 21]. Faraj and Yuan [9] presented a method for automatically optimizing the MPI collective subroutines. Liu *et al.* [28] used the hardware multicast in native Infini-Band to improve the performance of MPI broadcast operation. Systems such as

STAR-MPI (Self-Tuned Adaptive Routines for MPI collective operations) [10] and HP-MPI [47] have shown that the profile data can be used for optimizing MPI performance at link time or launch. However, given that the power, area, and latency constraints for off-chip versus on-chip communication architectures differ substantially, prior off-chip communication architectures are not directly suitable for on-chip usage.

A large number of recent studies have provided the message-passing software and hardware on the top of NoC-based multicore processor designs. A natural method of providing MPI functionality on multicore processors is to port the conventional MPI implementation, such as MPICH [33]. However, this approach is unsuitable for on-chip systems that have limited resources. A number of implementations have been ported for high-end embedded systems with large memories, such as MPI/Pro [41]. However, such implementations are also unsuitable for NoC systems because the required hardware and software overhead is still large. The code size for these conventional MPI implementations is over 40 MB for all layers [12].

In Ref. [13], a nonstandard message passing support is proposed for distributed shared memory architecture. In Ref. [36], an MPI-like microtask communication is applied to the Cell Broadband Engine processor. But a cell processor has a maximum of eight processing elements (PEs), which leads to architectural scalability problems. In Ref. [30], an SoC-MPI library is implemented on a Xilinx Virtex FPGA to explore different mappings upon several generic topologies. For the NoC-based multiprocessor SoC platform, the multiprocessor MPI [14] has been introduced, which can provide a flexible and efficient multiprocessor MPI. In Ref. [39], a custom MPI called rMPI that targeted embedded systems using MIT's Raw processor and an on-chip network was reported. A lightweight MPI for embedded heterogeneous systems was reported in Ref. [1]. The STORM system [7] implements a small set of MPI routines for essential point-to-point and collective communication in order to provide more programmability and portability for the applications of the platform.

Another interesting work similar to ours, presented in Ref. [42], focuses on the parallel programming of multicore multi-FPGA systems based on message passing. This work presents TMD-MPI as a subset of MPI, the definition and extension of the packet format to communication systems in different FPGAs, as well as the intra-FPGA using a simple NoC architecture. Unlike TMD-MPI, we develop a hardware implementation of several MPI functions targeting NoC-based multicore processor systems and provide a detailed NoC router and associated MU to support efficient message passing. In Ref. [43], the concept of embedding partial reconfiguration into the MPI programming model is introduced, which allows hardware designers to create reusable template bitstreams for multiple applications. A hardware implementation of the MPI-2 RMA communication library primitive in the FPGA platform is described in Ref. [16]. Peng [38] considered the NoC designs for low overhead broadcast and reduced transmission but did not consider the communication protocol.

Intel has recently released an experimental processor, called the Single-chip Cloud Computer (SCC) [49]. The 48-core SCC explores the message passing model, which provides an on-chip low-latency memory buffer called the message passing buffer; the message passing buffer is physically distributed across the tiles. Such designs can eliminate the "coherency wall" between cores existing in conventional shared memory architectures. In Ref. [6], a hybrid approach that combines shared memory and message passing in a single general-purpose chip multiprocessor architecture is proposed, and allows efficient executions of applications developed with both parallel programming approaches.

There are also several research works to improve the performance metrics of underlying packet-switched NoCs by integrating a second switching mechanism. The reconfigurable NoC proposed in Ref. [48] reduces the hop count by physical bypass paths. Though the virtual bus on-chip network (VBON) also has some similarities with this reconfigurable NoC, we reduce the hop count by virtually bypassing the intermediate routers and further consider the multicast problem. In Ref. [24], asynchronous bypass channels are proposed at intermediate nodes, thus avoiding the synchronization delay. A new class of network topologies and associated routing algorithms is also proposed to complement the router design in this work. Another approach is the express topology which employs long links between nonlocal routers to reduce the effective network diameter [20, 27]. Though the express topology can reduce the unicast latency of NoC communication, it does not consider the multicast or broadcast communication pattern, which is the main focus of collective MPI hardware implementations.

The support for multicast communications in NoCs may be implemented in software or hardware. The software-based approaches [5] rely on unicast-based message passing mechanisms to provide multicast communication. Implementing the required functionality partially or fully in hardware has been found to improve the performance of the multicast operations such as the connection-oriented multicast scheme in wormhole-switched NoCs [29], XHiNoC multicast router [45], and virtual circuit tree multicasting [8]. In this chapter, we propose the use of a VB as an attractive network for support-efficient MPI hardware implementation. The proposed VBON design with specialized MUs is discussed in detail to improve performance of MPI communications.

9.3 MOTIVATION
9.3.1 MPI ADAPTION IN NoC DESIGNS

The challenge of effectively connecting and programming numerous cores for an NoC-based system has received significant attention from both academia and industry. A natural choice is a cache-coherent shared memory design based on previous symmetric multiprocessor architectures. These multicore processors likely have small private first-level (L1) or second-level (L2) caches but share a large last-level cache that is kept coherent with all L1 caches. However, as the number of cores increases, the protocol overhead would rapidly grow, leading to a "coherency wall" beyond which the overhead exceeds the value of adding cores [26]. To resolve this problem, message passing multicore architectures are introduced to eliminate cache coherence between cores.

The MPI has been proven to be a successful message passing framework for large-scale parallel computer systems [19]. It is known to be portable and extensible. It has numerous tools and parallel legacy codes to facilitate its use. The current MPI standard is large, containing over 200 function calls. However, several functions are essential to code a parallel application, with others facilitating its programming. The special capabilities of the MPI such as its standard interface, language-independent interface, and large user-base make it a potential and suitable solution for implementation in NoC systems. Previous studies [14, 39, 49] have successfully demonstrated the effectiveness of adapting the MPI into NoC-based multicore processors. This chapter is also based on the assumption that the MPI is a potential programming model candidate for future multicore processors.

9.3.2 OPTIMIZATIONS OF MPI FUNCTIONS

MPI communication functions can be classified as either point-to-point or collective functions. Point-to-point communication involves only two nodes: the sender and the receiver. Collective communication involves all the nodes in the application; typically, a root node coordinates the communication, whereas the remaining nodes merely participate in the collective operation. MPI communications can also be classified as synchronous or buffered functions. In synchronous communication, the sender and the receiver block the execution of the program until the information transfer is complete. Buffered communication enables the overlap of communication and computation but places the responsibility of avoiding data corruption in the memory on the programmer because data transmission occurs in the background, such that the transmission buffer can be overwritten by further computation. These types of MPI functions would be our target baseline communication operations.

The problem of current MPI implementations is that they are realized in software and do not consider the refinement of NoC designs for both point-to-point and collective functions, thus resulting in large communication latencies. In this work, we aim to accelerate the processing performance of MPI primitives in future massive multicore architectures through underlying hardware support. These multicore architectures usually present a mesh-type interconnect fabric. A number of factors have to be considered when improving the performance of MPI functions through hardware support, especially for collective functions. Our design includes two main hardware techniques for accelerating MPI primitives: the specialized NoC design and the optimized MU.

9.4 COMMUNICATION CUSTOMIZATION ARCHITECTURES

9.4.1 ARCHITECTURE OVERVIEW

Figure 9.1 shows a block diagram of the proposed implementation architecture with a baseline 8 × 8 mesh topology. We consider a multicore processor chip where each core has a private L1 cache and logically shares a large L2 cache. The L2 cache

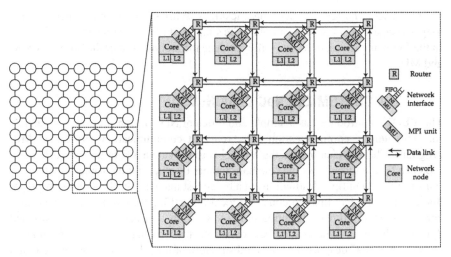

FIGURE 9.1

The architecture overview of the proposed design.

may be physically distributed on the chip, with one slice associated with each core (essentially forming a tiled chip). As shown in Figure 9.1, the underlying NoC design is the actual medium used to transfer messages, which can be designed with consideration for specialized features of MPI communications. Each node also has an MU between the core and the network interface (NI), which is used to execute corresponding instructions for MPI primitives.

The latency and bandwidth of the NoC are important factors that affect the efficiency of computations with numerous intercore dependencies. To support the MPI efficiently, good service for control messages must be provided. Control messages may be used to signal network barriers and changes in network configuration, as well as the onset or termination of computation. The process requires minimal bandwidth, but needs very low latency for broadcast (one-to-all or all-to-all) communication. However, the multihop feature and inefficient multicast (one-to-many) or broadcast (one-to-all) support in conventional NoCs have degraded the performance of such kinds of communications. To facilitate the efficient transmission of data and control messages, a customized network is needed. In the proposed design, a hierarchical on-chip network, called VBON, is introduced.

By directly executing the MPI primitives and interrupting service routines, the MU reduces the context switching overhead in the cores and accelerates software processing. The MU also performs the message buffer management as well as the fast buffer copying for the cores. The MU transfers messages to and from dynamically allocated message buffers in the memory to avoid buffer copying between system and user buffers. This process eliminates the need for the sending process to wait for the message buffer to be released by the communication channel. The MU also reserves a set of buffers for incoming messages. With use of the above methods, the long message transmission protocol can be simplified to reduce transmission latency. In the following subsections, we will introduce the architecture of the proposed NoC and MU.

9.4.2 THE CUSTOMIZED NoC DESIGN: VBON

The VBON is introduced as a simple and efficient NoC design offering low latency for short unicast communication and broadcast communication services in MPI programs. The key idea behind the VBON is to provide a VB in the network. Such VBs can be used to bypass intermediate routers by skipping the router pipelines or transmitting to multiple destinations in a broadcast manner. Unlike a set of proposals that employ physical links to construct local buses [32, 34, 40], the bus mentioned here is a specialized packet routing in the VBON. Its transaction link is constructed dynamically for communication requests from the existing point-to-point links of conventional NoC designs. The detailed structure of the VBON design can be found in Chapter 4.

9.4.3 THE MPI PRIMITIVE IMPLEMENTATION: MU

The MPI supports both the point-to-point and collective communication functions. Given the popularity of send-and-receive-based message-passing systems, *MPI_Send* and *MPI_Receive* are implemented for point-to-point communication functions. All other MPI communication functions can be realized by these two primitives. However, the collective communication based on these two primitives would be inefficient because the NoC, rather than an underlying transparent communication layer, should be involved in the transactions of collective communication to improve the performance. Furthermore, collective communications are always the performance bottlenecks for data-parallel applications [31]. In this chapter, we support three collective operations—*MPI_Bcast*, *MPI_Barrier*, and *MPI_Reduce*—which represent three kinds of communication patterns. An efficient implementation of these collective primitives is crucial for the performance improvement. *MPI_Bcast* is a commonly used collective function in parallel applications. In this process, the root broadcasts its data to all processes. Almost all collective communications would incorporate the broadcast communication. *MPI_Barrier* blocks the calling process until all the other processes have also called it. *MPI_Barrier* can return to any process only after all the processes have entered the call. *MPI_Reduce* operations collect data from all the processes using an associative operator, such as addition, maximum value, or even user-defined data operator. The final results are placed in the root. The following section specifies the detailed design of the proposed MU for the implementation of these MPI operations.

9.4.3.1 The architecture of the MU

The MU architecture is shown in Figure 9.2. The MU provides hardware support to address the communication protocol used in the MPI implementation. The primary

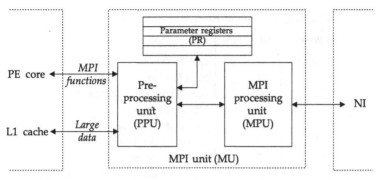

FIGURE 9.2

Block diagram of the MU.

functionality of the MU is to serve as a middle layer between the processor core and the interface of the NoC. The MU will receive the message and send requests from the PE core and will handle various messages from other processor cores. Two sources trigger the MU: the local processor core and the NI. The local processor core may request the MU to perform the MPI primitive functions, and then the associated communication data are transferred through this interface. Another source is the NI, which may request the MU to receive messages from the on-chip network and then perform corresponding operations for handling the received messages.

The MU consists of three key components: the preprocessing unit (PPU), parameter registers (PRs), and the MPU. The PPU is used to translate the instructions from the processor core into control signals and to generate the message passing parameters for data transfer that are temporarily stored in the PRs. Another important task of the PPU is to exchange data with the CPU cache to read or write the communication data. The PRs include several registers in the MU. Table 9.1 describes these registers. When MPI functions are performed, the MU first receives the message parameters from the PE core and then updates these registers.

Similarly to conventional MPI implementations [33], we use four parameters to identify a message: the source/destination *rank*, the message *tag*, and the *communicator*. The rank of a process is a unique number that identifies this particular process in a parallel application. The tag is an integer number that identifies a message. The meaning of a tag is entirely user definable. The rank of a process and the message tags are scoped to a communicator, which is a communication context that enables the MPI to communicate with a group of processes selectively within a particular context. As listed in Table 9.1, the *OPCODE* register specifies the operations for message communications; the *PID*, *RID*, *TAG*, and *COM* registers represent the four parameters for identifying the message; the *MSGSIZE* register indicates the length of the message.

The MPU is the key component of the MU that performs the actual operations for MPI primitive functions. Table 9.2 lists the set of primitive functions implemented

Table 9.1 Parameter Registers in the MU

Register	Bit Width	Description
OPCODE	8	Operation code
PID	32	Rank ID of local processor core
RID	32	Rank ID of remote processor core
TAG	32	Message tag
COM	64	Message communicator
MODE	8	Communication mode
ADDR	32	Memory address of network data
DATA	128	Communication data from register
MSGLEN	32	Message length

by the MU with their MPI equivalent primitives in the middle column. Any other communication-related MPI functions can be implemented using these hardware-supported primitives. Such primitives are classified into two categories. The first category includes the first five MU operations that are related to communication functionalities. The second category includes the last five MU operations that are related to the MU context handling.

The *Send* and *Receive* operations both have four parameters, where data can be specified by either the memory address *srcaddr* or the register value *reg*. When the data are small and can be represented in a register, only their value is directly sent to the MU because such data are usually generated by a processor core and reside in the register. In this case, the direct move operation can reduce the latency. When the data are large, only the starting memory address of the data is sent to the MU, and then the MU will request the cache controller to load the data to create the network message. The collective operations have similar considerations for fetching data. The *Broadcast* operation sends the data block to all other processor nodes, so the destination nodes do not need to be specified. The *Barrier* operation is used to synchronize all the processor nodes, which do not have any parameter. The *Reduce* operation combines the elements of the data block of each processor node using a specified operation and then returns the results to the root processor node. This operation has six parameters: specifying the source data, destination data, data length, tag, reduce operation, and root node. To accelerate the reduce operation in each node, the MPU also implements a reduction function unit (RFU) to perform the reduce operations. Table 9.3 lists the reduction operations supported by the RFU. The operation data types can be integer numbers, floating-point numbers, or both according to different application scenarios. For the experiments in this chapter, both integer and floating-point numbers are supported. To reduce the hardware cost of the RFU, we do not implement area-consuming reduce operations, such as the product, specified in the MPI. The RFU is realized on the basis of an adder.

The execution of these functions is performed in two separate pipelines: the send and receive pipelines. The send pipeline is used for active operations such as *Send* and *Broadcast*. The receive pipeline is used for passive operations such as *Receive*.

Table 9.2 Hardware-Implemented Primitive Functions

MU Operation	Primitives	Description
Send(srcaddr|reg, len, dpid, tag)	*MPI_Send*	Send a *len* size message with *tag* to target processor node *dpid*, where data is read from memory address *srcaddr* or register *reg*
Receive(recaddr|reg, len, spid, tag)	*MPI_Receive*	Receive a *len* size message with *tag* from processor node *spid*, where data is stored into memory address *recaddr* or register *reg*
Broadcast(srcaddr| reg, len, tag)	*MPI_Bcast*	Broadcast a *len* size message with *tag* to a group of cores in the communicator, where data is read from memory address *srcaddr* or register *reg*
Barrier()	*MPI_Barrier*	Synchronize all processor nodes, each calling it will be blocked until all the nodes have called it
Reduce(srcaddr|srcreg, dstaddr|dstreg, len, tag, op, rpid)	*MPI_Reduce*	Collect data from all the processor nodes, each sends a *len* size message with *tag* from memory address *srcaddr* or register *srcreg* to the root node *rpid* with memory address *dstaddr* or register *dstreg* for the reduce operation *op*, which is specified in Table 9.3
Init(com, spid, size)	*MPI_Initial*	Initialize the MU context including communicator *com*, Rank ID, *spid*, number of group cores *size*, and default *hybrid* communication mode
End()	*MPI_Finalize*	Clear the MU context
SetMode(mode)	–	Set the communication mode for MU operations with three types of mode supported: *synchronous*, *buffered*, and *hybrid* (default mode)
ContextSave(caddr)	–	Save the register and buffer data into memory addressed by *caddr*
ContextRestore(caddr)	–	Restore the register and buffer data from memory addressed by *caddr*

9.4.3.2 MPI processing unit

Figure 9.3 shows a simplified block diagram of the MPU. The MPU is capable of handling unexpected messages and dividing large messages into smaller packets. The MPU comprises six main components organized in two separate pipelines: the message packetizing and building units for the send pipeline; the packet reception, expectation, and response units for the receive pipeline; the collective logic unit for

Table 9.3 The Reduction Operations Supported by the RFU

Operations	Operation Code	Description
MAX	0000	Maximum operation
MIN	0001	Minimum operation
SUM	0010	Sum
LAND	0011	Logical and
BAND	0100	Bit-wise and
LOR	0101	Logical or
BOR	0110	Bit-wise or operation
LXOR	0111	Logical exclusive or operation
BXOR	1001	Bit-wise exclusive or operation
USER	1000	User-defined operation

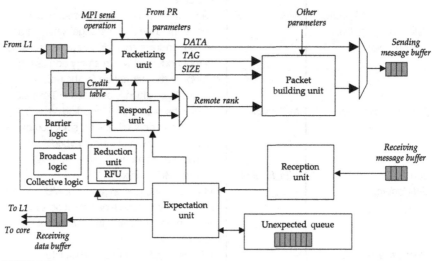

FIGURE 9.3

The block diagram of the MPU.

handling three collective operations. Moreover, it also has an internal buffer to store unexpected messages. Thus, sending and receiving packets can occur simultaneously. We will discuss these components through the sending and receiving procedures.

When the local processor core issues the sending operation, the message parameters related to the MPI communication are initially sent to the PR. The local processor core sends the data to the MPU through the cache controller. For the collective packet sending, the collective logic unit is also involved to generate the corresponding parameters. These parameters, as well as the data, are first sent to the packetizing unit to determine whether a message is larger than the maximum packet size. If such is the case, the packet size field in the packet header is modified accordingly. The large message is divided into equal-sized packets as long as the data exist. The packetized data, along with other message information, such as the packet size, message tag, *PID*, and *RID*, are sent to the building unit to create the packet.

The receiving procedure of the MPU is described as follows. The incoming packets are first decoded by the reception unit to obtain the message information for executing the receiving operations. The information is also provided to other units in the MPU on the basis of the received values. Such information is sent into the expectation unit to determine whether a message is expected. An unexpected message is a message which has been received by the MU for which a receive operation has not been posted (i.e., the program has not called a receive function like *MPI_Recv*). If the message is unexpected, the message packet will be stored in the message queue. The expectation unit also determines whether packets from a previous unexpected message are in the envelope memory queue. For the packets already determined as expected, the respond unit and collective logic unit may be involved to send respond packets according to the information acquired from incoming packets.

9.4.3.3 *The collective operation implementation*
The collective operations are the key functionalities implemented in the MU. In this subsection, we will discuss how the MU is used to support these collective operations, including *Broadcast*, *Barrier*, and *Reduce* operations. Figure 9.3 contains a diagram of the collective logic, which is capable of handling packets related to collective communications. It accepts the collective operations from the PPU, and generates the control signals for the packetizing unit to form packets. Moreover, the expected receiving packet information will also be sent to the collective logic to generate the respond information.

The broadcast operation is the most basic primitive among the three operations. When the local processor core issues the *Broadcast* instruction, the group information is first sent to the MU. Broadcast messages can be transmitted within a group (the multicast communication pattern). Thus, target group addresses are specified by the *bitstring* field in the packet. The status of each bit indicates whether the visited node is a target of the multicast. The broadcast unit should check this *bitstring* and generate packets accordingly. For other collective operations, this *bitstring* is also involved to indicate the group members of the communications. The packets are then transmitted through row/column VBs. Figure 9.4a shows the design method of broadcast operations based on the VB. For a 4 × 4 mesh configuration, only the latency of a maximum of two VB transactions is needed to broadcast data to all other cores.

The barrier operation is used to synchronize processes among parallel applications. Given that multicore parallel applications tend to exploit fine-grained parallelism, such application can be highly sensitive to the barrier performance [46]. To support fast barrier operations, we design a VB-based synchronization barrier. The design concept of this implementation is shown in Figure 9.4b. This design uses a master-slave barrier, as shown at the top of Figure 9.4b, for a configuration with 16

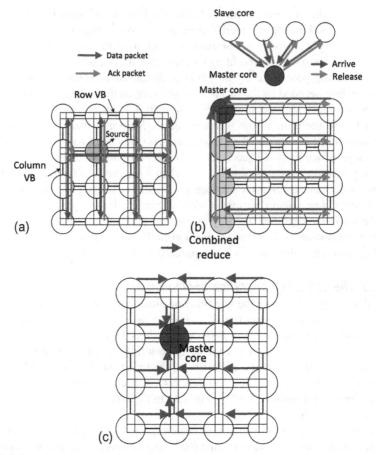

FIGURE 9.4

Proposed collective communication operation designs. (a) VB-based broadcast operation. (b) VB-based master-slave barrier. (c) Distributed reduce operation.

cores. The design employs a centralized approach where a master core is responsible for locking and releasing slave cores. The barrier can be divided into two phases. In the first phase, each core waits for other cores to arrive at the same phase, whereas in the second phase, all cores have arrived at the barrier, and a release command is given by the master core.

To support such a barrier efficiently, we add special barrier signal lines for each VB to transmit the signals required by the synchronization process. They are wires that first connect horizontal cores, then connect the horizontal master cores to form vertical transmission. The barrier signals are transmitted through the request and acknowledge flow control. Without loss of generality, we describe the simultaneous executions by all the cores on a 16-core mesh layout, as shown at the bottom of

Figure 9.4b. Each cycle of four barrier signals can be transmitted to the master. In the first arrival phase, the horizontal barrier arriving signal is transmitted through the barrier signal lines to the horizontal master. These cores then wait until the horizontal master sends a signal to resume execution. When the horizontal master receives all the barrier arriving signals along the horizontal barrier lines, it sends these signals to the master core through the vertical barrier signal line. The barrier operation only enters into the second release phase when the master core has received all the barriers from its horizontal master core. In this phase, the master core sends the release signals to its slave cores through the VB in the opposite direction. Finally, when the slave cores receive the release signals, they will resume the execution to complete the barrier operation.

The last collective operation is the reduce operation, which combines the partial results of a group of cores into a single final result. Considering that the reduce operation accounts for a significant portion of the execution time of MPI applications, the efficient implementation of the reduce operation is thus beneficial to overall performance [25]. Figure 9.4c shows a block diagram of the routing path for the reduce operation. This path is based on simple XY routing, that is, the reduce path first goes through X then through Y. Unlike the conventional centralized reduce operation, the proposed method distributes the reduce operations to all cores along the reduce path, as shown in Figure 9.4c. The reduce hardware in each core will receive the data from a neighboring core, after which it will dynamically determine the reduce data and will operate on the data locally only if all the expected data are received. After the reduce operation, the reduce hardware will send the result to the upper core for further reduction. To accelerate the reduce operation, such an operation is executed by the RFU according to the operation type and the data type, as described in Table 9.3. Performing the reduce operation can be classified into two cases. The first case is when the operation can be supported by the RFU. In such a case, all the data can be reduced normally in a distributed manner. The second case is when complex operations, such as product in the MPI, cannot be supported by the RFU. We use the centralized reduce method, where all the data are sent to the root, to perform the reduce operation.

9.4.3.4 Communication protocols

One of the key functionalities of MPI hardware implementation is to support the MPI communication modes. The MPI has two important communication modes: the synchronous mode and the buffered mode, as shown in Figure 9.5a and b respectively. The buffered mode enables the completion of the MPI send operation before the response packet is received. This process is based on the assumption that all the expected and unexpected packets can be buffered in the target MU. By contrast, the synchronous mode will send a request before sending data to the receiver. After the sender receives the ready packet, the actual data transfer will commence. Although this mode incurs a higher message overhead than the buffered mode,

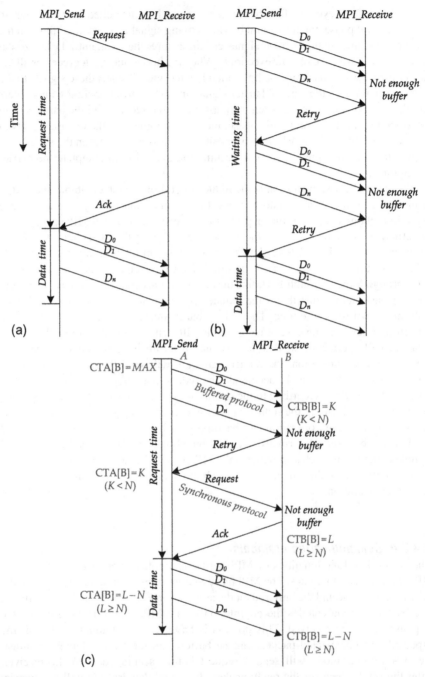

FIGURE 9.5

Supported MPI communication modes. (a) Synchronous protocol. (b) Buffered protocol. (c) Hybrid protocol.

it demands less buffer space and ensures that the target MU can have sufficient buffer spaces to store messages.

When the MPI message has already been sent out and the target processor core does not have sufficient buffers to receive this message, the retry mechanism should be triggered to maintain correctness. However, this mechanism would significantly degrade the performance. The buffered mode does not have to wait for a response message before triggering the retry mechanism. By contrast, the synchronous mode has to wait for the response message but would not activate the retry mechanism. To maximize the utilization rate of the buffered mode and minimize the number of retry operations due to overflowing the receive buffers, we propose a new optimized hardware mechanism for credit-based MPI control flow called the hybrid communication mode. The credit can be defined as the number of receive buffers available at the target processor core. The example packet format is shown in Figure 9.6.

This mechanism is described as follows. Each MPU maintains a credit table listing the credit values for all the nodes, and their initial values are set as the maximum. The processor core initially performs the send operation in the buffered mode and then decreases the credit value accordingly. If the target node has insufficient space for buffering the message, it will trigger the retry procedure and place its newest credit value in the corresponding message. The sending core will send the message in the synchronous mode if the received credit value n is less than the mode threshold p. If n equals zero, then the sending core will block the send operation until n is larger than the block threshold q. If n continues to increase to p, messages will again be sent in the buffered mode. Figure 9.5c shows an example of this mode. In sender A, the initial credit table value for receiver B ($CTA[B]$) is MAX, so it will trigger the buffered protocol to send the data with size N. Since the buffer count in receiver B ($CTB[B] = K$) is smaller than the message size, it will perform the retry operation with the credit

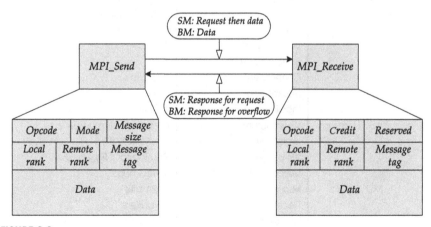

FIGURE 9.6

The MPI communication packet format.

table value (K) to notify sender A. When sender A receives this message, it will update its credit table and trigger the synchronous protocol to send the data again. So the hybrid protocol will achieve better performance than the buffered protocol in total.

9.5 EVALUATION

In this section, we present a detailed evaluation of the proposed communication architecture with a baseline 2D mesh topology. We will describe the evaluation method, followed by the results using synthetic and real application traffic patterns. Thereafter, the effect of the proposed method on hardware costs is discussed.

9.5.1 METHODOLOGY

To evaluate the proposed communication design, we implemented its architecture using a SystemC-based cycle-level NoC simulator augmented with the MU. It is modified from the NIRGAM simulator [17]. The simulator models a detailed pipeline structure for the NoC router and the MU. We can change various network configurations, such as the network size, topology, buffer size, routing algorithm, and traffic pattern. Table 9.4 lists the network configurations across all experiments. For send and receive operation scenarios, once the hop count of the point-to-point message is larger than four (threshold in a 4×4 mesh) or eight (threshold in an 8×8 mesh), the message is transmitted by VBs. For collective operation scenarios, all multicast messages are transmitted through VBs.

Table 9.4 Communication Design Configurations

Designs	Parameter	Measure
Basic design	Topology	8×8 mesh
	Basic routing	Dimensional ordered XY
	Ports	5
	VCs per port	4
	Buffers per VC	16
	Channel width/flit size	128 bits
	Packet size	8 flits
	Pipeline frequency	1 GHz
VBON design	VB structure	Row and column VBs
	Hierarchy	$2\times2, 4\times4, 8\times8$
	VB usage	Routing for *MPI_Bcast*
	Maximum VB length	4 NoC hops
MU design	Maximum message size	1024 bits
	Receiving buffer	32 buffers
	Unexpected buffer	32 buffers
	Credit threshold (p,q)	(4,2) entries

We compare the characteristics of the proposed communication architecture based on the VBON scheme against the conventional NoC design. The basic router represents conventional NoC designs. The basic router originally has a four-stage router pipeline. To shorten the pipeline, the basic router uses lookahead routing [15] and the speculative method [37].

The synthetic traffic patterns used in this research are the round-trip and uniform random traffic patterns. Each simulation runs for 1×10^6 cycles. To obtain stable performance results, the initial 1×10^5 cycles are used for simulation warm-up to enable the network to reach its steady state. The following 9×10^5 cycles are then used for analysis. When destinations are chosen randomly, we repeat the simulation run five times and then determine the average of the values obtained in each run. The time for initializing the MU is not counted for the message transmission.

We also studied the proposed approach using real application communication traffic. Traces for the baseline conventional implementations were obtained on a full-system multicore simulator, M5 [4]. We collect the message passing and memory access requests from the full-system simulator, then we extract the MPI functions and network messages from them to generate the NoC application traffic. The target multicore system is modeled with the Alpha instruction set architecture, which is the stablest instruction set architecture supported in the M5 simulator. Each core is modeled with two-way 16-kB L1 ICache, two-way 32-kB DCache, and 1-MB L2 cache. For the MPI applications, the cache is configured without coherence protocols just like the Intel SCC processor [49]. We also integrate Orion [50] to estimate the NoC energy and Cacti [35] to estimate the cache energy, thus obtaining the power metric for the cache-NoC system. We use the NAS Parallel Benchmarks (NPB 2.4) suite as application traffic to evaluate the proposed design. The applications used to perform the experiments are a subset of the A class NPB, a well-known, allegedly representative set of application workloads often used to assess the performance of parallel computers. These applications include three kernels, namely conjugate gradient (CG), integer sort (IS), and discrete 3D fast Fourier transform (FT), as well as two pseudoapplications, block tridiagonal solver (BT) and scalar pentadiagonal solver (SP). For the baseline hardware implementation with only point-to-point MPI communication support, other communication types, such as the collective communication, are performed through the basic point-to-point MPI communications. For the proposed hardware implementation, other communication operations are performed through the supported point-to-point or collective communications.

9.5.2 EXPERIMENTAL RESULTS

9.5.2.1 The effect of point-to-point communication: Bandwidth

We first discuss the performance of point-to-point communications in the proposed design. We record the time taken for a number of round-trip message transfers. The randomly generated messages (i.e., destinations of unicast messages at each node are selected randomly) are sent to the target node by *MPI_Send* instructions. When the target node receives this message, this node will first perform the *MPI_Receive*

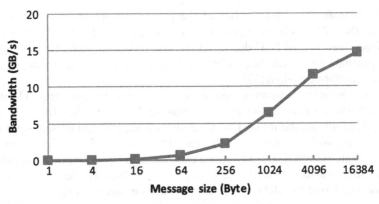

FIGURE 9.7

Bandwidth results of the proposed design with different message sizes.

instruction and then return the message to the source node immediately without any change through the *MPI_Send* instruction. This round-trip test can help determine the network bandwidth of the communication system. Such a bandwidth is considered as the average peak performance on a link channel. In the experiment, the simulation of MPI instructions, such as *MPI_Send* and *MPI_Receive*, will be triggered by benchmarks, such that the L1 cache controller will access the data and interconnect with the MU. Assuming that the maximum capacity of the L1 cache in the multicore processor is 32 kB, the maximum length of the message triggered by MPI primitive instructions should be set to 16 kB (for send and receive operations).

Figure 9.7 illustrates the bandwidth results of the point-to-point communication. When the size of the message is more than 1 kB, the bandwidth of the communication system could reach more than 5 GB/s. Compared with software-based MPI implementations, such as TMD-MPI [42] with 10 MB/s, the bandwidth of the proposed design exhibits a qualitative leap. For the hardware implementation in Ref. [44], a bandwidth of 531.2 MB/s is obtained. The proposed approach exhibits improved performance, which adequately demonstrates the potential benefit of supporting the parallel programming model by using a special hardware mechanism.

9.5.2.2 The effect of collective communication: Broadcast operations

One of the key features of the proposed communication architecture is the effective hardware support for the *MPI_Bcast* primitive. In conventional MPI implementations, *MPI_Bcast* is typically implemented using software with a tree-based algorithm. Such implementations exploit point-to-point communication operations. Thus, the number of hops to reach leaf nodes increases with the total number of nodes (typically in a logarithmic manner); the latency of *MPI_Bcast* also increases. As demonstrated in Figure 9.8a, the latency of *MPI_Bcast* evidently increases with the number of nodes.

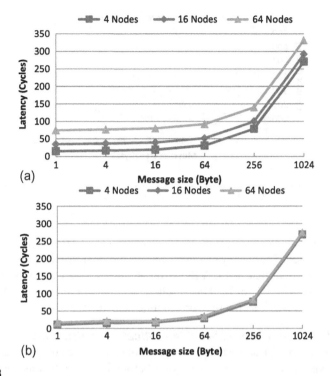

FIGURE 9.8

MPI_Bcast latency results based on different designs. (a) Conventional NoC design result. (b) Proposed communication design result.

When we implement the *MPI_Bcast* primitive based on the VBON network, this relationship changes significantly. Figure 9.8b shows the performance results of *MPI_Bcast* based on the VBON with different numbers of processor cores. Figure 9.8b shows small increments in the latency of *MPI_Bcast* as the number of processor cores increases. This is because 4-core or 16-core systems have a latency of only two VB transactions and 64-core systems have a latency of only four VB transactions for broadcast operation. So the latency increases by two VB transactions from 16-core to 64-core systems. Since each VB transaction takes only a few cycles, the increased latency is not much. This feature is useful for achieving high performance when parallel applications involve a large number of processor cores. Furthermore, the low latency of the short-message *MPI_Bcast* primitive facilitates the synchronization of tasks among different processor cores.

To show the broadcast performance under increasing contending traffic, Figure 9.9 illustrates the average latency of *MPI_Bcast* operations for the two different MPI implementations. A fixed number of *MPI_Bcast* messages are generated during one period. That is, only after the transmission of all the messages is completed can the messages be generated again. Because of increased congestion, the latencies of the

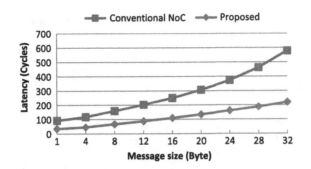

FIGURE 9.9

MPI_Bcast latency results under increasing contending traffic.

two implementations increase as the number of concurrent *MPI_Bcast* transmission messages increases. The latency increment on conventional NoCs (by about two time) is more obvious than that of the VBON (by about one time) when the number of generated messages is 1 to 32. This result shows that the support for *MPI_Bcast* operation in the VBON is effective for handling contending traffic.

9.5.2.3 The effect of collective communication: Barrier operations

To determine the performance impact of barrier operation, we compare our network-optimized barrier with two other hardware barrier implementations based on the VBON. The first implementation employs a centralized sense-reversal barrier, where each core increments a centralized shared counter as it reaches the barrier and spins until such a counter indicates that all cores are present. The second implementation employs a binary combination tree barrier, where several shared counters are distributed in a binary tree fashion. Thus, all cores are divided into groups assigned to each leaf (variable) of the tree. Each core increments its leaf and spins. Once the last core arrives in the group, this core continues up the tree to update the parent and so on until it reaches the root. The release phase employs a similar process but in the opposite direction (toward the leaves).

Figure 9.10 illustrates a comparison of the results of barrier performance with varying core numbers. The conventional centralized barrier has the longest operation latency since each core needs to communicate with the centralized core to complete the barrier operation, which would become the performance bottleneck. The tree barrier mitigates this situation by distributing the barrier combination in a binary tree fashion. However, this process also suffers from multihop network latency. The resulting latency reduction is approximately 23% for a 64-core configuration compared with the conventional barrier. To reduce further the latency of barrier operation, the proposed barrier uses a specialized network to transmit the barrier message. This feature reduces the latency of the tree barrier by approximately 79% for a 64-core configuration. Considering that multicore applications can be highly

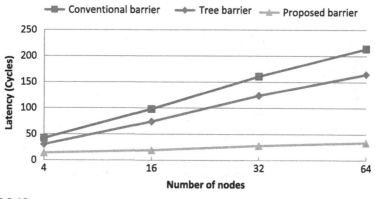

FIGURE 9.10

MPI_Barrier performance with different core counts.

sensitive to barrier latency, such a reduction can result in significant performance improvement.

9.5.2.4 The effect of collective communication: Reduce operation

For the implementation of the reduce operation, we also compare the proposed method with the conventional centralized method, in which the actual reduction is performed solely by the root core. Figure 9.11a shows a comparison of results in terms of reduce operation performance with varying core count. To test the feasibility of the reduce operation, the data size of the reduce operations is set to 1. Thus, the latency results of reduce operations are related only to the number of cores and the corresponding reduction methods. The figure shows that when the number of cores is relatively small (approximately 4-16 cores), the two methods have comparable latencies since the network latency and contention introduced by the centralized method with small-scale cores do not significantly affect the reduce operation. However, as the number of cores increases, the proposed method exhibits significant performance advantage over the conventional centralized method. For a 64-core configuration, the proposed method reduces the latency by approximately 46%, which is beneficial to the overall application performance.

Figure 9.11b shows a comparison of the results in terms of reduce operation performance with varying message size for a 64-core configuration. This reduce operation is performed by reducing intermediate arrays in each core. The performance benefit of the proposed reduce method becomes more evident as the message size increases from 1 to 1024 bytes. This result highlights the advantage of distributing the reduce operation of data into each core. The advantage stems from two reasons. First, the proposed method does not require each core to communicate with the root, but requires it to communicate only with its neighboring cores. This process reduces not only the network traffic but also the network contention. Second, numerous data

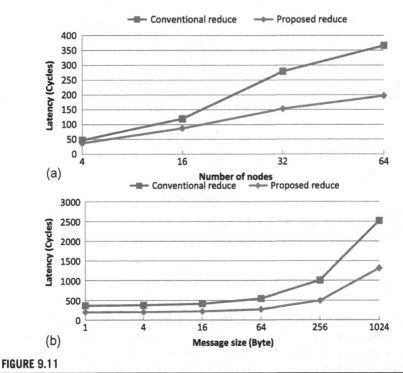

(a)

(b)

FIGURE 9.11

MPI_Reduce performance results for different scenarios. (a) Result obtained by varying the core count. (b) Result obtained by varying the message size.

reduce operations can be performed simultaneously along the communication path. This process minimizes the reduce operations solely through the root core.

9.5.2.5 The effect of application communication: Performance

We run the application traffic to evaluate the proposed design and then use the message delay as the performance metric for various communication designs. These benchmarks were chosen for their large number of message transmissions. Figure 9.12a shows the performance comparison results for different benchmarks with a 64-core configuration. We also have two other MPI support designs. The first conventional MPI support design provides hardware support for point-to-point MPI communication (including *MPI_Send* and *MPI_Receive*) based on conventional NoC architectures, and the second VBON MPI support design is based on the proposed VBON architecture. We can see that the VBON MPI support outperforms the conventional MPI support by 8% on average. This improvement can be attributed to the fact that the VBON can reduce the latency when transferring long-latency and collective messages using the VB. When collective MPI support is added to the second design, the proposed design can achieve significant performance improvement

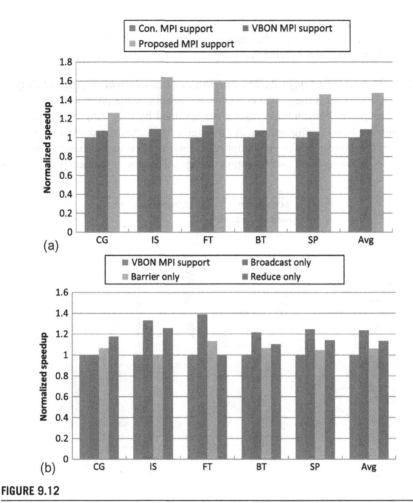

FIGURE 9.12

The impact comparison for application performance with different scenarios. (a) Comparison result for different MPI implementations. (b) Comparison result for different collective communication supports.

of 48% on average. This finding further shows that the hardware support for collective communication in MPI implementation would be an effective method.

To present clearly the characteristics of collective operation support with regard to these benchmarks, Figure 9.12b shows the performance speedups for the benchmarks obtained in the baseline VBON point-to-point MPI implementation architecture with different collective communication supports. We can see that all the benchmarks except CG can benefit from the broadcast support, whose performance improvement can be up to 23% on average. The other two collective communication supports have less performance improvement, 6% and 13% respectively. This is because these two collective operations have small portions in terms of execution time. Nevertheless, the evaluation results show that the hardware support for collective communications is beneficial for accelerating MPI applications.

9.5.2.6 The effect of application communication: Power and scalability

Figure 9.13a illustrates the power consumption results of the proposed NoC-cache system. The power consumption results are obtained by executing the application traffic with a 64-core configuration. As the figure indicates, the conventional NoC design has the largest power consumption, which is used as the baseline design. The VBON point-to-point MPI support reduces the power consumption by an

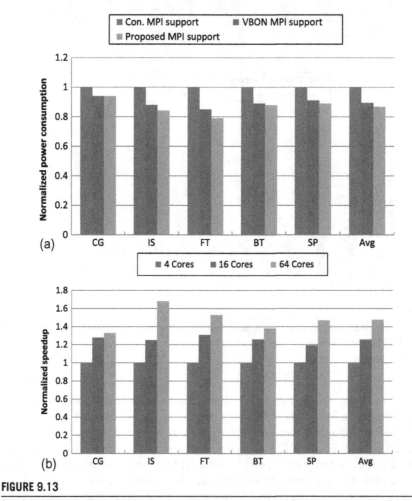

FIGURE 9.13

The impact comparison for application with power and scalability measurements. (a) Power result: 64-core configuration. (b) Scalability result: varying core count.

average of about 11%. This is mainly due to a reduction in the buffer and crossbar power consumption in bypassing routers. The proposed communication design can further reduce the power consumption, by about 3%. In summary, the proposed communication design clearly outperforms the other two NoC designs across all benchmarks not only in the performance but also in the power consumption.

To show the scalability of the proposed design, Figure 9.13b illustrates the computation results for the speedup over the conventional MPI support design with 4-core to 64-core configurations. The 4-core design is used as the baseline design, whose performance speedup is normalized to 1. That is, we compare the speedup of each configuration of the core count with the baseline MPI support design. The speedup of the 64-core configuration outperforms the speedup of 4-core and 16-core configurations for all benchmarks. For the 16-core system, about 25% more speedup can be obtained over the 4-core system; for the 64-core system, about 47% more speedup can be obtained over the 4-core system. This is because as the number of cores increases, reducing the MPI communication delay through hardware support becomes more important. The evaluation result successfully demonstrates that the proposed design has good scalability as the number of cores increases.

9.5.2.7 Implementation overheads

To determine the implementation overheads, the communication architectures, including the specially designed routers and MUs, were described in HDL and synthesized with a Taiwan Semiconductor Manufacturing Company 90 nm technology under typical operating conditions, 1.0 V and 25 °C. The area of the selected router and MU hardware was measured at their maximum supported frequencies and calculated by Synopsys Design Compiler. For comparability, the conventional NoC and VBON routers were equipped with the same buffer size in each direction. Approximately 6% area overhead for the VBON router is observed in the experiment.

Table 9.5 shows the area overhead breakdown of the MU implementation, excluding the RFU. In this observation, the MPU occupies most of the router area, which includes various kinds of buffers and execution logic. The area of the MU is approximately $0.11\,\text{mm}^2$. This area calculation does not take the RFU into account because it may change a lot according to different functionalities. For the configuration of supporting the fixed-point and floating-point reduce operations implemented in this chapter, the area of the RFU is approximately $0.08\,\text{mm}^2$. So the sum of the router overhead and the MU overhead is about $0.21\,\text{mm}^2$. For the evaluated processor core with area above $10\,\text{mm}^2$, such an area overhead is affordable since the performance improvement of real applications is large.

Table 9.5 Implementation Overheads for the MU

	PR	PPU	MPU	Total
Area ($10^{-3}\,\text{mm}^2$)	9	13	91	113
Ratio (%)	8.0	11.5	80.5	100

9.6 CHAPTER SUMMARY

Recent trends in NoC-based multicore architectures have considered message passing as a technology to enable efficient parallel processing for improved suitability, scalability, and performance. In this chapter, we presented a communication architecture that aims at accelerating basic MPI primitives by exploiting all hardware features of multicore processors. We first proposed the VBON, an underlying customized NoC that incorporates buses into NoCs, to achieve high performance for both point-to-point and collective data transfers. Furthermore, an optimized MU based on the VBON was designed to relieve the processor core from handling the message passing protocols as well as to reduce software processing overheads. As demonstrated in the experiment, the proposed design can significantly improve the overall performance of the communication system.

REFERENCES

[1] A. Agbaria, D.-I. Kang, K. Singh, LMPI: MPI for heterogeneous embedded distributed systems, in: Proceedings of the International Conference on Parallel and Distributed Systems (ICPADS), 2006, pp. 79–86. 286, 288

[2] Arteris, Network on Chip (NoC) Interconnect IPs, http://www.arteris.com/, cited 2011. 286

[3] K. Asanovic, R. Bodik, J. Demmel, T. Keaveny, K. Keutzer, J. Kubiatowicz, N. Morgan, D. Patterson, K. Sen, J. Wawrzynek, D. Wessel, K. Yelick, A view of the parallel computing landscape, Commun. ACM 52 (2009) 56–67. 286

[4] N.L. Binkert, R.G. Dreslinski, L.R. Hsu, K.T. Lim, A.G. Saidi, S.K. Reinhardt, The M5 simulator: modeling networked systems, IEEE Micro 26 (4) (2006) 52–60. 303

[5] E.A. Carara, F.G. Moraes, Deadlock-free multicast routing algorithm for wormhole-switched mesh networks-on-chip, in: Proceedings of the Symposium on VLSI (ISVLSI), 2008, pp. 341–346. 289

[6] M.R. Casu, M.R. Roch, S.V. Tota, M. Zamboni, A NoC-based hybrid message-passing/shared-memory approach to CMP design, Microprocess. Microsyst. 35 (2) (2011) 261–273. 289

[7] U.S. Da Costa, I.S. De Medeiros Júnior, M.V.M. Oliveira, Specification and verification of a MPI implementation for a MP-SoC, in: Proceedings of the International Colloquium Conference on Theoretical Aspects of Computing (ICTAC), 2010, pp. 168–183. 288

[8] N. Enright Jerger, L.-S. Peh, M. Lipasti, Virtual circuit tree multicasting: a case for on-chip hardware multicast support, in: Proceedings of the International Symposium on Computer Architecture (ISCA), 2008, pp. 229–240. 289

[9] A. Faraj, X. Yuan, Automatic generation and tuning of MPI collective communication routines, in: Proceedings of the International Conference on Supercomputing (ICS), 2005, pp. 393–402. 287

[10] A. Faraj, X. Yuan, D. Lowenthal, STAR-MPI: self tuned adaptive routines for MPI collective operations, in: Proceedings of the International Conference on Supercomputing (ICS), 2006, pp. 199–208. 288

[11] K. Feind, K. McMahon, An ultrahigh performance MPI implementation on SGI cc-NUMA Altix systems, Comput. Meth. Sci. Tech. 12 (2006) 67–70. 287

[12] E. Fernandez-Alonso, D. Castells-Rufas, J. Joven, J. Carrabina, Survey of NoC and programming models proposals for MPSoC, Int. J. Comput. Sci. Issues 9 (2012) 22–32. 288

[13] P. Francesco, P. Antonio, P. Marchal, Flexible hardware/software support for message passing on a distributed shared memory architecture, in: Proceedings of the Design, Automation & Test in Europe Conference & Exhibition (DATE), 2005, pp. 736–741. 288

[14] F. Fu, S. Sun, X. Hu, J. Song, J. Wang, M. Yu, MMPI: a flexible and efficient multiprocessor message passing interface for NoC-based MPSoC, in: Proceedings of the International SOC Conference (SOCC), 2010, pp. 359–362. 288, 290

[15] M. Galles, Spider: a high-speed network interconnect. IEEE Micro 17 (1) (1997) 34–39. 303

[16] R.C. Gamom Ngounou Ewo, E. Kiegaing, M. Mbouenda, H.B. Fotsin, B. Granado, Hardware MPI-2 functions for multi-processing reconfigurable system on chip, in: Proceedings of the International Parallel and Distributed Processing Symposium Workshops & PhD Forum (IPDPSW), 2013, pp. 273–280. 288

[17] M.S. Gaur, B.M. Al-Hashimi, V. Laxmi, R. Navaneeth, N. Choudhary, L. Jain, M. Ahmed, K.K. Paliwal, Varsha, Rekha, Vineetha, NIRGAM: a simulator for NoC interconnect routing and application modeling, in: Proceedings of the Design, Automation & Test in Europe Conference & Exhibition (DATE), 2007. 302

[18] K. Goossens, J. Dielissen, A. Radulescu, AEthereal network on chip: concepts, architectures, implementations, IEEE Des. Test Comput. 22 (5) (2005) 414–421. 286

[19] W. Gropp, E. Lusk, A. Skjellum, Using MPI: Portable Parallel Programming with the Message-Passing Interface, MIT Press, Cambridge, MA, USA, 1999. 290

[20] B. Grot, J. Hestness, S.W. Keckler, O. Mutlu, Express cube topologies for on-chip interconnects, in: Proceedings of the International Symposium on High-Performance Computer Architecture (HPCA), 2009, pp. 163–174. 289

[21] T. Hoefler, C. Siebert, W. Rehm, A practically constant-time MPI broadcast algorithm for large-scale InfiniBand clusters with multicast, in: Proceedings of the International Parallel and Distributed Processing Symposium (IPDPS), 2007, pp. 1–8. 287

[22] Y. Hoskote, S. Vangal, A. Singh, N. Borkar, S. Borkar, A 5-GHz mesh interconnect for a teraflops processor, IEEE Micro 27 (5) (2007) 51–61. 286

[23] L. Huang, Z. Wang, N. Xiao, Accelerating NoC-based MPI primitives via communication architecture customization, in: Proceedings of the International Conference on Application-Specific Systems, Architectures and Processors (ASAP), 2012, pp. 141–148. 285

[24] T.N.K. Jain, P. Gratz, A. Sprintson, G. Choi, Asynchronous bypass channels: improving performance for multi-synchronous NoCs, in: Proceedings of the International Symposium on Networks-on-Chip (NOCS), 2010, pp. 51–58. 289

[25] A. Kohler, M. Radetzki, Optimized reduce for mesh-based NoC multiprocessors, in: Proceedings of the International Parallel and Distributed Processing Symposium Workshops & PhD Forum (IPDPSW), 2012, pp. 904–913. 299

[26] R. Kumar, T.G. Mattson, G. Pokam, R.F.V. der Wijngaart, The case for message passing on many-core chips, in: M. Hbner, J. Becker (Eds.), Multiprocessor System-on-Chip, Springer, Berlin, Germany, 2011, pp. 115–123. 289

[27] A. Kumar, L.-S. Peh, P. Kundu, N.K. Jha, Express virtual channels: towards the ideal interconnection fabric, in: Proceedings of the International Symposium on Computer Architecture (ISCA), 2007, pp. 150–161. 289

[28] J. Liu, A.R. Mamidala, D.K. Panda, Fast and scalable MPI-level broadcast using InfiniBand's hardware multicast support, in: Proceedings of the International Parallel and Distributed Processing Symposium (IPDPS), 2004, pp. 1–10. 287

[29] Z. Lu, B. Yin, A. Jantsch, Connection-oriented multicasting in wormhole-switched networks on chip, in: Proceedings of the International Symposium on Emerging VLSI Technologies and Architectures (ISVLSI), 2006, pp. 205–210. 289

[30] P. Mahr, C. Lörchner, H. Ishebabi, C. Bobda, SoC-MPI: a flexible message passing library for multiprocessor systems-on-chips, in: Proceedings of the International Conference on Reconfigurable Computing and FPGAs (ReConFig), 2008, pp. 187–192. 286, 288

[31] A.R. Mamidala, R. Kumar, D. De, D.K. Panda, MPI collectives on modern multicore clusters: performance optimizations and communication characteristics, in: Proceedings of the International Symposium on Cluster Computing and the Grid (CCGRID), 2008, pp. 130–137. 292

[32] R. Manevich, I. Walter, I. Cidon, A. Kolodny, Best of both worlds: a bus enhanced NoC (BENoC), in: Proceedings of the International Symposium on Networks-on-Chip (NOCS), 2009, pp. 173–182. 292

[33] MPICH, High-Performance Portable MPI, http://www.mcs.anl.gov/mpi/mpich, cited 2012. 288, 293

[34] N. Muralimanohar, R. Balasubramonian, Interconnect design considerations for large NUCA caches, in: Proceedings of the International Symposium on Computer Architecture (ISCA), 2007, pp. 369–380. 292

[35] N. Muralimanohar, R. Balasubramonian, N. Jouppi, CACTI 6.0: a tool to model large caches, Technical Report HPL-2009-85, HP Laboratories, April 2009. 303

[36] M. Ohara, H. Inoue, Y. Sohda, H. Komatsu, T. Nakatani, MPI Microtask for programming the Cell Broadband Engine™ processor, IBM Syst. J. 45 (2006) 85–102. 288

[37] L.-S. Peh, W. Dally, A delay model and speculative architecture for pipelined routers, in: Proceedings of the International Symposium on High-Performance Computer Architecture (HPCA), 2001, pp. 255–266. 303

[38] Y. Peng, M. Saldana, P. Chow, Hardware support for broadcast and reduce in MPSoC, in: Proceedings of the International Conference on Field Programmable Logic and Applications (FPL), 2011, pp. 144–150. 287, 288

[39] J. Psota, A. Agarwal, rMPI: message passing on multicore processors with on-chip interconnect, in: Proceedings of the International Conference on High Performance Embedded Architectures and Compilers (HiPEAC), 2008, pp. 22–37. 286, 288, 290

[40] T.D. Richardson, C. Nicopoulos, D. Park, V. Narayanan, Y. Xie, C. Das, V. Degalahal, A hybrid SoC interconnect with dynamic TDMA-based transaction-less buses and on-chip networks, in: Proceedings of the International Conference on VLSI Design (VLSID), 2006, pp. 657–664. 292

[41] RunTime Computing Solutions, MPI/Pro, http://www.runtimecomputing.com/products/mpipro/, cited 2013. 288

[42] M. Saldaña, P. Chow, TMD-MPI: an MPI implementation for multiple processors across multiple FPGAs, in: Proceedings of the International Conference on Field Programmable Logic and Applications (FPL), 2006, pp. 1–6. 286, 288, 304

[43] M. Saldaña, A. Patel, H.J. Liu, P. Chow, Using partial reconfiguration and message passing to enable FPGA-based generic computing platforms, Int. J. Reconfig. Comput. 2012 (2012) 3:3–3:3. 288

[44] M. Saldaña, A. Patel, C. Madill, D. Nunes, D. Wang, P. Chow, R. Wittig, H. Styles, A. Putnam, MPI as a programming model for high-performance reconfigurable computers, ACM Trans. Reconfig. Tech. Syst. 3 (2010) 22:1–22:29. 287, 304

[45] F.A. Samman, T. Hollstein, M. Glesner. Multicast parallel pipeline router architecture for network-on-chip. in: Proceedings of the Design, Automation & Test in Europe Conference & Exhibition (DATE), 2008, pp. 1396–1401. 289

[46] J. Sartori, R. Kumar, Low-Overhead, high-speed multi-core barrier synchronization, in: Proceedings of the International Conference on High Performance Embedded Architectures and Compilers (HiPEAC), 2010, pp. 18–34. 297

[47] D.G. Solt, A profile-based approach for topology aware MPI rank placement, in: Invited Presentation to HPCC, 2007. 288

[48] M.B. Stensgaard, J. Sparsø, ReNoC: a network-on-chip architecture with reconfigurable topology, in: Proceedings of the International Symposium on Networks-on-Chip (NOCS), 2008, pp. 55–64. 289

[49] I.A.C. Ureña, M. Riepen, M. Konow, RCKMPI—lightweight MPI implementation for Intel's single-chip cloud computer (SCC), in: Proceedings of the European MPI Users' Group Conference on Recent Advances in the Message Passing Interface (EuroMPI), 2011, pp. 208–217. 288, 290, 303

[50] H.-S. Wang, X. Zhu, L.-S. Peh, S. Malik, Orion: a power-performance simulator for interconnection networks, in: Proceedings of the International Symposium on Microarchitecture (MICRO), 2002, pp. 294–305. 303

[51] D. Wentzlaff, P. Griffin, H. Hoffmann, L. Bao, B. Edwards, C. Ramey, M. Mattina, C.-C. Miao, J.F. Brown III, A. Agarwa, On-chip interconnection architecture of the TILE processor, IEEE Micro 27 (5) (2007) 15–31. 286

[52] J.A. Williams, I. Syed, J. Wu, N.W. Bergmann, A reconfigurable cluster-on-chip architecture with MPI communication layer, in: Proceedings of the Symposium on Field-Programmable Custom Computing Machines (FCCM), 2006, pp. 350–352. 286

Message passing interface communication protocol optimizations†

CHAPTER OUTLINE

†Part of this research was first published in *ACM Transactions on Architecture and Code Optimization* [14]. This work has also been selected to be presented at the 9th International Conference on High Performance and Embedded Architecture and Compilers (HiPEAC-2014).

10.1 INTRODUCTION

Multicore architectures with network-on-chip (NoC) connecting cores have been pervasively recognized as the de facto design for the efficient utilization of the ever-increasing density of transistors on a chip. Recent proposals including the 64-core TILE64 processor from Tilera [31], Intel's 80-core Teraflops chip [12], Arteris's NoC interconnect IPs [1], and NXP-Philips's AEtheral NoC [8] have successfully demonstrated the potential effectiveness of NoC designs.

To reduce the gap between the multicore programming model and the underlying NoC-based hardware, a new NoC design that incorporates parallel programming models, such as the message passing interface (MPI) [10], is regarded as a promising option. Numerous applications have been ported to or developed for the MPI standard, thus making performance optimization a necessity for multicore architectures [19]. The software overhead constitutes a very large percentage of message latency. This issue will increase in severity when high-speed on-chip parallel communication channels are used to transmit messages. To accelerate the software processing time, the hardware support features of NoC designs require further exploration.

To provide efficient support for the MPI to boost the performance of parallel applications, we exploit the on-chip hardware available in NoC-based multicore architectures. Existing solutions [18, 21, 24, 25, 30, 32] mainly focus on the MPI adaption on NoC-based multicore processors such as the minimal MPI function selection and the MPI software stack modification. However, special techniques that can take advantage of the fine-grained, low-latency features of NoC hardware have not been fully exploited. In this chapter, we focus on the special communication protocol for the further performance optimization of MPI functionality.

One of the key functionalities of hardware implementation is the support of MPI communication protocols. Synchronous and buffered protocols are the two dominant classes of communication protocols for message passing. In a synchronous system, the sender and the receiver block the execution of the program until the information transfer is complete. In a buffered system, the sender can complete the transmission without the corresponding receive operation being executed. Thus, buffered protocols can achieve lower latencies than synchronous protocols on data transmission by eliminating acknowledgment delay.

By buffering to avoid the acknowledgment delay for subsequent data transmission, the buffered protocol outperforms the synchronous protocol when buffers are plentiful, but the latter outperforms the former when buffers are limited, such that the retry mechanism is activated. Designing a single communication protocol to provide high performance for numerous system configurations and workloads is difficult. We advocate an adaptive approach to address this challenge. This adaptive scheme is desirable for two reasons. First, considering the trend toward NoC-based multicore processors, a single protocol must suffice for multiple hardware configurations and applications. Second, statically choosing between a buffered protocol and a synchronous protocol is undesirable owing to the varying behavior of different workloads and the time-varying behavior within a work-load. Further, a given workload's demand on the buffer size varies dynamically over time.

Thus, an adaptive hybrid protocol that provides robust performance is preferable to a static selection of either a buffered or a synchronous protocol. Our contribution is an adaptive communication mechanism (ADCM) that performs buffered communication if buffers are plentiful but performs synchronous communication if buffers are limited. The ADCM adapts dynamically by determining the communication protocol on a per-request basis to provide robust performance. The ADCM attempts to combine the advantages of both buffered and synchronous communication modes and achieves better throughput and performance. Simulations of various workloads show that the proposed communication mechanism can be effectively used in future NoC designs.

10.2 BACKGROUND

We first briefly discuss two dominant classes of communication protocols, the buffered and synchronous protocols. Then, we present the limitations of existing protocols. Finally, we introduce the related work for optimizing MPI communications.

10.2.1 COMMUNICATION PROTOCOLS IN MPI

In this section, we describe two basic communication protocols, known as synchronous and buffered protocols in the MPI, which also serve as the base cases. Figure 10.1 illustrates the process of the two basic communication protocols. The buffered protocol is asynchronous since it enables the completion of a send operation without the corresponding receive operation being executed. Thus, the overlap of communication and computation is possible. However, this protocol assumes sufficient memory at the receiver to store all the expected or unexpected messages, which could be of the order of kilobytes or even megabytes depending on the size of the messages; otherwise, buffer overflows will occur. The allocation of substantial memory for buffering may lead to wasted memory in cases where the buffer is underutilized. Thus, the programmer must avoid the corruption of the data, given that data transmission occurs in the background, and the transmission buffer can be overwritten by further computation. In an embedded system with limited resources, this protocol may not scale well.

By contrast, the synchronous protocol requires the producer first to initiate a request to the receiver. This request is called the message envelope and includes the details of the message to be transmitted. When the receiver is ready, it will reply with a clear-to-send packet to the producer. Once the producer receives the clear-to-send packet, the actual transfer of data will begin. This protocol incurs a higher message overhead compared with the buffered protocol because of the

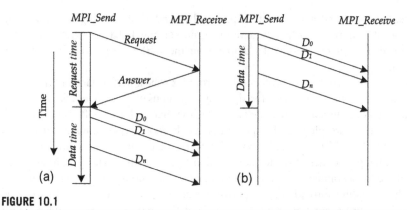

FIGURE 10.1

Basic MPI communication protocols. (a) Synchronous mode. (b) Buffered mode.

synchronization process. However, less memory is required, and buffer overflows are less likely to occur because the protocol needs only to store message envelopes, which are 8 bytes long, in the event of unexpected messages.

10.2.2 EXISTING PROBLEMS

The buffered protocol is generally found to outperform the synchronous protocol when the receiving buffers are sufficient. However, using a single buffered protocol alone would result in numerous problems. Before discussing these problems, we assume that the on-chip network is adequately robust to perform any message transmission operation to the target receivers correctly. That is, we do not need to receive the acknowledgment message to ensure correct NoC transmissions. This condition is also the case for real NoC-based processor designs, which differ from the large multiprocessor systems.

10.2.2.1 Correctness problems

For MPI applications, messages out of order or unexpected are common [9]. The unexpected messages would have to be buffered, and a large number of unexpected small messages or a small number of large messages may result in buffer overflows. Such buffer overflows would give rise to a correctness problem because the transmission data are not received correctly. Figure 10.2 illustrates this correctness problem. In NoC-based systems with limited memory spaces, buffering messages would be inadequate because it would limit the scalability of NoC designs. Thus, a programmer must prevent the corruption of the data. The incorrect use of the buffered protocol would result in incorrect MPI executions, which poses a programming burden. To mitigate this problem, the retry mechanism could be used. When the receiver does not have sufficient buffers to store the unexpected messages, the retry mechanism would be triggered to resend the lost messages.

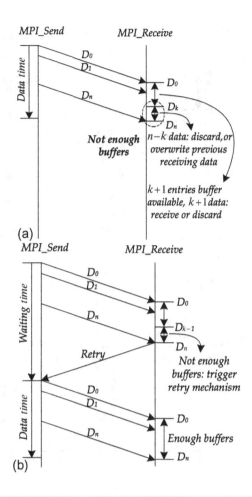

FIGURE 10.2

Correctness problems. (a) Buffered mode: incorrect execution. (b) Buffered mode: correct execution.

10.2.2.2 Retry problems

Although the retry mechanism can maintain the correctness of the buffered protocol, the utilization of the retry mechanism in the data transmission still faces a number of problems. First, we need to know when to trigger the retry operation. Different ways of triggering the retry mechanism will yield different performance results. Figure 10.3 illustrates four different timing methods for the retry operation. The first method, as shown in Figure 10.3a, is to trigger the retry procedure after all the data have been received, even if these data are discarded. This method is simple but may delay the transmission time when a receiving buffer is already available. The second method, as shown in Figure 10.3b, is to trigger the retry procedure immediately when

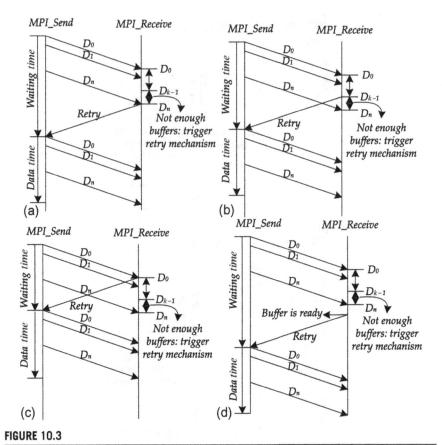

FIGURE 10.3

Retry problems: timing exploration. (a) Retry after. (b) Retry immediately. (c) Retry before. (d) Retry ready.

buffers overflow. This method is better than the first one because it can send the retry message in advance. The third method, as shown in Figure 10.3c, is to trigger the retry procedure immediately after the first packet of the message has been received, when the buffer is insufficient. This method appears to be an optimal approach that can trigger the retry procedure as soon as possible. However, this method may introduce additional data transmissions when buffers are not ready. The last method, as shown in Figure 10.3d, is to trigger the retry procedure only when the receiver has sufficient buffers to complete the data transmission.

Determining what to retry is also an important problem. Different degrees of data granularity for retry also yield different performance results. Figure 10.4 illustrates three basic methods for retry. The first method, as shown in Figure 10.4a, is called retry whole, which sends the retry packet to resend the whole message. With use of this method, the receiver does not need to record any data received from the previous

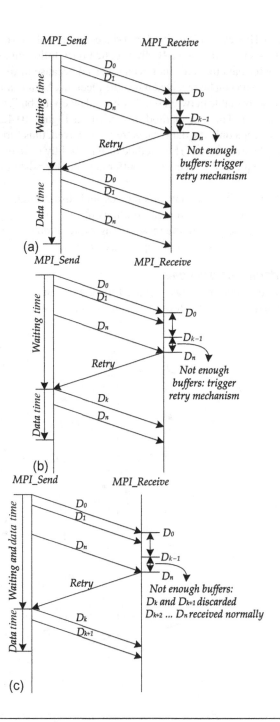

FIGURE 10.4

Retry problems: data granularity exploration. (a) Retry whole. (b) Retry following. (c) Retry discarded.

data transmission. However, this method introduces additional data transmission. The second method, as shown in Figure 10.4b, is called retry following, which sends the retry packet only for data that have not been received because of the lack of buffers. That is, part of the data can be received when the receiving buffers are available. This method utilizes the available network bandwidth and receiving buffer more efficiently than the first method. The last method, as shown in Figure 10.4c, is called retry discarded, which sends only the data that were not received, thus further reducing the size of the data to be transferred. This method differs slightly from the retry following method in that it requires the receiver to buffer any message data once the buffer is available.

These methods each have their advantages and disadvantages. To achieve an optimized retry mechanism, further explorations of these methods are required. In this chapter, we propose to combine these timing exploration methods with the data granularity exploration to establish an optimized approach.

10.2.2.3 Performance problems

The buffered protocol does not always outperform the synchronous protocol. In some cases, the buffered protocol is slower than the synchronous protocol. Figure 10.5

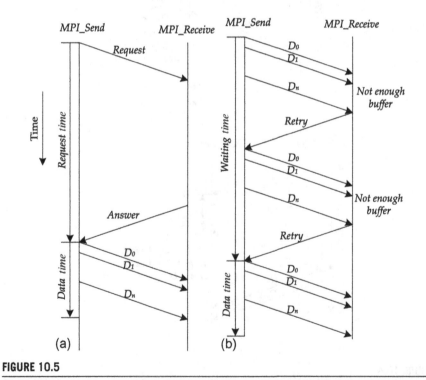

FIGURE 10.5

Performance problems. (a) Synchronous protocol. (b) Buffered protocol.

illustrates such a case, where the buffers are not ready to receive messages immediately. After three attempts at message transmission, the sender successfully sends its data to the receiver. However, the synchronous protocol needs to send data only once, after which the receiver can trigger the retry procedure immediately when buffers are available. The buffered protocol has two other disadvantages. First, this protocol may introduce a large amount of network traffic, which would increase network contention and consequently result in performance degradation. Second, the buffered protocol should send the confirmation message to the sender to release the sending buffer after the activation of the retry mechanism. This process may also degrade performance because of the additional message transmission and delayed sending buffer release.

10.2.3 **RELATED WORK**

To improve the performance of MPI applications, there have been a large number of works on software optimization and hardware support. For multiprocessor systems, the MPI optimization is an extensively investigated domain along various directions, including optimizing implementations on a generic architecture such as a cluster [11] or on a specific machine [5].

Ogawa and Matsuoka [20] used compiler modifications to optimize the MPI. The compiler recognizes the MPI calls in a program, performs a static analysis to determine which arguments are static, and then creates specialized MPI functions for that program. Faraj and Yuan [3] presented a method for automatically optimizing the MPI collective subroutines. Karwande *et al.* [15] presented a method for compiled communication, which applies more aggressive optimizations to communications with information that is known at compile time. Liu [16] and Hoefler [11] used hardware multicast in native InfiniBand to improve the performance of MPI broadcast operation. However, these approaches cannot be used directly in NoC infrastructures under different constraints. For on-chip multicore systems, Peng [23] considered the NoC designs for low-overhead broadcast and reduced transmission but did not consider the communication protocol.

The communication protocol has been investigated, and applications were found to have the tendency to consume time on traversing message queues, thus resulting in an increase in performance gap [9]. Thus, a unique hardware structure is extended to accelerate the list traversal and matching [27]. An active research area is the use of reconfiguration to improve the application performance such as adaptive MPI [13] and reconfigurable iterative MPI [17]. Venkata *et al.* [28, 29] showed how reconfiguration can be used to improve bandwidth availability. They also used profile data for fine-grained run-time reconfiguration and provided a framework that can be used to implement other similar reconfigurations. Other systems such as STAR-MPI [4] and HP-MPI [26] have shown that the profile data can be used for optimizing the MPI performance at link time or launch.

Unlike the aforementioned optimization techniques, this chapter focuses on multicore architectures and utilizes the advantages of on-chip hardware resources. Given that power, area, and latency constraints for off-chip versus on-chip communication architectures differ substantially, prior off-chip communication architectures are not directly suitable for on-chip usage. Thus, in this chapter, we propose a new communication mechanism for accelerating MPI functions using an adaptive implementation technique.

10.3 **MOTIVATION**

After considering the aforementioned problems of conventional protocols, we attempt to establish an ideal communication protocol that can address these issues. Figure 10.6 illustrates the protocol that ideally has prior knowledge of the buffer usage and is capable of performing the corresponding operations. The figure has three main timelines: the first is D_0 arrival time AT_{D_0}, which is the arrival time of first communication data D_0 at the target node; the second is D_n arrival time AT_{D_n}, which is the arrival time of the last communication data D_n at the target node; and the last is D_0(retry) arrival time $\text{AT}_{D_0(\text{R})}$, which is the arrival time of the first communication data D_0 for the first retry operation at the target node. These three timelines will help determine the appropriate operation for different cases.

Figure 10.6a shows an ideal case where the receiving buffer is sufficiently large ($n + 1$ free buffers are available before AT_{D_0}). We simply use the buffered communication protocol to deal with this transaction. Figure 10.6b shows a case where the receiving buffer is limited ($n + 1$ free buffers are only available before AT_{D_n}). That is, the receiving buffer is not ready to receive the data, but will be ready soon. In such a case, the sender can initially send the data using the buffered protocol, such that the receiver will receive the data using available receiving buffers until no buffers are left. Once buffers become available, the data will be received. After $n + 1$ buffers are ready, the receiver will trigger the retry mechanism for discarded data using the retry discarded method illustrated in Figure 10.4c. Given that this process can ensure that the data will be successfully received, the *Confirm* message is unnecessary. This approach outperforms the two basic protocols and will have less network traffic than the buffered protocol. Figure 10.6c shows a case with a more limited number of free buffers wherein only $n + 1$ free buffers are available between AT_{D_n} and $\text{AT}_{D_0(\text{R})}$. In such a case, we use the retry following method. That is, the receiver triggers the retry operation immediately after receiving data D_0 for the following $n + 1 - k$ data. Compared with the retry discarded method, this approach may send some data twice. However, this method is capable of immediately triggering the retry mechanism. Figure 10.6d shows the last case, where the free buffers are limited and will be ready only after $\text{AT}_{D_0(\text{R})}$. A simple solution for this issue is to use the synchronous protocol. Although the buffered protocol may achieve better performance than the synchronous protocol in some cases, it may introduce a considerable amount of network traffic that will likely degrade the overall performance.

The proposed ideal protocol improves the two basic protocols in terms of performance and network traffic. However, it is based on the knowledge of accurate current and future buffer usage information. This requirement is impossible for practical

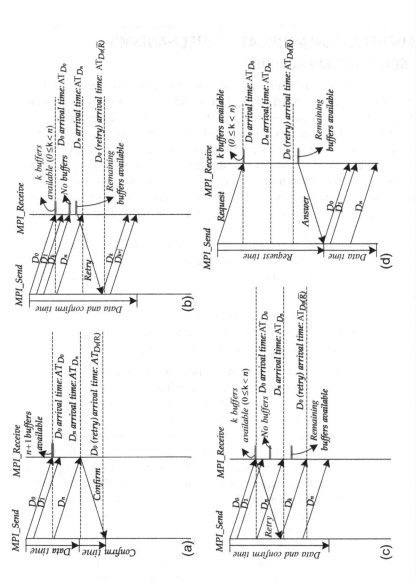

FIGURE 10.6

The ideal communication protocol that optimally addresses different cases. (a) $n + 1$ free buffers available before D_0 arrives. (b) $n + 1$ free buffers available between D_0 and D_n arriving. (c) $n + 1$ free buffers available between D_n and D_0 (retry) arriving. (d) $n + 1$ free buffers available after D_0 (retry) arrives.

implementations but sets a goal that we can attempt to achieve. In the following sections, we will introduce the ADCM, which enables the hardware implementation of the proposed ideal protocol with a number of simplifications.

10.4 ADAPTIVE COMMUNICATION MECHANISMS
10.4.1 GOALS AND APPROACHES

The goal of the ADCM is to minimize the average data transmission latency. Given an infinite buffer, using the buffered protocol would achieve this goal by eliminating all waiting time for receiver acknowledgment. However, a finite buffer may result in buffer overflows and retry delays that outweigh the benefit of eliminating waiting time. Nevertheless, mean retry delays dominate only when the receiving buffer on the target NoC node is highly occupied. The mechanism we propose for the ADCM uses feedback to keep the buffer utilization below a critical level and thus mitigate retry delays. Our mechanism uses a combined local-remote estimation of buffer utilization to keep utilization below a prespecified threshold by dynamically adjusting the probability of buffered transmission.

Our adaptive communication implementation uses a simple mechanism to estimate the buffer utilization and correspondingly adjust the communication protocol. The mechanism can be discussed from two aspects: the sender and the receiver. The sender operations comprise three parts:

(1) estimating the buffer utilization for the target receiving node,
(2) determining whether to trigger the buffered communication, and
(3) reacting according to the messages from the receiving node.

First, the sender uses the buffer utilization information along with the receiving message as a local estimate of the target buffer utilization. Although this static and somewhat obsolete information does not capture future buffer utilization, it is easy to obtain and correlates strongly with future buffer utilization because of the buffered nature of the requests that are most likely to cause buffer occupation. Each core uses a simple, signed, saturating utilization counter to calculate whether the buffer utilization is above or below a static threshold. When the counter is sampled, a positive value indicates that the buffers used are more than the threshold, and a negative value means that the buffers used are less than the threshold. The counter is initially reset to the maximum value. Second, a generated message is transmitted through the buffered or synchronous protocol with a probability proportional to the policy counter. The node sends the messages using the synchronous protocol if the policy counter is smaller than the random number; otherwise, the node uses the buffered protocol. Finally, the sender will react according to the feedback message. If buffered communication packets are generated, then a message for the retry operation may be received until a confirmation message is received. If synchronous communication packets are generated, then the data are sent after an answer packet has been received.

The receiver operations comprise two parts:

(1) estimating its own buffer utilization and
(2) performing the corresponding receive operations.

This adaptivity is reflected in the time of triggering the retry mechanism and the granularity of data retry. First, the receiver estimates the buffer utilization according to the number of free buffers and future utilization trends. Second, the receiver generates a message according to different timing and levels of data granularity. Given that the time of this retry can be regarded as the answer of the target node, the sender may not need to wait for the confirmation message.

Although the ADCM is broadly inspired by the credit-based flow control often used in NoCs, solely using the sender-to-receiver credit counts at a large granularity (perhaps 1 kB per credit) to manage flow control in the MPI communication is difficult. This is because the MPI communication information, such as the message size, is transparent to the NoC credit-based flow, which knows only the packet size. Thus, this flow control logic would be implemented in an upper layer such as an MPI engine (ME), as described in the following section.

10.4.2 BASELINE MPI-ACCELERATED NoC DESIGNS

This section describes the baseline communication architecture of processors. This design aims at accelerating the processing performance of MPI primitives in future massive multicore architectures by way of underlying hardware support. These multicore architectures usually present a mesh-type interconnect fabric. No hardware cache coherence exists for the first-level (L1) and second-level (L2) caches, similarly to the Intel SCC processor. A number of factors have to be considered in the performance improvement of MPI primitives through the hardware support. The basic design discussed in this chapter includes two main hardware techniques for the acceleration of MPI primitives: the NoC design and the ME [25]. Figure 10.7 shows a block diagram of the baseline implementation architecture with a 4 × 4 mesh topology. The underlying NoC design is the actual medium used to transfer messages, which can be designed with consideration for MPI communication. Each node also has an ME between the core and the network interface (NI); this ME is used to execute corresponding instructions for MPI primitives.

By directly executing the MPI primitives and interrupt service routines, the ME reduces the context switching overheads in the cores and can accelerate the software processing. The ME also performs the message buffer management for the cores as well as the fast buffer copying. This engine transfers messages to and from dynamically allocated message buffers in the memory to avoid the buffer copying between system and user buffers. This process also eliminates the need for the sending process to wait for the release of the message buffer by the communication channel. The ME also reserves a set of buffers for the incoming messages. Using the above methods, the long-message transmission protocol can be simplified, consequently reducing the transmission latency.

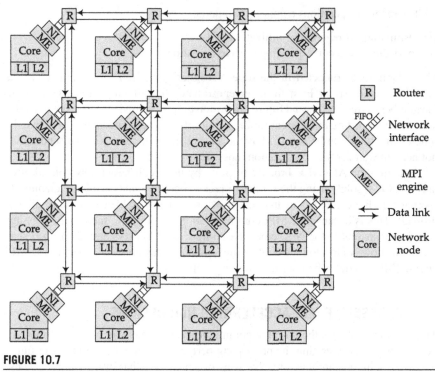

FIGURE 10.7

Block diagram of baseline communication architecture.

The ME architecture is shown in Figure 10.8. This design provides the hardware support to address the communication protocol used in the MPI implementation. Primary functionalities include serving as a middle layer between the processor core and the interface of the NoC. The ME receives the message send requests from the PE core and handles various messages from other processor cores. Two sources can trigger the ME to work: the local processor core and the NI. The local processor core may request the MPI unit perform MPI primitive functions, such that the associated communication data are transferred through this interface. Another source is the NI, which may request the ME to receive messages from the on-chip network and perform corresponding operations for handling the received messages.

The ME generally comprises three key components: the preprocessing unit (PPU), parameter registers (PRs), and the MPI processing unit (MPU). The PPU is used to translate the instructions from the processor core into control signals and to generate the message passing parameters used for transfer of data that are temporarily stored in the PRs. Another important task of the PPU is to exchange data with the CPU cache for reading or writing communication data. The PRs include several registers in the ME. When MPI functions are performed, the MPI unit first receives the message parameters from the PE core and then updates these registers. The MPU is the key

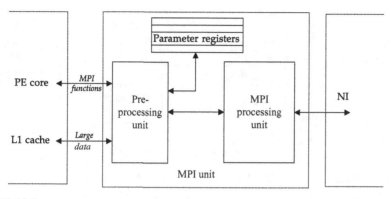

FIGURE 10.8

Block diagram of the ME.

component of the MPI unit that performs the actual operations for MPI primitive functions, as shown in Figure 10.9a. The execution of these functions is generally performed in two separate pipelines: the send and receive pipelines. The send pipeline is used for active operations such as *MPI_Send.* The receive pipeline is used for passive operations such as *MPI_Receive.*

10.4.3 ADCM ARCHITECTURAL SUPPORT

In this subsection, we describe the proposed ADCM architectural support based on the MPI-accelerated NoC designs.

10.4.3.1 ADCM hardware

To support the adaptive communication protocol, hardware modifications are primarily applied to the MPU. The conventional MPU is capable of handling unexpected messages and dividing large messages into smaller packets, as shown in Figure 10.9a. The MPU generally comprises five main components organized in two separate pipelines: message packetizing and building units for the send pipeline as well as packet reception, expectation, and response units for the receive pipeline. The MPU also has an internal buffer to store the unexpected messages. Thus, sending and receiving can be performed simultaneously. Figure 10.9b shows a block diagram of the proposed modified MPU. Two small logical units with associated registers are added to the MPU; they are shaded in Figure 10.9b. For the sending pipeline, the sending policy unit (SPU) is added. To support the sending policy generation, a receiving buffer credit (RBC) table is also added. The SPU generates the selection results according to the RBC table. The SPU comprises two subcomponents: the sending policy generator, which implements the adaptive sending protocol algorithm (buffered or synchronous protocol), and the preparation unit, which generates the corresponding header data according to the selection result and then sends these data to the packetizing unit.

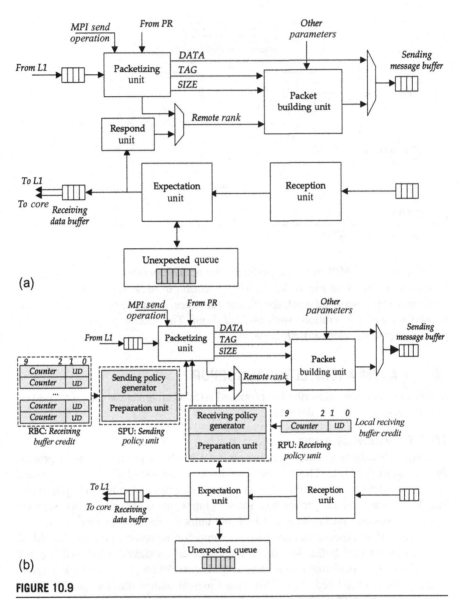

FIGURE 10.9

Block diagram of the MPU. (a) Conventional MPU. (b) Modified MPU for the ADCM.

The RBC table is the basis for the selection of different protocols. This table includes two fields for each target NoC node: *Counter* and *UD*. The 2-bit *UD* field is used to indicate the trend of the receiving buffer number, such that "00" indicates that the buffer number is unchanged, "01" indicates that the buffer number is increased, and "10" indicates that the buffer number is decreased. The *UD* field is an important parameter for protocol selection because it can predict available buffers more accurately. The 8-bit *Counter* field stores the number of receiving buffers available for other nodes, and their initial values are set as the maximum. The update of the RBC value is triggered by the response packet from the corresponding node. The first two bits can be used to specify the buffer number in different angularities, that is, "00" for 0-1 kB, "01" for 1-4 kB, "10" for 4-8 kB, and "11" for 8-32 kB. The last six bits are used to indicate the buffer number in that granularity. This process is sufficiently accurate for storing the buffer number information and for reducing the hardware overhead of the RBC table. Each core has an RBC for every possible NoC node. The value of *Counter* is directly calculated from the received value for the send operation, and the value of *UD* is calculated from records of previous buffer number changes. These two parameters serve as the available information for the estimation algorithm of buffer usage.

For the receiving pipeline, the receiving policy unit (RPU) is added. The RPU also includes two subcomponents: the receiving policy generator, which implements the adaptive receiving protocol algorithm according to the received packets and local receiving buffers, and the preparation unit, which generates corresponding header data according to the receiving policy and sends these data to the packetizing unit. These header data specify the type of the response packets, the RBC data, and so on. To access the buffer utilization information, a local RBC register is configured. This register records the number of free buffers and the utilization trend of buffers, the fields of which have the same meaning as those of the SPU. The *UD* field is set according to the comparison result of the current number and previous stored number of free receiving buffers. Notably, the receiving buffers are counted as an unexpected queue, as shown in Figure 10.9. Evidently, the expected packets can be handled promptly by the processor core and will not occupy the configured free buffer.

10.4.3.2 Adaptive algorithm implementation

The implementation of the adaptive algorithm is an important component of the ADCM approach for determining how to integrate the buffered and synchronous protocols. Our main goal is to improve the performance of MPI functions. However, the hardware complexity is also an important metric in our design. Thus, the proposed algorithm should facilitate an optimal tradeoff between the performance and the hardware complexity. This algorithm can be implemented in two parts from the sender/receiver perspective: the adaptive sending and receiving algorithms.

Figure 10.10 shows the mainframe of these two algorithms. The first part is the adaptive sending algorithm, as shown in Figure 10.10a. This algorithm is relatively simple since it only needs to select the communication protocol from two options for sending messages. This algorithm accepts two inputs: the RBC table information and the data to be sent. The algorithm compares the RBC value of the receiving node with the length of sending data flits. If the estimated number of receiving buffers is adequate to store all the data, the buffered protocol is simply chosen for message transmission. If the estimated number of free receiving buffers is insufficient for

```
1: // Note: RBC: receiving buffer credit table for all other nodes
2: //       Data: data to be sent
3: void adaptive_sending (RBC, SendingData) {
4:    if (RBC[dst][9:2] > len(SendingData) )              // Enough free buffers
5:       Buffered_sending();                              // Send using buffered protocol
6:    if (RBC[dst][9:2] > T_s_threshold && RBC[dst][1:0]==`01` )  // buffer increase
7:       Buffered_sending();                              // Send using buffered protocol
8:    else                                    // Other cases: little buffer, buffer decrease
9:       Synchronous_sending();                           // Send using synchronous protocol
10: }
```
(a)

```
1:  // Note: LocalRBC: Local receiving buffer credit
2:  //       ReceivingData: data received
3:  void adaptive_receiving (LocalRBC, ReceivingData) {
4:     if ( is_confirm_received )                         // Confirm signal received
5:        release_sendingbuffer();                        // Release the sending buffers
6:     elseif ( is_synchronous_protocol )                 // Synchronous data received
7:        ReceiveNormally();                              // Send answer or Receive Data
8:     if ( is_buffered_protocol ) {                      // This packet is in buffered protocol
9:        if ( is_retry_operation ) {                     // This packet has retry data
10:          if (is_speculative_retry ) {                 // The retry is speculative
11:             if ( is_enough_buffer) {                   // Enough buffers
12:                ReceiveNormally();                      // Receive normally
13:                Confirm();                              // Send confirm signals
14:             }
15:             else {                                     //Not enough buffers
16:                Receive_with_FreeBuffer();    // Receive with free buffers
17:                TriggerRetryFollowing_when_BufferReady();    // Retry
18:             }
19:          }
20:          else                                // Not speculative retry operation
21:             ReceiveNormally();    // Receive normally with enough buffers
22:       }
23:       else if ( LocalRBC[9:2] > len (ReceivingData) ) {    // Enough buffers
24:          ReceiveNormally();                            // Receive with enough buffers
25:          Confirm();                                    // Send confirm signals
26:       }
27:       elseif ( LocalRBC[9:2] > T_r_threshold && LocalRBC[1:0]==`01` ) {
28:          Receive_with_FreeBuffer();                    // Receive with free buffers
29:          TriggerSpeculative_RetryFollowing();          // Speculative retry
30:       }
31:       else {                                           // Other cases
32:          Receive_with_FreeBuffer();                    // Receive with free buffers
33:          TriggerRetryFollowing_when_BufferReady();  // Retry following
34:       }
35: }
```
(b)

FIGURE 10.10

The adaptive algorithm for the ADCM. (a) Adaptive sending protocol algorithm. (b) Adaptive receiving protocol algorithm.

storing all the data but is likely to increase, the buffered protocol is also selected. This process is based on the following assumption: the buffer can be free after a short time when data are received. Nevertheless, a minimal number of current free buffers to make more free buffers is required for this case, which we call $T_{s\_threshold}$. Generally, we can set this value as half of the sending data size, as used in this chapter. For other cases such as when the number of free buffers is less than $T_{s\_threshold}$ or the estimated number of receiving free buffers is likely to decrease, the synchronous protocol is chosen.

Figure 10.10b shows the adaptive receiving protocol algorithm. Similarly to the sending algorithm, the receiving algorithm also accepts two data sources (*localRBC* and received data) as input. This algorithm is implemented in the following two steps:

(1) analyze the received data and determine the next step of communication and
(2) send the response message to the data sender according to the decision of step 1.

Considering that various types of messages could be received, this procedure is more complex for handling different messages. Line 4 verifies whether the confirmation message is received. Receipt of the confirmation message indicates that the data transmission is successfully completed. The response operation is very simple in that the receiver only needs to release its sending buffers for the storage of a specific transmission. Line 6 verifies whether the message is received using the synchronous protocol. Such messages are of two types: the request message and the data message. To deal with the request message, the receiver can send the answer message only after it has already sufficient free buffers that can be reserved for this transmission. Receipt of the data message indicates that the receiving free buffer is sufficient for the synchronous protocol. Thus, the receiver can receive the data without consideration of free buffers.

Line 8 verifies whether the message is received using the buffered protocol. Various cases should be considered to maximize the utilization of the buffered protocol to improve performance. Line 9 initially verifies the retry operation cases. The algorithm supports two different retry operations: the *ready retry following* and the *speculative retry following* operations. The *ready retry following* requires the receiver to trigger the retry operation only after the free buffers are ready to receive. If the free buffers are inadequate, buffers will be reserved only when they are free. To reduce the transmission delay further, the *speculative retry following* operation is also integrated. For such a type of retry operation, the receiver does not need to wait until the free buffer is ready before triggering the retry operation upon finding that the number of free buffers is insufficient for receiving the message data. Line 10 verifies whether the retry mechanism is speculative, whereas lines 10–21 do the respond operations accordingly. Notably, if the free buffers remain insufficient for receiving the *speculative retry following* message data, the *ready retry following* mechanism is triggered in the following transmission.

Lines 23-34 deal with the normal first-time data message reception (no retry operation) using the buffered protocol, which can be handled in three cases:

(1) the number of receiving free buffers is adequate,
(2) the number of free buffers is more than $T_{r\_threshold}$ and is likely to increase, and
(3) other cases that do not belong to the two aforementioned cases.

The first case is very simple in that the message only has to be received normally, after which the confirm signal is sent to the sender. The second case is related to the *speculative retry following* operation, which is used when current free buffers are insufficient for storing the received data, but the RPU predicts that the number can be adequate when the data transmission is again processed by the retry operation. Thus, the RPU speculatively triggers the retry operation immediately after it finds insufficient free buffers. The last case is related to the *ready retry following* operation, which is the worst case. The synchronous protocol is likely to be used when the free buffers will be ready only after a long time. Such a policy can save the traffic workload if the *speculative retry following* operation is triggered and is likely to improve the performance.

On the basis of the description of the adaptive algorithm for the ADCM, we can see that the integration of buffered and synchronous protocols combines the best features of the two protocols to achieve the ideal communication when the buffer prediction is correct. Otherwise, optimal performance is achieved by dynamically adjusting the response operations.

10.4.3.3 The packet format

Given that the communication protocol is implemented by the hardware, the conventional packet format of the network should be modified to provide the receivers with more information to perform adaptive operations. Figure 10.11 illustrates the packet format of the ADCM. For comparison, a conventional NoC packet format is also shown in the figure. We explain the packet fields as follows. In the conventional 69-bit packet format shown in Figure 10.11a the packet comprises the header flit followed by payload flits. The *Type* and *ID* (identity) bits are two additional 3-bit heads. The source and target addresses of the packet are included in the header flit. Passing a communication segment of the NoC, each packet has the same local identity number (ID tag) for differentiation from other packets. The local ID tag of the data flits of one packet will vary over different communication segments to provide a scalable concept.

To support the adaptive protocol for NoC transmission, additional bits should be integrated into the conventional packet format, called the *ADCM*, which is shaded in Figure 10.11b. The specific bits of *ADCM* are listed in the figure, and it comprises five control fields: *P*, *Type*, *R*, *S*, and *RBC*. The control bit *P* indicates whether the transmission is under the buffered or synchronous protocol. *P* also determines the following steps for handling the message: this bit will be set by the sender according to different buffer usage scenarios. The *Type* field specifies the packet type for the ADCM: *data*, which is the actual packet for sending data that are located in the following data flits; *request*, which is the control packet for requesting data receiving under the synchronous protocol and does not contain actual data of the message; *answer*, which corresponds to the *request* packet under the synchronous protocol; *retry*, which is the control packet for requesting a retry operation under the buffered protocol; and *confirm*, which is used to confirm to the sender that the receiver has

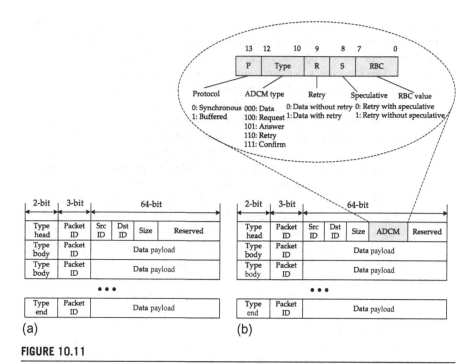

FIGURE 10.11

The packet format. (a) Conventional NoC packet format. (b) ADCM packet format.

received the message. The R field indicates whether the received data is in the retry operation. The S field is the speculative bit used to indicate whether the retry operation for sending data is speculative. If speculative retry is observed, the receiver will send the confirmation packet to the sender to complete the data transmission. The RBC field is used to provide the $localRBC$ for the sender to estimate future buffer usage information.

This *ADCM* packet format will not introduce a burden for packet encoding and transmission delay because additional flits will not be required. Furthermore, the ADCM provides adequate information for the sender and receiver to determine the adaptive operation.

10.4.4 COMPARISON WITH THE IDEAL PROTOCOL

To reduce the hardware complexity and establish a practical design, the ADCM approach does not perform the same dynamic operations as the ideal protocol discussed in Section 10.3. The ADCM differs from the ideal protocol mainly in the following aspects:

- The buffer usage information. In the ADCM, the current and future buffer usage information is estimated on the basis of previous buffer usage information, which is inaccurate since the buffers can considerably change depending on the node receiving the data. Such inaccuracy may cause the ADCM to perform operations that are less optimal compared with those performed by the ideal protocol. This inaccuracy is mainly attributed to the long network latency of information transmission and the estimation algorithm. To improve the estimation accuracy, we can perform optimizations from two aspects: the hardware support for efficient buffer usage transmission and the accurate buffer usage estimation of local buffers such as some hints from programs. These optimizations are beyond the scope of this chapter and will be our future work.

- The retry granularity. In the ADCM, the retry following method is used. This process may introduce more network traffic compared with the retry discarded method used in the ideal protocol. However, the retry following method does not need to record the address of discarded data and does not possess a complex control logic to send/receive these data. Furthermore, the retry following method can trigger the retry mechanism immediately after finding that the free buffers are insufficient. This will improve the performance in some cases.

- The retry timing. The selection of retry timing methods is based on the estimation of buffer usage information. Appropriate retry timing will facilitate performance improvement. In the ADCM, the *speculative retry following* method (i.e., retry immediately method in Figure 10.3b) is used when estimates show sufficient free buffers. Otherwise, the *ready retry following* method in Figure 10.3b is used. Given that the buffer usage estimation information in the ADCM is inaccurate, the ADCM may make a wrong decision and result in poorer performance than the ideal protocol in some cases.

10.5 EVALUATION

10.5.1 METHODOLOGY

To evaluate the proposed communication design, we implemented its architecture using a SystemC-based cycle-level NoC simulator augmented with the ADCM, which is modified from a NIRGAM simulator [7]. The ADCM is based on the ME architecture. The simulator models a detailed pipeline structure for the NoC router and the ME. We can change various network configurations, such as the network size, topology, buffer size, routing algorithm, and traffic pattern. Table 10.1 lists the NoC design configurations in this study. There are 32 receiving and 32 unexpected buffers in the NoC design, each having a 128-bit size. We compare the characteristics of the proposed communication architecture based on the ADCM scheme against the characteristics of the conventional ME design. The basic router is representative of the conventional NoC design, which originally had a four-stage router pipeline. The first stage is the buffer write (BW), and the routing computation (RC) occurs in the second stage. In the third stage, virtual channel allocation (VA) and switch allocation (SA) are performed. In the fourth stage, the flit traverses the switch (ST). Each pipeline stage takes one cycle, followed by one cycle to perform the link traversal (LT) to the

Table 10.1 Communication Design Configurations

Designs	Parameter	Measure
Processor design	Frequency	1 GHz
	Cores	Alpha 21,264
	L1 instruction cache	2-way 16 kB
	L1 data cache	2-way 32 kB
	L2 cache	16-way 1 MB
NoC design	Topology	4 × 4 mesh
	Basic routing	Dimensional ordered XY
	Ports	5
	VCs per port	4
	Buffers per VC	16
	Channel width/flit size	128 bits
	Packet size	8 flits
	MPU receiving buffers	32
	MPU unexpected buffers	32

next router. To shorten the pipeline, the basic router uses lookahead routing [6] and the speculative method [22].

Each simulation experiment is run until the network reaches the steady state. The time for initializing the ME is not considered part of message transmission. For the sake of comprehensive study, numerous validation experiments were performed for several combinations of workload types and network sizes. In the following section, the capability of the proposed communication design is assessed for different traffic patterns, including synthetic traffic and real application traffic.

The synthetic traffic patterns used in this research are the round-trip, uniform random, and hotspot traffic patterns to achieve a more specific evaluation for different traffic patterns. Each simulation runs for 1×10^6 cycles. To obtain stable performance results, the initial 1×10^5 cycles are used for simulation warm-up, and the following 9×10^5 cycles are used for analysis. When destinations are chosen randomly, we repeat the simulation run five times and obtain the average of the values obtained in each run. The MPI communication packet data size is also chosen randomly, and ranges from 128 bits (one flit) to 1024 bits (eight flits).

We also studied the ADCM approach using real application communication traffic. Traces for the baseline conventional implementations were obtained using a full-system multicore simulator, M5 [2]. We model our target multicore systems with the Alpha instruction set architecture, which is the stablest instruction set architecture supported in the M5 simulator. Each core is modeled with a two-issue in-order SPARC processor with two-way 16-kB L1 ICache, two-way 32-kB DCache, and 1-MB L2 Cache. We use the NAS Parallel Benchmarks (NPB 2.4) suite as the application traffic to evaluate the ADCM design. The applications used to perform the experiments are a subset of the A class NPB, a well-known, allegedly representative set of application workloads often used to assess the performance of parallel computers. These applications include three kernels—conjugate gradient (CG), integer sort (IS), discrete 3D fast Fourier transform (FT)—as well as two pseudoapplications—block tridiagonal solver (BT) and scalar pentadiagonal solver (SP). Considering that the current baseline hardware implementation of MPI functions only supports point-to-point MPI communication, other communication types such as collective communication are performed through these basic MPI communications.

10.5.2 SYNTHETIC TRAFFIC RESULTS

The synthetic traffic represents different communication patterns that facilitate the evaluation of the ADCM approach. In the following text, we analyze the experimental results from the aspects of bandwidth, traffic, and delay. Higher MPI bandwidth indicates that the communication design can achieve higher execution performance, which has been adopted as the evaluation metric in numerous experiments for MPI implementations.

10.5.2.1 Round-trip traffic pattern

We first record the time taken for a number of round-trip message transfers. The randomly generated message (i.e., the destinations of unicast messages at each node are selected randomly) is sent to the target node by the *MPI_Send* instruction. When the target node receives this message, it will first perform the *MPI_Receive* instruction and then return the message back to the source node immediately without changing anything by the *MPI_Send* instruction. Considering that receivers pre-post *MPI_Receive* instructions, this round-trip test can help us to obtain the maximum network bandwidth of the communication system. Assuming that the maximum capacity of the L1 cache in the multicore processor is 32 kB, the maximum length of the message triggered by MPI primitive instructions should be set to 16 kB (for send and receive operations).

Figure 10.12 shows the bandwidth results for different protocols under a round-trip traffic pattern with message size ranging from 1 to 16,384 bytes. As the message size increases, the communication bandwidth rapidly increases, such that more time can be used for the real data transmission. It can be seen from the figure that the buffered protocol achieves significantly higher bandwidth than the synchronous protocol primarily because the buffered protocol does not need the handshaking process or retry operations for its pre-post receives. As expected, the proposed ADCM approach achieves the same bandwidth as the buffered protocol. With sufficient buffers and pre-post receive operations, the buffered protocol exhibits a special case of the ADCM.

To understand further the buffered protocol's disadvantages and the ADCM's advantages, Figure 10.13 illustrates how the bandwidth varies in terms of the percentage of pre-post receive operations with a 4-kB message. Pre-post receive indicates that such receive operations are pre-posted before the time messages arrive.

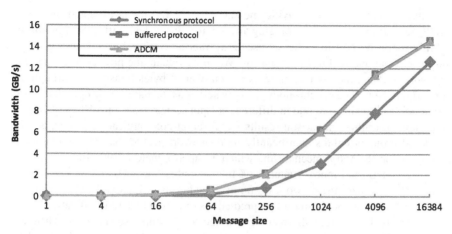

FIGURE 10.12

The bandwidth comparison under round-trip traffic.

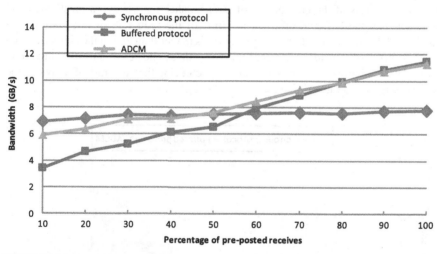

FIGURE 10.13

The bandwidth comparison for pre-post receive ratio variation.

The synchronous protocol outperforms the buffered protocol when 58% or less receives are pre-posted. The ADCM approach achieves better communication bandwidth by dynamically performing corresponding protocol behavior on the basis of the buffer usage and communication demand.

10.5.2.2 Hotspot traffic pattern

For the hotspot traffic pattern, we set one or two network nodes as the hotspot nodes to which other nodes send data messages with a greater probability. To simulate the unexpected cases of receiving messages, *MPI_Receive* is executed later than *MPI_Send* (uniform random ranging from 1 to 256 cycles) with 30% possibility. Unlike for the round-trip traffic, we use the average network traffic and message delay in the hotspot and real traffic scenarios. We first evaluate the protocols through the network traffic, which is identified as the number of bytes transmitted through the underlying NoC. The synchronous protocol serves as the baseline design; that is, the protocol is normalized to one in different hotspot scenarios. Figure 10.14 illustrates the network traffic comparison results for different communication protocols. The buffered protocol entails significantly more network traffic than the synchronous protocol for the retry mechanism, which is approximately 29% for the single-hotspot scenario and approximately 43% for the double-hotspot scenario. The ADCM approach minimizes the network traffic overhead. This traffic overhead is primarily attributed to the wrong buffer usage predictions or partial data retry operations.

Figure 10.15 illustrates the average message delay comparison results for different protocols. We define the message delay as the time interval between the send and receive operations for each data unit. The ADCM approach has the least message delay compared with the two other protocols. The ADCM outperforms the synchronous protocol in terms of the elimination of the handshaking process in most cases and is better than the buffered protocol in terms of the reduction of retry delay. The message delay in the double-hotspot scenario is longer than that in the single-hotspot scenario primarily because of the increased retry delay and network traffic contention. The ADCM minimizes this negative impact, achieving the best tradeoff between the message delay and the traffic load.

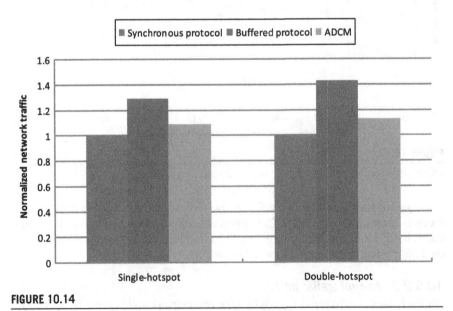

FIGURE 10.14

The network traffic comparison under hotspot traffic.

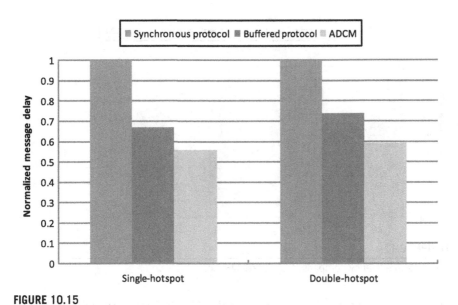

FIGURE 10.15

The message delay comparison under uniform random traffic.

10.5.3 REAL APPLICATION RESULTS

We run the application traffic to evaluate the ADCM. These application benchmarks were selected for their large number of message transmissions and unexpected receives. Considering that the on-chip buffers are limited for preserving these data, a pure buffered protocol is unsuitable. Figure 10.16 shows the network traffic

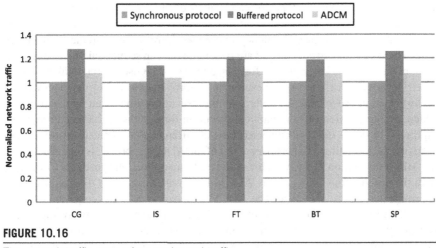

FIGURE 10.16

The network traffic comparison under real traffic.

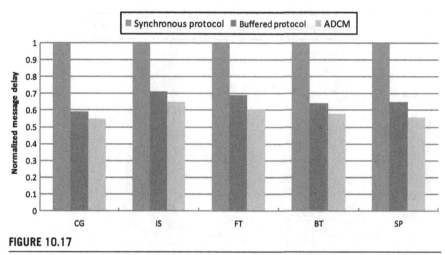

FIGURE 10.17

The message latency comparison under real traffic.

comparison results for different benchmarks. The buffered protocol required an average of approximately 22% more network traffic than the synchronous protocol. This network traffic overhead is decreased to 7% when using the ADCM.

Figure 10.17 shows the message delay comparison results for different benchmarks. The buffered protocol has 34% lower message delay than the synchronous protocol, whereas the ADCM has 42% lower message delay than the synchronous protocol on average. This reduced message delay can facilitate the application performance improvement. The CG benchmark achieves the lowest message delay reduction because of its short message size and relatively low number of unexpected receives. In conclusion, the real traffic results demonstrate that the proposed ADCM approach achieves the best performance metric with minimized network traffic overhead.

The above memory access latency and network traffic measurements are more related to instructions per cycle and are beneficial only in sequential processing. In a parallel environment, the valid performance metric to be used is the total execution time. Nevertheless, these two measurement improvements would result in overall performance improvement. To determine the execution time of an application, we added the information on computation execution time to the real application trace. Figure 10.18 shows the performance results of different communication protocols in terms of execution time. The primary effect of the ADCM approach is the reduction in the MPI transmission delay, which would result in application speedups. Compared with the synchronous protocol, the ADCM has greater execution time reduction of approximately 30% on average. The ADCM also has a performance advantage over the buffered protocol of approximately 4% execution time reduction on average. That is, using the ADCM, the programmer can avoid the burden of handling the buffers in the MPI application to achieve even better performance than the buffered protocol.

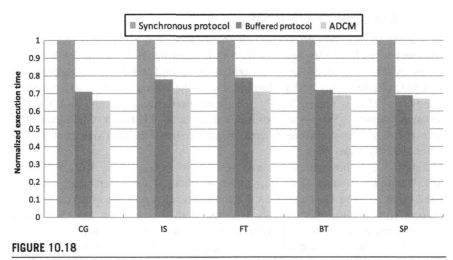

FIGURE 10.18

The execution time comparison under real traffic.

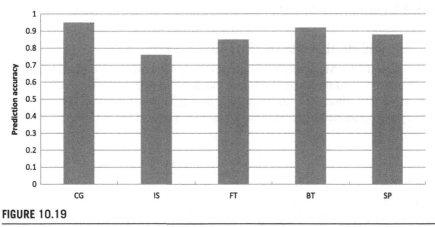

FIGURE 10.19

The ADCM prediction accuracy for the test benchmarks.

Figure 10.19 illustrates the ADCM prediction accuracy for the tested benchmarks. The prediction accuracy is defined as the appropriate operation ratio for the receiving buffer usage. This result can help us understand the effectiveness of the prediction mechanism. The figure shows that the ADCM achieves an average of approximately 87% prediction accuracy for the NPB benchmarks. That is, in most cases, the ADCM can make the best use of the receiving buffer. A wrong prediction may trigger unnecessary retry operations or other operations used in synchronous protocol, which is why the ADCM also has a minor network traffic overhead.

10.5.4 SENSITIVITY ANALYSIS

To analyze further the performance of the ADCM, this subsection presents the performance results of varying the ADCM design parameters. Figure 10.20 shows the sensitivity results for receiving buffer sizes of 16, 32, and 64 kB. We find that a buffer with 64 kB achieves the best performance with the largest memory consumption. The 32-kB buffer is approximately 38% better performing than the 16-kB buffer but approximately 15% worse performing than the 64-kB buffer. Considering the memory resource for multicore systems, we set the size of the hardware management buffer to 32 kB in this work.

Figure 10.21 shows the sensitivity results for $T_{s\_threshold}$ of 40%, 50%, and 60%. The optimal value of $T_{s\_threshold}$ can be changed for different applications. If the value is set to too large, then the communication will act as a synchronous protocol with minimal performance benefits. If the value is set too small, then the communication will act as a buffered protocol with numerous retry operations, resulting in performance degradation. $T_{s\_threshold}$ of 50% has an evident performance advantage over $T_{s\_threshold}$ of 40% by an average of 19%. However, this case does not hold true in the comparison with $T_{s\_threshold}$ of 60%. This finding demonstrates the application-specific features of $T_{s\_threshold}$. In this chapter, we choose 50% as a reasonable configuration value for the evaluation.

10.5.5 THE HARDWARE OVERHEAD

The proposed ADCM approach is implemented on the basis of a conventional MPU. The hardware realization overheads introduced include the ADCM control logic and RBC table. To estimate the hardware cost, we implemented the ADCM hardware in Verilog and performed logic synthesis by using Synopsys Design Compiler to obtain

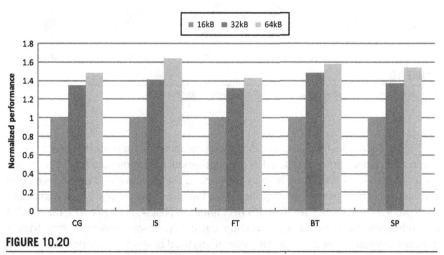

FIGURE 10.20

Comparison results for different receiving buffer sizes.

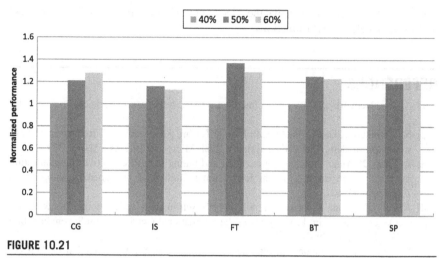

FIGURE 10.21

Comparison results for different $T_{s\_threshold}$ values.

Table 10.2 The Hardware Estimation for the ME in the Proposed ADCM

	PR	PPU	MPU	Total
Area (gate count)	975	1325	9570	94
Ratio (%)	8.2	11.2	80.8	100

the area information. We used Taiwan Semiconductor Manufacturing Company 90-nm CMOS generic process technology for logic synthesis. The hardware overhead for the entire proposed ME hardware, including the PR, PPU, and MPU, is listed in Table 10.2. For the impact of the proposed modifications, approximately 3% area overhead for the ADCM is observed. This small hardware overhead of the ADCM brings significant network traffic and message delay reduction, and the ADCM is demonstrated to be an effective communication protocol.

10.6 CHAPTER SUMMARY

In this chapter, the ADCM for accelerating MPI functions on NoC-based multicore architectures was proposed. The ADCM integrates two conventional communication protocols, namely the buffered and synchronous protocols, and behaves adaptively according to the application and NoC configurations. The ADCM exhibits behaviors similar to the buffered communication when a sufficient number of buffers are available in the receiver, but exhibits behaviors similar to the synchronous protocol when the receiver has limited buffers. It combines the advantages of the buffered and synchronous communication protocols to achieve better throughput and performance.

The promising results confirm that the ADCM could be effectively used in future NoC designs to accelerate MPI functions.

REFERENCES

[1] Arteris, Network on Chip (NoC) Interconnect IPs, http://www.arteris.com/, cited 2011. 318

[2] N.L. Binkert, R.G. Dreslinski, L.R. Hsu, K.T. Lim, A.G. Saidi, S.K. Reinhardt, The M5 simulator: modeling networked systems, IEEE Micro 26 (4) (2006) 52–60. 339

[3] A. Faraj, X. Yuan, Automatic generation and tuning of MPI collective communication routines, in: Proceedings of the International Conference on Supercomputing (ICS), 2005, pp. 393–402. 325

[4] A. Faraj, X. Yuan, D. Lowenthal, STAR-MPI: self tuned adaptive routines for MPI collective operations, in: Proceedings of the International Conference on Supercomputing (ICS), 2006, pp. 199–208. 325

[5] K. Feind, K. McMahon, An ultrahigh performance MPI implementation on SGI cc-NUMA Altix systems, Comput. Meth. Sci. Tech. 12 (2006) 67–70, June 2006. 325

[6] M. Galles, Spider: a high-speed network interconnect, IEEE Micro 17 (1) (1997) 34–39. 339

[7] M.S. Gaur, B.M. Al-Hashimi, V. Laxmi, N.R. Choudhary, L. Jain, M. Ahmed, K.K. Paliwal, Varsha, Rekha, Vineetha, NIRGAM: a simulator for NoC interconnect routing and application modeling, in: Proceedings of the Design, Automation & Test in Europe Conference & Exhibition (DATE), 2007. 338

[8] K. Goossens, J. Dielissen, A. Radulescu, AEthereal network on chip: concepts, architectures, and implementations, IEEE Des. Test Comput. 22 (5) (2005) 414–421. 318

[9] S. Goudy, A preliminary analysis of the MPI queue characteristics of several applications, in: Proceedings of the International Conference on Parallel Processing (ICPP), 2005, pp. 175–183. 320, 325

[10] W. Gropp, E. Lusk, A. Skjellum, Using MPI: Portable Parallel Programming with the Message-Passing Interface, MIT Press, Cambridge, MA, USA, 1999. 318

[11] T. Hoefler, C. Siebert, W. Rehm, A practically constant-time MPI broadcast algorithm for large-scale InfiniBand clusters with multicast, in: Proceedings of the International Parallel and Distributed Processing Symposium (IPDPS), 2007, pp. 1–8. 325

[12] Y. Hoskote, S. Vangal, A. Singh, N. Borkar, S. Borkar, A 5-GHz mesh interconnect for a teraflops processor, IEEE Micro 27 (5) (2007) 51–61. 318

[13] C. Huang, G. Zheng, L. Kalé, S. Kumar, Performance evaluation of adaptive MPI, in: Proceedings of the Symposium on Principles and Practice of Parallel Programming (PPoPP), 2006, pp. 12–21. 325

[14] L. Huang, Z. Wang, N. Xiao, Y. Wang, Q. Dou, Adaptive communication mechanism for accelerating MPI functions in NoC-based multicore processors, ACM Trans. Archit. Code Optim. 10 (3) (2013) 18:1–18:25. 317

[15] A. Karwande, X. Yuan, D.K. Lowenthal, An MPI prototype for compiled communication on Ethernet switched clusters, J. Parallel Distrib. Comput. 65(10) (2005) 1123–1133. 325

[16] J. Liu, A.R. Mamidala, D.K. Panda, Fast and scalable MPI-level broadcast using InfiniBand's hardware multicast support, in: Proceedings of the International Parallel and Distributed Processing Symposium (IPDPS), 2004, pp. 1–10. 325

[17] K.E. Maghraoui, B.K. Szymanski, C.A. Varela, An architecture for reconfigurable iterative MPI applications in dynamic environments, in: Proceedings of the International Conference on Parallel Processing and Applied Mathematics (PPAM), 2005, pp. 258–271. 325

[18] P. Mahr, C. Lörchner, H. Ishebabi, C. Bobda, SoC-MPI: a flexible message passing library for multiprocessor systems-on-chips, in: Proceedings of the International Conference on Reconfigurable Computing and FPGAs (ReConFig), 2008, pp. 187–192. 318

[19] J.J. Murillo, HW-SW Components for Parallel Embedded Computing on NoC-based MPSOCS, PhD Thesis, Barcelona, Spain, 2009. 318

[20] H. Ogawa, S. Matsuoka, OMPI: optimizing MPI programs using partial evaluation, in: Proceedings of the ACM/IEEE Conference on Supercomputing (SC), 1996, pp. 1–15. 325

[21] M. Ohara, H. Inoue, Y. Sohda, H. Komatsu, T. Nakatani, MPI microtask for programming the Cell Broadband Engine^{TM} processor, IBM Syst. J. 45 (2006) 85–102. 318

[22] L.-S. Peh, W. Dally, A delay model and speculative architecture for pipelined routers, in: Proceedings of the International Symposium on High-Performance Computer Architecture (HPCA), 2001, pp. 255–266. 339

[23] Y. Peng, M. Saldana, P. Chow, Hardware support for broadcast and reduce in MPSoC, in: Proceedings of the International Conference on Field Programmable Logic and Applications (FPL), 2011, pp. 144–150. 325

[24] J. Psota, A. Agarwal, rMPI: message passing on multicore processors with on-chip interconnect, in: Proceedings of the International Conference on High Performance Embedded Architectures and Compilers (HiPEAC), 2008, pp. 22–37. 318

[25] M. Saldaña, P. Chow, TMD-MPI: an MPI implementation for multiple processors across multiple FPGAs, in: Proceedings of the International Conference on Field Programmable Logic and Applications (FPL), 2006, pp. 1–6. 318, 329

[26] D.G. Solt, A profile-based approach for topology aware MPI rank placement, in: Invited Presentation to HPCC, 2007. 325

[27] K.D. Underwood, K.S. Hemmert, A. Rodrigues, R. Murphy, R. Brightwell, A hardware acceleration unit for MPI queue processing, in: Proceedings of the International Parallel and Distributed Processing Symposium (IPDPS), 2005, pp. 96.2–96.11. 325

[28] M.G. Venkata, P.G. Bridges, MPI/CTP: a reconfigurable MPI for HPC applications, in: Proceedings of the European PVM/MPI User's Group Conference on Recent Advances in Parallel Virtual Machine and Message Passing Interface (EuroPVM/MPI), 2006, pp. 96–104. 325

[29] M.G. Venkata, P.G. Bridges, P.M. Widener, Using application communication characteristics to drive dynamic MPI reconfiguration, in: Proceedings of the International Parallel and Distributed Processing Symposium (IPDPS), 2009, pp. 1–6. 325

[30] Verari Systems, Inc., MPI Software, http://www.mpi-softtech.com/, 2012. 318

[31] D. Wentzlaff, P. Griffin, H. Hoffmann, L. Bao, B. Edwards, C. Ramey, M. Mattina, C.-C. Miao, J.F. Brown III, A. Agarwa, On-chip interconnection architecture of the TILE processor, IEEE Micro 27 (5) (2007) 15–31. 318

[32] J.A. Williams, I. Syed, J. Wu, N.W. Bergmann, A reconfigurable cluster-on-chip architecture with MPI communication layer, in: Proceedings of the Symposium on Field-Programmable Custom Computing Machines (FCCM), 2006, pp. 350–352. 318

Epilogue

Conclusions and future work 11

CHAPTER OUTLINE

11.1 CONCLUSIONS

The advancement of semiconductor technology and the severe design challenges for single-core processors are together driving computer architecture rapidly into the many-core era. Although the community has already made significant breakthroughs, the development of efficient many-core processors still faces several challenges, including high-level parallel programming paradigms, intermediate-level communication structure, and low-level logic implementations. Communication-centric cross-layer optimizations can not only increase the performance for the communication layer, but can also efficiently mitigate the challenges for both the programming paradigm layer and the logic implementation layer. On the basis of this insight, this book has explored the network-on-chip (NoC) design space in a bottom-up, coherent, and uniform fashion, from low-level router, buffer, and topology implementations, to network-level routing and flow control designs, to co-optimizations of the NoC and high-level programming paradigms.

The main content of this book was presented in Part II, on logic implementations, Part III, on routing and flow control, and Part IV, on programming paradigms. Part I contains the Prologue and Part V contains the Epilogue. Part II, consisting of Chapters 2–4, tackled the logic implementations of the NoC router architecture, buffer structure, and topology. More specifically, in Chapter 2, we designed a single-cycle router with wing channels to reduce the communication latency. The wing channels forward incoming packets to free ports immediately with the inspection of switch allocation results to achieve single-cycle per-hop delay. Also, the packets traversing wing channels fill in the free time slots of the crossbar to improve the network throughput. Chapter 3 first introduced a dynamically allocated virtual channel (VC) design to share buffers among VCs of the same port, and then a hierarchical bit-line buffer-based structure to share buffers among different ports was designed. Both the dynamically allocated VC and the hierarchical bit-line buffer adaptively avoid network congestion according to network traffic and buffer occupations. Chapter 4 presented a hierarchy topology which combines the packet-switched network with

Networks-on-Chip. http://dx.doi.org/10.1016/B978-0-12-800979-6.00011-1

the transaction-based bus structure. The proposed virtual bus on-chip network dynamically configures point-to-point links of conventional NoCs into virtual bus structures. This topology efficiently supports both unicast and multicast/broadcast communications.

Part III, including Chapters 5–7, shifted the attention to a higher level of abstraction, the routing and flow control of the NoCs. On the basis of a holistic approach, Chapter 5 delved into the design of routing algorithms for workload consolidation. The proposed destination-based selection strategy achieves both high adaptivity and dynamic isolation for multiple concurrent applications. Chapter 6 explored efficient flow control mechanisms to maximize the utilization of limited buffer resources for fully adaptive routing algorithms. It presented two novel flow control designs. First, whole packet forwarding (WPF) reallocates a nonempty VC if the VC has enough free buffers for an entire packet. We proved that WPF does not induce deadlock, and that it is an important extension to several deadlock avoidance theories. Second, we extended Duato's theory to apply aggressive VC reallocation on escape VCs without deadlock. Chapter 7 continued our exploration of deadlock-free flow control in torus NoCs. The flit bubble flow control theory presented achieves deadlock freedom by maintaining one free flit-size buffer slot inside the ring. The two implementations support both high frequencies and efficient buffer utilization.

Part IV covered co-optimizations of the NoC and programming paradigms in three chapters, Chapters 8–10. In Chapter 8, we optimized the NoC design for shared memory programming paradigms. We provided hardware implementations for collective communications, including multicast and reduction ones, for cache-coherent protocols to prevent these communications from becoming system bottlenecks. In Chapter 9, we customized the NoC for message passing programming paradigms. The NoC designed provides special and low-cost hardware implementations for message passing interface (MPI) communication primitives. Since most other MPI functions can be built upon these hardware-implemented primitives, this design greatly and efficiently improves the performance for MPI communication. Chapter 10 studied supporting adaptive MPI communication protocols in NoCs. The proposed adaptive communication mechanism combines the advantages of both buffered and synchronous communication modes to enhance throughput and latency; it performs similarly to the buffered mode with large free receiving buffers, while it changes to the synchronous mode with limited buffers.

In summary, in this book we have applied the communication-centric cross-layer method in a bottom-up fashion. The research presented here has addressed a multitude of pressing concerns spanning a wide spectrum of design topics. In the lower logic implementation layer, the exploration of low-latency router architectures was followed by the design of efficient dynamic VC structures. The study of this layer ended with an NoC topology enhanced with virtual bus structures. The exploration of the intermediate network routing and flow control layer first focused on routing algorithms for workload consolidation, and then delved into flow control designs for fully adaptive routing and deadlock-free torus NoCs. For the co-design of the NoC and the upper programming paradigm layer, both the mainstream shared memory paradigm and message passing paradigms were addressed by providing customized and special communication hardware.

11.2 FUTURE WORK

The content of this book opens several interesting avenues for future research. The low-latency router architectures with wing channels proposed in Chapter 2 can be enhanced with priority arbitrations to support critical packets or traffic flow more efficiently; reserving express wing channels for critical traffic can mitigate the performance bottlenecks due to the communication latency. The dynamic VC structures proposed in Chapter 3 can be leveraged to design the buffer structure in the network interface. The virtual bus on-chip network in Chapter 4 can be extended to support power gating techniques to shut down all routers. The reconfigurable bus links act as the backup connected network.

The idea of the destination-based selection strategy in Chapter 5 can be used in the design of an injection control mechanism. If the packet destination is integrated into the injection control procedure, the network can maintain the performance in a more robust fashion. The WPF theory presented in Chapter 6 can be leveraged to enhance the turn-model-based routing algorithm by allowing the short packets to cross the prohibited turns. Combining the flit bubble flow control in Chapter 7 and Duato's theory can result in the design of efficient fully adaptive routing algorithms for torus NoCs.

The message combination framework presented in Chapter 8 can be extended to support more general cache coherence protocols, where the requesting node collects the acknowledgments. The customized hardware implementation for MPI primitives studied in Chapter 9 can be improved by integrating the special support for latency-critical short messages. The adaptive communication mechanism proposed in Chapter 10 may be improved with novel mechanisms which deliver timelier end-point buffer status.

Although this book offers a thorough exploration of the NoC design space, several emerging technology trends or techniques indicate numerous new topics for future research in the NoC field. In addition to reducing the dynamic power consumption, reducing the static power consumption is becoming more and more important. Power gating the network components, such as the routers, buffers, allocators, or crossbars, is the general way to optimize the static power consumption. The effect of wake-up delay and maintaining connectivity for networks are two important issues. The deadlock-free flow control mechanisms presented in Chapters 6 and 7 can be used to provide deadlock freedom for partially powered down networks.

The synchronization procedure easily becomes a system bottleneck for many-core processors with shared memory programming paradigms. Exploring the NoC structure to design a fast communication structure for synchronization signals can mitigate this challenge. After about 20 years of developments, hardware transactional memory is becoming a reality. Providing customized NoC features for transactional memory programming paradigms is also an interesting research direction. The heterogeneous architecture, including both CPUs and graphics processing units, is a widely accepted method to address the power consumption problems. The latency-critical CPUs and throughput-oriented graphics processing units have different requirements for the communication structure. Deploying isolation mechanisms in the NoCs to support these kinds of communications is important for the efficiency of heterogeneous architectures.

In general, using communication-centric cross-layer methods to integrate low-level circuit and logic knowledge and high-level programming paradigms and application knowledge into the NoC designs will provide everlasting vigor to the NoC and computer architecture community; research into NoCs has countless opportunities and challenges.

Index

Note: Page numbers followed by *f* indicate figures and *t* indicate tables.

Printed in the United States
By Bookmasters